PURINE METABOLISM IN MAN—III

Biochemical, Immunological, and Cancer Research

ADVANCES IN EXPERIMENTAL MEDICINE AND BIOLOGY

Recent Volumes in this Series

Volume 120A
KININS – II: Biochemistry, Pathophysiology, and Clinical Aspects
Edited by Setsuro Fujii, Hiroshi Moriya, and Tomoji Suzuki

Volume 120B
KININS – II: Systemic Proteases and Cellular Function
Edited by Setsuro Fujii, Hiroshi Moriya, and Tomoji Suzuki

Volume 121A
MACROPHAGES AND LYMPHOCYTES: Nature, Functions, and Interaction, Part A
Edited by Mario R. Escobar and Herman Friedman

Volume 121B
MACROPHAGES AND LYMPHOCYTES: Nature, Functions, and Interaction, Part B
Edited by Mario R. Escobar and Herman Friedman

Volume 122A
PURINE METABOLISM IN MAN – III: Clinical and Therapeutic Aspects
Edited by Aurelio Rapado, R. W. E. Watts, and Chris H. M. M. De Bruyn

Volume 122B
PURINE METABOLISM IN MAN – III: Biochemical, Immunological, and Cancer Research
Edited by Aurelio Rapado, R. W. E. Watts, and Chris H. M. M. De Bruyn

Volume 123
GABA – Biochemistry and CNS Functions
Edited by Paul Mandel and Francis V. DeFeudis

Volume 124
THE ENDOCRINE PANCREAS AND JUVENILE DIABETES
Edited by David M. Klachko, Ralph R. Anderson, Thomas W. Burns, and Harold V. Werner

Volume 125
STRUCTURE AND FUNCTION OF GANGLIOSIDES
Edited by Lars Svennerholm, Paul Mandel, Henri Dreyfus, and Paul-Francis Urban

Volume 126
BIOLOGICAL EFFECTS OF ALCOHOL
Edited by Henri Begleiter

Volume 127
LITHIUM EFFECTS ON GRANULOPOIESIS AND IMMUNE FUNCTION
Edited by Arthur H. Rossof and William A. Robinson

PURINE METABOLISM IN MAN–III

Biochemical, Immunological, and Cancer Research

Edited by

Aurelio Rapado
Fundación Jiménez Díaz
Madrid, Spain

R.W.E. Watts
M.R.C. Clinical Research Centre
Harrow, England

and

Chris H.M.M. De Bruyn
Department of Human Genetics
University of Nijmegen Faculty of Medicine
Nijmegen, The Netherlands

PLENUM PRESS • NEW YORK AND LONDON

Library of Congress Cataloging in Publication Data

International Symposium on Purine Metabolism in Man, 3d, Madrid, 1979.
Purine metabolism in man, III.

(Advances in experimental medicine and biology; v. 122A-122B)
Includes index.
CONTENTS: [1] Clinical and therapeutic aspects. – [2] Biochemical, immunological and cancer research.
1. Purine metabolism – Congresses. 2. Hyperuricemia – Congresses. 3. Immunopathology – Congresses. 4. Cancer – Congresses. I. Rapado, A. II. Watts, R. W. E. III. De Bruyn, C. H. M. M. IV. Title. V. Series. [DNLM: 1. Purine-pyrimidine metabolism, Inborn errors – Congresses. 2. Purines – Metabolism – Congresses. W3 IN922NM 3d 1979p/WD205.5P8 I61 1979p]
QP801.P8I56 1979 612'.0157 79-22555

ISBN 978-1-4684-8561-5 ISBN 978-1-4684-8559-2 (eBook)
DOI 10.1007/978-1-4684-8559-2

Proceedings of the second half of the Third International Symposium on Purine Metabolism in Man, held in Madrid, Spain, June 11–15, 1979

Softcover reprint of the hardcover 1st edition 1980

A Division of Plenum Publishing Corporation
227 West 17th Street, New York, N.Y. 10011

Preface

These volumes contain the papers which were presented at the Third International Symposium on Purine Metabolism in Man held in Madrid (Spain) in June, 1979. The previous meetings in the series were held in Tel Aviv (Israel) and in Baden (Austria) in 1973 and 1976, respectively. The proceedings were also published by Plenum.

Knowledge of the pathophysiology of the purines has developed greatly since the 1950's when it was mainly related to clinical gout, and it is now relevant to many fields of Medicine and Biology. These volumes include papers reporting new work on clinical gout and urolithiasis as well as on some of the subjects which have featured prominently in the previous volumes, including: regulatory aspects of the intermediary metabolism of purines and related compounds, enzymology, methodology, and the results of mutations which affect purine metabolism. However, there have been many new developments during the last three years and the scope of the communications reflects not only increasing depth of knowledge, but also a widening of the field. This publication has clinical and fundamental implications for internal medicine, pediatrics, urology, biochemistry, immunology, genetics, and oncology.

It is interesting to compare the scope of this volume with that of its predecessors. The main emphasis has shifted from the study of gout and the dissection of metabolic pathways to encompass investigations in the fields of oncology, immunology, and lymphocyte physiology. There are pointers to possible implications in relation to cardiology and neuromuscular diseases, which may well prove to be growing points for the future. In spite of considerable work on the mechanism of urinary stone formation, the inter-relationship between uric acid and calcium oxalate urolithiasis remains obscure.

It is no longer logical to discuss clinically related purine research without including comparable work in the less studied field of pyrimidine metabolism. Some such studies were reported at the Madrid meeting, and this development will be formally encouraged in the future.

The use of some animal and single cell models as tools with complexity intermediate between man and the single or multi-enzyme systems represents another new development in this area of clinical investigation.

We acknowledge the support which we received from the distinguished members of the scientific community who served on the Organizing and Scientific Committees, as well as their contributions to the high standards of the material presented.

We also thank the "Fundacion Jimenez Diaz" and the Autonomous University of Madrid, both of whom sponsored the meeting, the Department of Cultural Relations in the Ministry of Foreign Affairs, the Madrid City Council and the Wellcome Research Laboratories (England) for their financial support, and Plenum Publishing Corporation (U.S.A.) for their assistance in the publication of the proceedings. The meeting would not have been possible without the cheerful and spirited help of Maria Luisa San Roman and Mireya Usano, and our special thanks are due to them.

A. Rapado
R.W.E. Watts
C.H.M.M. de Bruyn

Contents of Part B

IV. LYMPHOCYTE PURINE METABOLISM RESEARCH

Contents of Part A

IV. STUDIES ON XANTHINURIA

V. THERAPEUTICAL ASPECTS OF DISORDERS IN PURINE METABOLISM

VI. MUTATIONS AFFECTING PURINE METABOLISM

A. Phosphoribosyltransferases

B. Nucleoside Phosphoribosylating Enzymes

DE NOVO PURINE SYNTHESIS IN CULTURED HUMAN FIBROBLASTS

R.B. Gordon, L. Thompson, L.A. Johnson and B.T. Emmerson

University of Queensland Department of Medicine
Princess Alexandra Hospital, Ipswich Road
Brisbane. 4102, Australia

In mammalian cells, purine nucleotides may be formed via (a) de novo synthesis and (b) salvage pathways. Both these pathways utilise PRPP and the concentration of this substrate is thought to be important in regulating the rate of purine synthesis (1, 2). Individuals with a deficiency of the purine salvage enzyme hypoxanthine phosphoribosyltransferase (HPRT), have increased production of uric acid (3, 4). Lymphocyte and fibroblast cultures of HPRT-deficient ($HPRT^-$) cells have been reported as having accelerated rates of purine de novo synthesis associated with elevated concentrations of PRPP (5, 6).

In studies of $HPRT^-$ and $HPRT^+$ fibroblast cells we have found increased concentrations of PRPP in $HPRT^-$ cells. However, the rate of purine de novo synthesis in both types of cells was dependent on the assay conditions. A marked differentiation in rates could only be established when the assay was carried out in medium containing hypoxanthine, which inhibited the $HPRT^+$ cells.

MATERIALS AND METHODS

Fibroblast cells were grown in RPMI 1640 medium containing 20 mM Hepes (pH 7.4), penicillin (100 U/ml), streptomycin (100 U/ml) and 10% bovine fetal calf serum. Rates of purine de novo synthesis were assayed by measuring (1) the incorporation of ^{14}C-formate into α-N-formylglycinamide ribonucleotide (FGAR) in the presence of 0.3 mM azaserine which inhibits further metabolism of FGAR, and (2) the incorporation of ^{14}C-formate into cellular purines isolated as the silver complexes (7). PRPP in cell extracts was assayed by the conversion of ^{14}C-adenine to ^{14}C-AMP in the presence of purified APRT (6, 8).

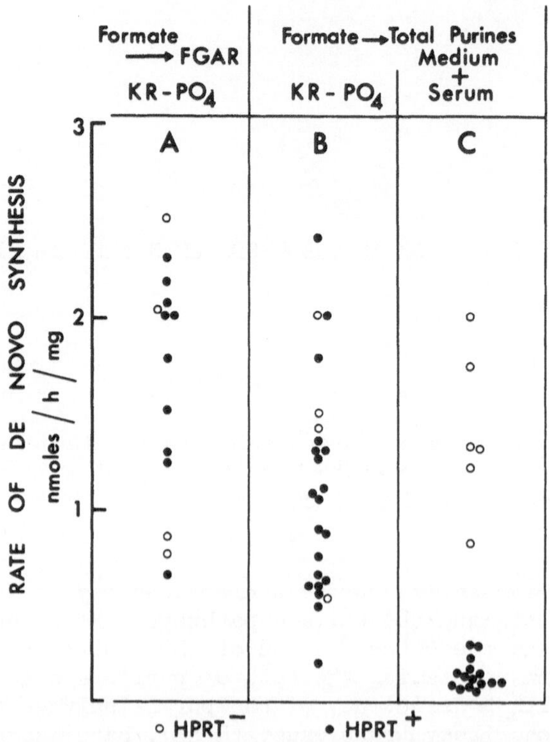

Figure 1. Comparison of assays for the rate of purine de novo synthesis. (A) incorporation of ^{14}C-formate into FGAR: (B), (C) incorporation of ^{14}C-formate into cellular purines plus excreted purines. Assays were carried out in Krebs-Ringer phosphate buffer (A) and (B), or in RPMI-1640 with 10% undialysed fetal calf serum (C) (from Gordon et al, 1979).

RESULTS

Figure 1 (A and B) shows the lack of differentiation between HPRT$^+$ and HPRT- cells when rates of purine de novo synthesis were measured in KRP buffer. However, differentiation was observed when the assay was performed in tissue culture medium containing 10% fetal calf serum (Figure 1C). It has been reported that calf serum contains hypoxanthine in sufficient quantities to markedly inhibit purine de novo synthesis in HPRT$^+$ lymphoblast cells (9). We have assayed several batches of bovine fetal calf serum by HPLC and found hypoxanthine at a concentration of 10-15 µM. This means that the hypoxanthine in our culture medium used for fibroblast growth and assays was 1-1.5 µM. To test the effect of hypoxanthine we measured the rates of purine de novo synthesis in medium containing

Table 1. Cells were grown in RPMI-1640 medium with 10% undialysed fetal calf serum. Twenty-four hours prior to assay this medium was replaced with fresh medium containing 10% undialysed or dialysed serum. The incorporation of ^{14}C-formate into purines and ^{3}H-glycine into PCA-insoluble material was measured as described (8).

DE NOVO PURINE SYNTHESIS IN MEDIUM CONTAINING
DIALYSED AND UNDIALYSED FOETAL CALF SERUM
(n mole/mg protein/hr; 1 hour incubation)

	UNDIALYSED FOETAL CALF SERUM				DIALYSED FOETAL CALF SERUM			
Cells	^{14}C-Formate incorporated into purines			$[^{3}H]$-glycine into $HClO_4$ insoluble fraction	^{14}C-Formate incorporated into purines			$[^{3}H]$-glycine into $HClO_4$ insoluble fraction
	Cells	Incubation Medium	Total		Cells	Incubation Medium	Total	
$HGPRT^{+}$-1	.07	.018	.09	.83	1.07	.078	1.15	1.22
$HGPRT^{+}$-2	.06	.016	.08	.82	1.00	.036	1.04	1.15
$HGPRT^{-}$	1.02	.40	1.42	.92	1.14	.53	1.67	1.29

undialysed and dialysed serum. Purine bases in the serum can be completely removed by dialysis (9). Dialysed serum greatly affected the rate of ^{14}C-formate incorporated into total purines in $HPRT^{+}$ cells (13-fold stimulation) but not in HPRT- cells (Table 1). Although with dialysed serum the rates of intracellular purine synthesis were similar in $HPRT^{+}$ and $HPRT^{-}$ cells, more purines were excreted into the medium in the case of $HPRT^{-}$ cells (32%) than $HPRT^{+}$ cells (6%). Rates of ^{3}H-glycine incorporation into PCA-insoluble material were similar for both media indicating no effect of dialysed serum per se on growth or metabolism of fibroblast cells.

The inhibition of purine de novo synthesis by purine bases in the medium was concentration dependent. Considerable inhibition was observed at quite low concentrations of the bases (70% at 1.5 μM Hx; 70-80% at 1 μM Ad). $HPRT^{-}$ cells were only inhibited by adenine but not to quite the same degree as $HPRT^{+}$ cells. Adenine was slightly more inhibitory to $HPRT^{+}$ cells than hypoxanthine. These results indicate that the rate of purine de novo synthesis of fibroblast cells can be inhibited by hypoxanthine present in undialysed serum as has been shown for lymphoblasts (9).

Since PRPP is a substrate for purine de novo synthesis as well as purine salvage, both these pathways may compete for the common

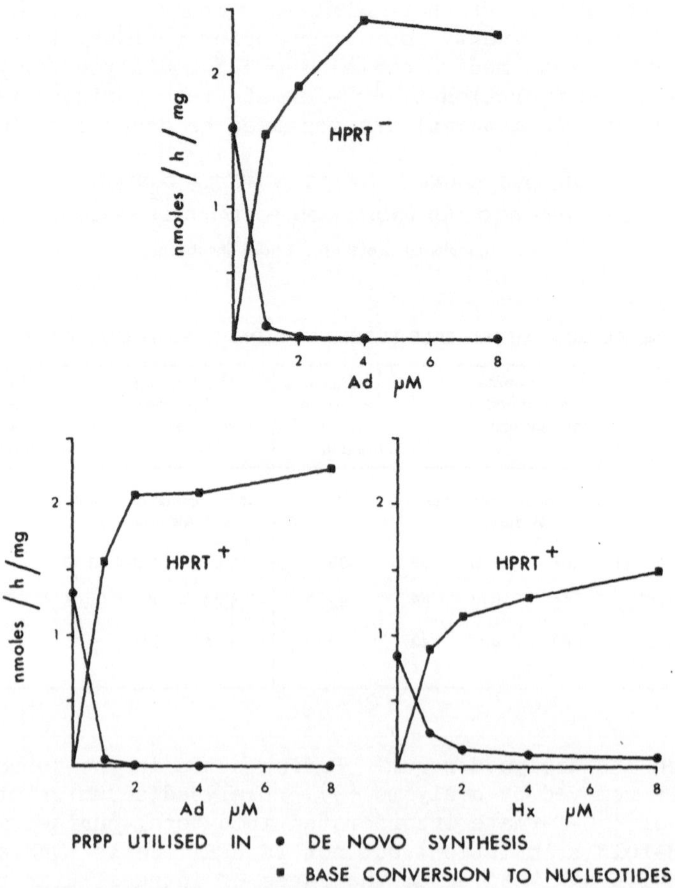

Figure 2. The relative utilisation of PRPP. The simultaneous incorporation of ^{14}C-formate and ^{3}H-purine base into cellular purines was assayed using a double-isotope technique (8).

substrate. As one molecule of PRPP is used in each pathway, the simultaneous incorporation of ^{14}C-formate and 3H-purine base into cellular purines can be taken as a measure of PRPP utilised in de novo synthesis and via the salvage pathway respectively. Figure 2 shows this relative utilisation of PRPP. Preferential utilisation of PRPP by the salvage pathways is apparent. There was a similar rate of PRPP utilisation in the salvage of adenine in HPRT$^+$ and HPRT$^-$ cells.

The inhibition of purine de novo synthesis by purine bases is accompanied by a decrease in the intracellular PRPP concentration (Table 2). However, the PRPP content of HPRT$^-$ cells grown in medium containing dialysed fetal calf serum was about three-fold that of

Table 2. Cell growth conditions were the same as for Table 1. Purine de novo synthesis and PRPP content were determined on parallel cultures.

EFFECT OF PURINE BASES ON CONCENTRATION OF PRPP

Cells	^{14}C Formate incorporation into purines (nmol/mg protein/hr) in absence of base	Purine base μM	Purine *de novo* synthesis %	PRPP content (nmol/mg protein)
		Adenine		
$HPRT^-$	1.65	0	100	.97
		1.0	44	.81
		1.5	25	.52
		2.0	13	.56
$HPRT^+$-1	1.57	0	100	.15
		1.5	12	.08
$HPRT^+$-2	0.84	0	100	.38
		1.5	12	.10
		Hypoxanthine		
$HPRT^+$-1	1.57	0	100	.15
		1.0	42	.11
		2.0	27	.12
$HPRT^+$-2	0.84	0	100	.38
		1.0	44	.07
		2.0	22	.07

$HPRT^+$ cells, but both cell types had similar rates of purine de novo synthesis (Table 3). This lack of a correlation between the rate of purine de novo synthesis and PRPP concentration may be interpreted in two ways: (a) either the PRPP concentration does not necessarily reflect the PRPP available for purine synthesis or (b) factors in addition to PRPP play a role in the control of purine synthesis.

The potential for purine de novo synthesis in $HPRT^+$ and $HPRT^-$ fibroblast cells appears similar. How well the fibroblast cell reflects the in vivo situation of other mammalian cells is not known. Since HPRT-deficient humans are clinically overproducers of purines, presumably normal individuals must continually control purine de novo synthesis. Such control could be by limitation of PRPP through its preferential utilisation in the HPRT reaction. Whether in vivo this limitation is from the salvage of endogenous purine bases formed from normal cellular catabolism or from

Table 3. *Cultures from 4 HPRT^{+}-normals and from 3 HPRT^{-} patients.

exogenous purine bases from other cells is yet to be determined. Since the relative activities of enzymes involved in the degradation and salvage of purines vary widely in different mammalian cells (10), the control of purine synthesis may well differ depending on the type of cell.

PRPP CONCENTRATION AND RATE OF PURINE DE NOVO SYNTHESIS

(measured in dialysed foetal calf serum)

Cells *		Intracellular concentration of PRPP (nmol/mg protein)	Purine de novo synthesis (nmol/mg protein/hr)
HGPRT^{+}	1	.41	.91
	2	.28	1.51
	3	.38	.84
	4	.15	1.57
HGPRT^{-}	1	.92	1.67
	2	.72	2.02
	3	.97	1.65

REFERENCES

1. Henderson J.F.(1972). Am Chem Soc monograph 170, Washington, D.C.
2. Bagnara A.S., Letter A.A. and Henderson J.F. (1974). Biochim Biophys Acta 374:259-270.
3. Lesch M. and Nyhan W.L. (1964). Am J Med 36:561-570.
4. Sorensen L.B. (1970). J Clin Invest 49:968-978.
5. Wood A.W., Becker M.A. and Seegmiller J.E. (1973). Biochem Genet 9:261-274.
6. Becker M.A. (1976). J Clin Invest 57:308-318.
7. Martin D.W. and Owens N.T. (1972). J Biol Chem 247:5477-5485.
8. Gordon R.B., Thompson L., Johnson L. and Emmerson B.T. (1979). Biochim Biophys Acta 562:162-176.
9. Hershfield M.S. and Seegmiller J.E. (1977). J Biol Chem 252: 6002-6010.
10. Shenoy T.S. and Clifford A.J. (1975). Biochim Biophys Acta 411:133-143.

COMPARATIVE METABOLISM OF A NEW ANTILEISHMANIAL AGENT, ALLOPURINOL RIBOSIDE, IN THE PARASITE AND THE HOST CELL

Donald J. Nelson*, Stephen W. LaFon*, Gertrude B. Elion*, J. Joseph Marr† and Randolph L. Berens†
*Wellcome Research Laboratories
Research Triangle Park, NC, 27709, USA
†Dept. of Medicine, St. Louis University
St. Louis, MO, 60314, USA

INTRODUCTION

Leishmania donovani and L. braziliensis are monocellular hemoflagellates which infect millions of people in tropical regions of the world. These parasitic organisms depend exclusively upon salvage of bases and nucleosides from the host mileu for their purine requirement[1]. Allopurinol (HPP) as well as 4-amino-pyrazolo(3,4-d)pyrimidine (4-APP) and oxipurinol (DHPP) are inhibitors of the growth of the promastigote form of these organisms in culture and adenine can reverse this effect[2]. Since HPP is a widely used therapeutic agent for the control of hyperuricemia in man and is remarkably non-toxic to mammalian cells, it was of interest to determine how the metabolism of HPP and allopurinol riboside (HPP-Rib), a normal metabolite of allopurinol in man[3], differs in the leishmaniae and mammalian host.

RESULTS AND DISCUSSION

Growth Inhibition - HPP-Rib unexpectedly displayed a very strong antileishmanial activity against promastigotes in culture. The inhibition of the growth of L. braziliensis and L. donovani by HPP-Rib was several hundred-fold greater than the effect of the base, HPP, whereas in L. mexicana, the effects of the two compounds were approximately equal (Table 1). Such differences may be related to dissimilar rates of metabolic transformation of the drugs in the various species.

Table 1. Comparison of the Growth Inhibition by HPP and HPP-Rib in Species of Leishmania.

	ED_{50}, micromolar	
	HPP	HPP-Rib
L. braziliensis	40	0.15
L. donovani	150	0.2
L. mexicana	25	15

The initial cell concentration was 3×10^5/ml and after 6 days the controls had reached a density of about 2×10^7 cells/ml. The concentration of drug which produced a 50% reduction in the growth rate = ED_{50}.

The growth inhibition by HPP-Rib was not antagonized by adenine in any species, which distinguished HPP-Rib from HPP[4]. This selective reversal by adenine of HPP, but not HPP-Rib, is probably related to the two different pathways by which the compounds are activated.

The growth of Detriot 98 and L cells (N.K. Cohn, unpublished) and P388D1 macrophage cells in culture was not influenced by HPP or HPP-Rib at concentrations as high as 200 µM, which attested to the general lack of toxicity of these compounds toward mammalian cells.

Metabolic Stability of the Ribosyl Linkage of HPP-Rib - Several lines of experimental evidence suggest that the riboside of allopurinol is much more stable to enzymatic cleavage than the analogous natural purine, inosine. Analysis of the medium from incubations of both L. braziliensis (Table 2) and L. donovani with HPP-Rib showed that the riboside was >99% intact at 24 hrs; by 4 days with L. donovani at a cell density of 5×10^6/ml, DHPP-Rib accounted for 7% of the metabolites and HPP and DHPP were not detected. Inosine was totally split to hypoxanthine in 30 min[1].

HPP-Rib given to mice was not a xanthine oxidase inhibitor[5] and no HPP or DHPP was found in the urine. In our present studies HPP-Rib was oxidized in the mouse to oxipurinol-1-ribonucleoside (DHPP-Rib) which accounted for 50 - 60% of the 50 mg/kg oral dose of HPP-Rib. This extensive oxidative metabolism in vivo is thought to be due largely to the action of aldehyde oxidase and has been reported to occur with the rabbit liver enzyme in vitro[6]. Thus, both mammalian cells and to a small extent, leishmaniae oxidize HPP-Rib without riboside cleavage.

Human purine nucleoside phosphorylase catalyses the formation of HPP-Rib from HPP[7]. With the calf thymus enzyme, inosine (at 0.3 mM) was split rapidly (74 μmol/min/mg) whereas HPP-Rib was cleaved to the base at a 10,000-fold slower rate (0.0052 μmol/min/mg). In complementary studies, human erythrocytes incubated with (6-^{14}C)HPP-Rib at 0.1 mM (with P_i=1mM) formed HPP at a rate of 0.76 nmol/hr/ml packed cells, whereas inosine was split to hypoxanthine at a rate 500-fold faster. The phosphorolytic reaction rate was very slow, so that the equilibrium was in favor of HPP-Rib formation.

Nucleotide Metabolites - HPP had been shown previously to be a precursor for large amounts of HPP-Rib-5'-P and 4-APP ribonucleotides in *Leishmania*[4]. Experiments with (6-^{14}C)HPP-Rib gave very similar results with *L. donovani*, *L. braziliensis* and *L. mexicana*. In Table 2, the data from a time course of incorporation in *L. braziliensis* are shown. The level of HPP-Rib-5'-P had steadily increased and by 24 hrs exceeded ATP by 4-fold. When *L. donovani* was examined, levels of the pyrazolopyrimidine ribonucleotides were similar to these at 24 hrs., but appeared at about a three times faster rate during the first four hours.

Table 2. Nucleotide Metabolites in *L. braziliensis* after Incubation with [6-^{14}C]HPP-Rib.

	Intracellular Conc., pmol/10^6cells		
	Control	+(6-^{14}C)HPP-Rib, 5 μg/ml	
		4 hrs	24 hrs
ADP	33	52	49
GDP	9	36	30
CTP	3.4	6	1.3
UTP	14	36	17
ATP	54	119	77
GTP	17	41	19
HPP-Rib-5'-P	-	131	291
APP-Rib-5'-P	-	0.7	3
APP-Rib-5'-DP	-	0.5	2
APP-Rib-5'-TP	-	0.5	1
RNA, dpm/mg	-	0	1225

Cells were incubated with (6-^{14}C) HPP-Rib, 11622 dpm/nmol, for 4 hrs with 66 x 10^6 cells/ml, and 24 hrs with 6.5 x 10^6 cells/ml. The results are from single experiments using methods previously reported[3].

In view of the stability of HPP-Rib to splitting, its greater inhibitory activity than HPP and non-reversal by adenine, it seemed possible that HPP-Rib was phosphorylated directly to the nucleotide. Studies were undertaken in L. donovani with (U-^{14}C-ribose)HPP-Rib to investigate this hypothesis. If the riboside had been split to HPP, the resulting ^{14}C-ribose fragment would be expected to enter the pentose pool and label the endogenous purine and pyrimidine nucleotides via PRPP and the HPP-Rib-5'-P formed from HPP would then have a much lower specific activity than the precursor HPP-Rib. The extent of the pool dilution occurring in such a transformation was estimated in a second experiment with (U-^{14}C-ribose)inosine at 9 μM. In this case inosine was completely split in 30 min and the ATP was labeled to an extent of 5.1%. In a parallel experiment 2-fluoroadenine was added as a probe to examine the labeling of the PRPP pool. From the specific activity of the 2-F-ATP, compared with the ^{14}C-inosine precursor, it was possible to calculate that the ^{14}C-ribose fragment suffered a 9-fold decrease in specific activity due to the mixing with endogenous pentose. In the experiment with (U-^{14}C-ribose)HPP-Rib at 14 μM, the specific activity of HPP-Rib-5'-P and the 4-APP ribonucleotides was nearly identical with the precursor, HPP-Rib, and no labeling of purines or pyrimidines occurred. These results strongly support a pathway for the direct phosphorylation of HPP-Rib in the parasite. The enzyme responsible for this has been characterized as a nucleoside phosphotransferase (EC 2.7.1.77)[8] which is distinct from the nucleoside kinases and from hypoxanthine phosphoribosyltransferase (EC 2.4.2.8) which catalyses the conversion of HPP to HPP-Rib-5'-P.

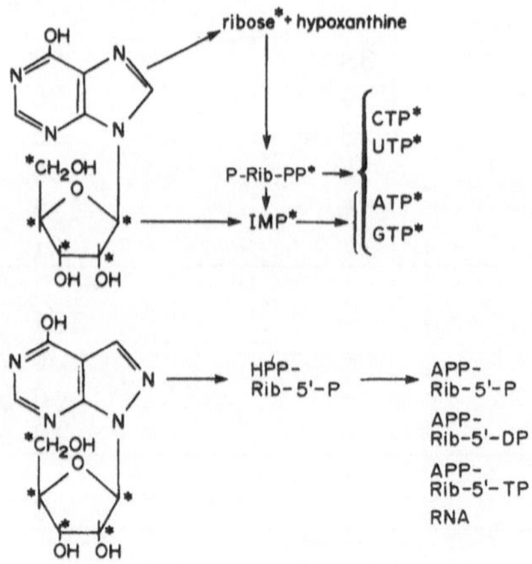

Metabolism of inosine and allopurinol riboside in *Leishmania donovani*

The formation of 4-APP ribonucleotides from HPP-Rib (Table 2) is peculiar to these homoflagellates and probably proceeds via an adenylosuccinate synthetase (EC 6.3.4.4)catalysed amination of HPP-Rib-5'-P. This enzyme in L. donovani appears to have a broader substrate specificity[9] than the corresponding mammalian enzyme, where HPP-Rib-5'-P was not a substrate, but was a weak inhibitor (Ki=1.2 mM)[10]. In previous studies, HPP was converted to micromolar levels of HPP-Rib-5'-P in rat liver, but there was no formation of 4-APP ribonucleotides[11].

The RNA of the parasites incubated with (6-^{14}C)HPP contained radioactivity (Table 2) which, in a sample hydrolysed to the bases, was totally associated with 4-APP; HPP was not present as such in the RNA and the endogenous purines were not labeled. The ratio of adenine to 4-APP in the RNA was 7200:1 in the 24 hr experiment with L. braziliensis, Table 2. In a similar experiment with L. donovani, the RNA had an adenine to 4-APP ratio of 1700:1. Although this is a low level of replacement of adenine by 4-APP in the RNA, it is conceivable that it is sufficient to alter the function of the RNA in these leishmaniae. The possibility of incorporation of pyrazolopyrimidines from ^{14}C-HPP into mammalian RNA was closely examined and found not to occur[12].

SUMMARY

HPP-Rib is a potent antileishmanial agent, which has been useful in defining new and unusual purine metabolizing pathways in leishmaniae. in comparison with those in the host. The ribosyl linkage both in the parasite and in the host is resistant to cleavage. In the parasite there is a selective and marked conversion of HPP-Rib to HPP-Rib-5'-P and 4-APP ribonucleotides as well as incorporation into RNA, which does not occur in the host. These findings with HPP-Rib suggest a new chemotherapeutic approach which may be exploited in the treatment of leishmaniasis.

REFERENCES

1. J. J. Marr, R. L. Berens, and D. J. Nelson, Biochem. Biophys. Acta 544:360 (1978).

2. J. J. Marr and R. L. Berens, J. Inf. Disease 136:724 (1977).

3. G. B. Elion, A. Kovensky and G. H. Hitchings, Biochem. Pharmacol. 15:863 (1966).

4. D. J. Nelson, C. J. L. Bugge, G. B. Elion, R. L. Berens and J. J. Marr, J. Biol. Chem. in press.

5. G. B. Elion, F. M. Benezra, I. Canellas, L. O. Carrington and G. H. Hitchings, Israel J. Chem. 6:787 (1968).

6. T. A. Krenitsky, S. M. Niel, G. B. Elion and G. H. Hitchings, Arch. Biochem. Biophys. 150:585 (1972).

7. T. A. Krenitsky, G. B. Elion, R. A. Strelitz and G. H. Hitchings, J. Biol. Chem. 242:2675 (1967).

8. T. A. Krenitsky, G. W. Koszalka, J. V. Tuttle, D. L. Adamczyk, G. B. Elion and J. J. Marr, These Proceedings.

9. T. Spector and T. Jones, J. Biol. Chem., in press.

10. T. Spector and R. L. Miller, Biochem. Biophys. Acta 445:509 (1976).

11. D. J. Nelson, C. J. L. Bugge, H. C. Krasny and G. B. Elion, Biochem. Pharmacol. 22:2003 (1973).

12. D. J. Nelson and G. B. Elion, Biochem. Pharmacol. 24:1235 (1975).

PURINE METABOLISM IN RAT SKELETAL MUSCLE

E.R. Tully and T.G. Sheehan

Department of Biochemistry
University College
Cork, Ireland

About 40 to 50% of man's weight is contributed by skeletal muscle,[1]. Muscle is a highly differentiated tissue, and as such is metabolically less complex than many other tissues. It has a contractile function in which ATP plays a major role. At rest there is a steady loss of some purines from muscle,[2], while during exercise release of inosine and hypoxanthine in particular, increases markedly,[3]. A mechanism or mechanisms must exist to replace the purine lost from the tissue.

Muscle has high activities of the purine-metabolising enzymes AMP deaminase (AMP aminohydrolase EC 3.5.4.6), 5'-nucleotidase (5'-ribonucleotide phosphohydrolase EC 3.1.3.5) and myokinase (ATP: AMP phosphotransferase EC 2.7.4.3). This suggests that muscle may be active in at least some aspects of purine metabolism.

A recent stimulus to research on this tissue has been the finding of five cases of a new disease who presented with muscular weakness or cramping after exercise. Muscle biopsies were normal when histologically examined, but gave no reaction to the AMP deaminase stain. This was confirmed by quantitative analysis of the homogenized tissue[4].

Although information is accumulating about purine metabolism in myocardium very little is known about this area in skeletal muscle.

The work reported here is an attempt to answer some of the questions concerning the ability of skeletal muscle to synthesize purine nucleotides <u>de novo</u> and/or from purine bases and nucleosides.

METHODS

After preliminary investigations the muscle selected for study was the extensor digitorum longus from the hindlimb of the rat. This is a small, long, very thin muscle, with tendons at both ends whcih facilitate dissection of the intact muscle.

Muscles (25-35 mg) were removed from male Wistar rats (70-90 g), anaesthetised with pentobarbitone. They were incubated in groups of four at 37°C in bicarbonate/saline,[5], containing 5 mM glucose, 1.5 g bovine serum albumen/1 and 0.2% silicine anti-foam emulsion. The medium was bubbled throughout the experiment with O_2/CO_2 (19:1). Muscles were held under 2 g tension, which improves viability over the experimental period.

For the investigation of purine biosynthesis de novo the incubation medium contained either 0.3 mM (1 - 14C) glycine (57 mCi/m mol) or 0.1 mM (1 - 14C) formate (61 mCi/m mol). For studies on the utilisation of purine bases and nucleosides, 20 μM of each labelled potential precursor was used. After 60 min. incubation, muscles were removed, washed quickly, frozen in liquid nitrogen, and extracted with 0.92 M $HClO_4$. After neutralisation, labelled nucleotides were separated by thin layer chromatography on poly (ethyleneimine) - cellulose by the procedure of Crabtree and Henderson ,[6]. Radioactivity was measured by liquid scintillation counting of the nucleotide areas removed from the thin layer plates.

RESULTS AND DISCUSSION

The results obtained with labelled glycine and formate are shown in table 1.

It can be seen that incubation with both labelled glycine and formate resulted in incorporation of label into purine nucleotides. Under the conditions employed, the incorporation with glycine was approximately seven times greater than with formate. With both precursors, adenine nucleotides were predominantly labelled, most of the label being recovered in ATP.

Azaserine, a well known inhibitor of purine nucleotide biosynthesis de novo in other tissues, reduced the rate of both glycine and formate incorporation dramatically. In the presence of this compound the rates were reduced by approximately 90%.

The rates of nucleotide synthesis were greater than those reported in other tissues,[7-8].

The results obtained with labelled purine bases and nucleosides are shown in table 2.

Table 1. Purine Nucleotide Biosynthesis from Glycine and Formate in Rat Skeletal Muscle.

Incorporation rates are expressed as nmol/h per g wet weight. The number of experimental results are shown in brackets after the rate values.

Nucleotide Precursor	Rate of incorporation	Distribution of Radioactivity (% of total radioactivity in nucleotide fraction) Adenine Nucleotides	Guanine Nucleotides	XMP+IMP
(1 - ^{14}C) Glycine	173 ± 12 (5)	71	14	15
(1 - ^{14}C) Formate	25 ± 7 (6)	78	11	11
(1 - ^{14}C) Glycine + 1 m M azaserine	15 (2)	72	16	12
(1 - ^{14}C) Formate + 1 m M azaserine	2.1 (1)	71	17	12

Table 2. Purine Nucleotide Biosynthesis from Bases and Nucleosides in Rat Skeletal Muscle.

Incorporation rates are expressed as nmol/h per g wet weight, and represent the mean ± S.D. of four experiments.

Nucleotide Precursor	Rate of incorporation	Distribution of Radioactivity (% of total radioactivity in nucleotide fractions) Adenine Nucleotides	Guanine Nucleotides	XMP+IMP
(8 - ^{14}C) Adenine	51 ± 8	96	3	1
(8 - ^{14}C) Hypoxanthine	15 ± 4	74	16	10
(8 - ^{14}C) Guanine	4 ± 1	36	54	10
(8 - ^{14}C) Xanthine	0	-	-	-
(1 - ^{14}C) Adenosine	67 ± 3	94	4	2
(1 - ^{14}C) Inosine	15 ± 4	82	12	6
(8 - ^{14}C) Guanosine	6 ± 1	22	72	6

Adenosine showed the greatest rate of incorporation into nucleotides. Despite this, almost 40% of the labelled adenosine was recovered as inosine, and 20% as hypoxanthine ,[9].

Under the conditions employed, deamination at some stage was greater than incorporation into nucleotides. Whether this occurred via adenosine deaminase after uptake of labelled adenosine, or via AMP deaminase after uptake and phosphorylation cannot be determined from the results.

Adenine nucleotides were predominantly labelled by ^{14}C-adenosine, most of the label being recovered in ATP. Of the bases adenine showed the highest rate of incorporation, 96% going into adenine nucleotides, once again mainly into ATP.

Inosine and hypoxanthine had the same incorporation rates, both being markedly lower than those of adenosine and adenine. Radioactivity was recovered principally in the adenine nucleotides.

0.5 mM Guanosine inhibited the formation of labelled hypoxanthine from (^{14}C) inosine by 68%. Under these conditions, nucleotide formation from inosine was inhibited by 52% ,[10]. This suggests that inosine was converted initially to hypoxanthine by nucleoside phosphorylase, and that the hypoxanthine thus produced was used for nucleotide synthesis.

Guanosine and guanine were incorporated into nucleotides at very low rates, most of the label being recovered in guanine nucleotides, chiefly GTP.

Xanthine was not utilized for nucleotide synthesis by rat skeletal muscle.

SUMMARY

1. Rat skeletal muscle can synthesize purine nucleotides de novo.

2. Nucleotide synthesis de novo was inhibited by approximately 90% in the presence of 1 mM azaserine.

3. Purine nucleotides can be synthesized via "salvage" pathways from purine bases and nucleosides. Under the conditions employed incorporation rates were adenosine adenine inosine/hypoxanthine guanosine guanine.

4. Xanthine was not utilized for nucleotide synthesis.

REFERENCES

1. C. Long, (ed.), Biochemists Handbook, 639 (Spon., London 1961)
2. E.L. Bockman, R.M. Berne, and R. Rubio, Am. J. Physiol., 230, 1531-1537 (1976).
3. A.W. Murray, Ann. Rev. Biochem., 40, 811-826 (1971).
4. W.N. Fishbein, V.W. Armbrustmacher, and J.L. Griffin, Science, 200, 545-548 (1978).
5. H.A. Krebs and K. Henseleit, Hoppe-Seylers Z. Physiol. Chem., 210, 33-66 (1932).
6. G.W. Crabtree and J.F. Henderson, Cancer Res., 31, 985-991 (1970).
7. P.C.L. Wong and J.F. Henderson, Biochem.J., 129, 1085-1094 (1972).
8. H.G. Zimmer, C. Trendelenburg, H. Kammermeier and E. Gerlach, Circ. Res., 32, 635-641 (1973).
9. T.G. Sheehan and E.R. Tully, unpublished work.
10. T.G. Sheehan and E.R. Tully, Biochem. Soc. Trans., 6, 1070-1072 (1978).

ALTERATIONS IN PURINE METABOLISM IN CULTURED FIBROBLASTS WITH HGPRT DEFICIENCY AND WITH PRPP SYNTHETASE SUPERACTIVITY

Esther Zoref-Shani and Oded Sperling

Tel-Aviv University Medical School, Department of Chemical Pathology, Tel-Shashomer, and Rogoff-Wellcome Medical Research Institute, Beilinson Medical Center Petah-Tikva, Israel.

Within the last decade, two different inborn enzyme abnormalities were identified as causing excessive purine production in man: deficiency of hypoxanthine-guanine phosphoribosyltransferase(HGPRT) whether partial (1) or complete (2), and superactivity of 5-phosphoribosyl-1-pyrophosphate (PRPP)synthetase, due to feedback-resistance (3-5) or various other molecular alterations. The finding of excessive de novo purine nucleotide synthesis in these enzyme abnormalities has contributed markedly to the understanding of the normal metabolism of purines in man, mainly in revealing the importance of HGPRT activity and of PRPP availability in the regulation of de novo purine synthesis.

In the present study we have investigated the differences in the rate of synthesis and in the metabolic fate of purine nucleotides in cultured fibroblasts obtained from purine overproducing patients, due to the above enzyme abnormalities. The metabolic-kinetic consequences of the enzyme mutations have been characterized in order to verify the postulated mechanisms of purine overproduction in the mutant cells and further improve our knowledge of the normal metabolism of purines in man.

Skin biopsies were obtained from control subjects, from a gouty patient (O.G.) affected with mutant feedback-resistant, physiologically-superactive PRPP synthetase (3-5), from a gouty patient (A.R.) with partial HGPRT deficiency (8) and from a Lesch-Nyhan patient (T.N.) with virtually complete HGPRT deficiency (9). Cultures were obtained and grown as described before (5). Purine synthesis de novo was gauged by the rate of (14-C) formate incorporation into cellular purines and into purines excreted by the cells into the medium (5,10). For the study of the distribution of label among the various purine derivatives, the precipitated purines were extracted in o.4 ml 0.1 n HCl at 100°C for 1 h and chromatographed on microcrystalline cellulose thin layer plates (10). The spots of adenine, hypoxanthine and guanine were identified under UV light, scrapped off and counted. The incorporation of labeled purine bases into cellular purine compounds was measured following 2 h incubation periods (37°C) of 2-3x10^6 cells, in monolayer cultures, in 10 ml of fresh growth medium containing 2.5 μCi of the labeled purine base (10).

The rate of purine synthesis de novo, was found to be accelerated in the mutant cells being in PRPP synthetase superactivity more than 10-fold the normal rate ($P<0.001$), and in the HGPRT deficient fibroblasts approximately 20-fold the normal rate (Table I). The relatively faster rate of de novo IMP synthesis in HGPRT deficiency in comparison to PRPP synthetase superactivity may be taken to reflect the differences in the mechanisms of purine overproduction operating in the two mutant cells.It has been suggested that in PRPP synthetase superactivity, purine synthesis is accelerated because of a

Table I. Incorporation of (14-C)formate into total and intracellular purines and distribution of label among intracellular purine derivatives in culutred fibroblasts from normal subjects and from patients affected with HGPRT deficiency and with superactivity of PRPP synthetase.

Fibroblast source	Incorporation into purines: Total (cpm/mg prot./3.5h)	Incorporation into purines: intracellular (% of total)	Distribution of label among cellular purine derivatives (% of total intracellular): Ad.	Hypox.	Gu.
Normal	13,048 ±7,616(7)	57.9 ±8.6(7)	68.3 ±3.6(19)	5.5 ±2.6(19)	25.9 ±3.4(19)
PRPP synthetase superactivity	156,450 ±47,691(6)	33.9 ±3.17(6)	77.8 ±3.9(10)	6.7 ±3.9(10)	15.4 ±2.7(10)
HGPRT deficiency complete	262,206 ±84,586(6)	69.3 ±5.23(6)	65.9 ±3.7(10)	3.7 ±2.1(10)	30.3 ±3.8(10)
partial	237,281 ±64,981(5)	55.4 ±12.1(6)	68.1 ±5.3(10)	4.3 ±1.8(10)	27.5 ±4.5(10)

single mechanism - the excessive formation of PRPP (3-5). On the other hand, in HGPRT deficiency two separate mechanisms have been suggested to cause the acceleration of purine synthesis (11), increased availability of PRPP, which is accumulating due to decreased consumption, and diminished feedback inhibition on the PRPP amidotransferase due to the decreased reutilization of guanine and hypoxanthine for nucleotide synthesis. In addition, the intensive salvage nucleotide synthesis in the cells with PRPP synthetase superactivity, in comparison to the marked reduction or even absence of this activity in the HGPRT deficient cells, results in the former cells in relatively greater inhibition of the PRPP amidotransferase resulting in a relative deceleration of the rate of purine synthesis de novo.

The mutant cells were found to differ in the rate of excretion of labeled newly de novo produced purines into the culture medium (Table I). In the HGPRT deficient cells, excretion of labeled purines into the incubation medium as measured in the present study, may be taken to represent almost quantatively nucleotide degradation. In contrast, in the normal cells and in the cells with PRPP synthetase superactivity, due to the operation in these cells of the salvage synthesis of nucleotides from hypoxanthine, the excretion of the latter to the medium does not represent accurately nucleotide degradation. The activity of this pathway affects the excretion of labeled hypoxanthine into the medium by two mechanisms. Due to the reutilization of hypoxanthine for nucleotide synthesis, only a proportion of the hypoxanthine deriving from nucleotide degradation will appear in the medium. On the other hand an artificial increase effect may be caused, by the presence in the incubation medium of exogenously supplied nonlabeled hypoxanthine, deriving from the fetal calf serum (12). The exogenous, non labeled hypoxanthine may compete in the cells with the endogenously produced labeled hypoxanthine, on the HGPRT catalyzed conversion of hypoxanthine to IMP, resulting in an artificial increase in the proportion of labeled hypoxanthine excreted into the medium. In the present study, the cells with superactivity of PRPP synthetase excreted into the incubation medium within the time of incubation, the greatest proportion of the newly formed purines (Table I). According to the above considerations, this finding is taken to indicate that in these mutant cells, the amount of IMP formed de novo is markedly exceeding the need for GMP and AMP synthesis and that, in addition, the reutilization of the preformed labeled hypoxanthine for nucleotide synthesis is markedly diminished by the presence of exogenous, non labeled hypoxanthine.

Differences were also found in the distribution of labeling, originating from the labeled newly de novo synthesized IMP, among the various cellular purine derivatives (Table I). In the cells with PRPP synthetase superactivity, the proportion of radioactivity in adenine derivatives was significantly greater than normal ($P<0.001$), and that in guanine derivatives markedly lower than normal ($P<0.001$). In the cells with the virtually complete HGPRT deficiency, the incorporation into guanine was greater than normal ($P<0.005$). The increased channeling of IMP to GMP in the HGPRT deficient cells was interpreted by us previously to reflect a shortage of GMP nucleotides in these cells (10). Nevertheless, no further evidence could be obtained to support this assumption. Furthermore, recent studies in our laboratory (13) suggest that the above differences in the metabolic fate of the newly formed IMP reflect the differences, which exist in these cells, in the rate of total (de novo and salvage) IMP synthesis. Increasing total IMP synthesis by incubating the cells at increasing Pi concentrations, resulted in that the proportion of labeling found in GMP decreased, whereas the proportion of labeling in the excreted purines increased. Accordingly, the higher proportion of labeling found in guanine nucleotides in the HGPRT deficient cells, indicate lower total IMP synthesis in these cells.

The results of the incorporation of labeled purine bases into the cellular purine nucleotides and the rate of interconversions between the various purine nucleotides during 2 h incubation periods are presented in Table II.

TABLE II. Distribution of label among the various cellular purine derivatives following incorporation of (14-C)purine bases in cultured fibroblasts from normal subjects and from patients affected with deficiency of hypoxanthine-guanine phosphoribosyltransferase and with superactivity of phosphoribosylpyrophosphate synthetase

Fibroblast source	Distribution of label among purine derivatives (in % of total)		
	Adenine	Hypoxanthine	Guanine
(14-C)Adenine incorporation			
Normal	88.7 ± 0.79	8.4 ± 1.05	2.7 ± 0.38
PP-rib-P synthetase superactivity	89.7 ± 2.60	7.3 ± 2.60	2.9 ± 0.15
HGPRT deficiency			
- virtually complete	93.4 ± 1.86	2.86± 1.20	3.71± 1.30
- partial	89.9 ± 1.30	5.86± 0.80	4.1 ± 0.64
(14-C)Hypoxanthine incorporation			
Normal	66.0 ± 1.35	4.6 ± 1.15	29.3 ± 1.68
PP-rib-P synthetase superactivity	78.4 ± 1.57	4.9 ± 0.42	16.5 ± 0.99
HGPRT deficiency			
- virtually complete	Too low values		
- partial	64.4 ± 3.18	3.75 ±1.34	31.7 ± 4.52
(14-C)Guanine incorporation			
Normal	0.5 ± 0	0.62± 0.09	98.4 ± 0.07
PP-rib-P synthetase superactivity	2.15± 0.35	0.99± 0.15	96.8 ± 0.28
HGPRT deficiency			
- virtually complete	Too low values		
- partial	1.2 ± 0	1.04 ± 0.51	97.7 ± 0.42

The results of the study of the incorporation of the purine bases into the cellular nucleotides and of the metabolic fate of the labeled nucleotides formed, are compatible with those obtained in previous studies in our laboratory (5) as well as in another laboratory (14). In addition, in many respects these results are also compatible with the results of (14-C)formate incorporation studies. The results clarify also the physiological significance of each of the pathways involved in the interconversion between the various nucleotides as well as of the routes of their degradation. Whereas there is only one known pathway for IMP degradation, through inosine to hypoxanthine, the degradation of AMP and of GMP may occur through two different pathways, through IMP or through the respective nucleosides adenosine and guanosine. The finding of increased proportion of conversion of (14-C)adenine into guanine derivatives in the cells with the virtually complete HGPRT deficiency (Table II), may be taken to indicate that in the cultured fibroblasts, AMP is converted to GMP through IMP, rather than through adenosine. This, since the latter pathway involves also the HGPRT catalyzed synthesis of IMP from hypoxanthine, a reaction which amounts in the cells with the virtually complete HGPRT deficiency to less than 1% of the normal (5). Evidence that indeed AMP degradation to adenosine has a minor role, if at all, in purine metabolism, was presented previously for a number of mammalian cells, including lymphoblasts (15,16). The finding in all cells, of almost no transfer of labeling from guanine nucleotides to IMP and to adenine nucleotides (Table II), clearly indicate that in the cultured human cells, guanine nucleotides are not utilized for synthesis of adenine nucleotides.

REFERENCES.

1. Kelley, W.N., Greene, M.L., Rosenbloom, F.M., Henderson, J.F. and Seegmiller, J.E. (1969). Ann. Intern. Med. 70, 155-206.
2. Seegmiller, J.E., Rosenbloom, F.M. and Kelley, W.N. (1967) Science 155, 1682-1684.
3. Sperling, O., Boer, P., Persky-Brosh, S., Kanarek, E. and de Vries, A. (1972). Europ. J. Clin. Biol. Res. 17, 703-706.
4. Sperling, O., Persky-Brosh, S., Boer, P. and de Vries, A. (1973). Biochem. Med. 7, 389-395.
5. Zoref, E., de Vries, A. and Sperling, O. (1975) J. Clin. Invest. 56, 1093-1099.
6. Becker, M.A., Meyer, L.J., Wood, A.W. and Seegmiller, J.E. (1973). Science (Wash. D.C.) 179, 1123-1126.
7. Becker, M.A., Meyer, L.J., Kostel, P.J. and Seegmiller, J.E. (1974). J. Clin. Invest. 53 4a.
8. Sperling, O., Frank, M., Ophir, R., Liberman, U.A., Adam, A. and de Vries, A. (1970). Europ. J. Clin. and Biol. Res. 15, 942-947.
9. Bashkin, P., Sperling, O., Schmidt, R. and Szeinberg, A. (1973) Israel J. Med. Sciences 9, 1553-1558.

10. Zoref-Shani, E., Sivan, O. and Sperling, O. (1978). BBA 521, 452-458.
11. Kelley, W.N. (1968) Fed. Proc. 27, 1047-1052.
12. Hershfield, M.S., Spector, E.B. and Seegmiller, J.E. (1977) Adv. Exp. Med. Biol. 76A, 303-313.
13. Zoref-Shani, E. and Sperling, O. in preparation.
14. Raivio, K.O. and Seegmiller, J.E. (1973) Biochim. Biophys. Acta 299, 273-282.
15. Brox, I.W. and Henderson, J.F. (1976). Can. J. Biochem. 54, 200-202.
16. Hershfield, M.S. and Seegmiller, J.E. (1977) J. Biol. Chem. 252, 6002-6010.

PURINE METABOLISM IN CULTURED CORONARY ENDOTHELIAL CELLS

S. Nees, A.L. Gerbes, B. Willershausen-Zönnchen and E. Gerlach

Physiologisches Institut der Universität München

Pettenkoferstr. 12, D-8000 München 2

Although it is well known that endothelial cells are involved in several biological processes such as transport[1], hemostasis[2], synthesis of collagen[3], histamine[4] and prostaglandins[5], our knowledge concerning intermediary metabolism of the endothelium is rather limited. In the course of studies on interrelationships between heart function and cardiac metabolism[6,7] we became interested in some features of purine metabolism of coronary endothelial cells. Our interest was initiated by the assumption that these cells might contribute to the production of vasoactive adenosine which is considered to play an important role in the metabolic regulation of coronary blood flow[8,9]. The studies - not possible of course to be performed under in vivo conditions - were carried out on cultured endothelial cells isolated from coronary vessels of guinea pig hearts as recently described[10].

MATERIALS AND METHODS

Culture Medium 199 (Seromed, München) containing penicilline (200 U/ml) and streptomycine (200 µg/ml) was supplemented with fetal calf serum (20%) and L-glutamine (2 mM). Column packings (totally porous silica) for High Pressure Liquid Chromatography (HPLC) were obtained from Macherey & Nagel, Düren. Nucleotides, nucleosides and bases for calibration were purchased from Boehringer Mannheim, all other materials of highest available purity from Merck, Darmstadt.

<u>Preparation of cells and cell culture</u>: Guinea pig hearts were cannulated through the aorta, and their coronary system, washed

free of blood, was filled with an isotonic buffer solution containing collagenase and trypsin (0.1% each). After an exposure of 20 min perfusion was started again and all endothelial cells which had been detached, were collected from the perfusate by centrifugation. Subsequently, the cells were washed with culture medium and seeded in culture dishes. Cultivation was performed at 37°C in a humidified air atmosphere containing 3% CO_2. Depending on the inoculum confluency was reached after 2 to 4 weeks. Contaminations with fibroblasts and smooth muscle cells were usually less than 2%. As judged by electron microscopy the cultivated endothelial cells revealed important morphological criteria of endothelial cells in vivo (clusters of free ribosomes, smooth and rough endoplasmic reticulum, clumps of coarse filaments, fine filaments and prominent microtubules in the cytoplasma).

Analysis of nucleotides, nucleosides and bases: Cultured endothelial cells were extracted with 0.4 N perchloric acid. Quantitation of the different purine compounds in the neutralized cell extracts was carried out by application of specially elaborated HPLC-techniques using weak anion exchange columns for the separation of the nucleotides and reverse phase columns for nucleosides and bases.

Determination of enzyme activities: Specific activities of enzymes involved in nucleotide metabolism were measured in a 20 000 g membrane preparation as well as in a soluble 200 000 g supernatant fraction of endothelial cells. Enzyme tests were performed using standard procedures, substrates and products were separated by HPLC.

RESULTS AND DISCUSSION

In Table 1 mean values from three individual series of analyses concerning contents of purine nucleotides, nucleosides and bases in non-growing confluent endothelial cell cultures are listed. For reasons of comparison respective data for normoxic myocardial tissue are also given. Obviously, endothelial cells contain extraordinarily high amounts of ATP, ADP and AMP. The sum of the adenine nucleotides (ΣATP, ADP, AMP) reaches with more than 15 µmoles/g a value which is about three times higher than the mean adenine nucleotide content of cardiac tissue. Another interesting feature of endothelial cells concerns their high levels of adenine nucleotide degradatives. The contents of adenosine, inosine, adenine and hypoxanthine are about 1 to 2 orders of magnitude higher than the respective values for the myocardium. In contrast to the high levels of adenine nucleotides guanine nucleotides are present in endothelial cells only in small quantities, which are similar to those in myocardial and other tissues.

Table 1: Content of adenine nucleotides and their dephosphorylated degradatives in confluent coronary endothelial cells and in myocardial tissue of guinea pigs. Mean values from three individual series of analyses of 11 culture dishes each.

	Endothelial cells nmoles/g	Myocardium nmoles/g
ATP	11 960	4 280
ADP	2 760	1 050
AMP	630	160
$\sum$'ATP, ADP, AMP	15 350	5 490
Adenosine	87	2
Inosine	100	1.2
Adenine	60	0.5
Hypoxanthine	50	0.9
GTP	233	200
GDP	157	100
GMP	37	23
Guanosine	*)	*)
Guanine	*)	*)

*) not detectable

Additional experiments revealed that growth state of the cultures did not profoundly influence the total content of adenine nucleotides. Furthermore, incubation of confluent cell cultures in purine-free media for three days did not result in any detectable reduction of the adenine nucleotide content. On the other hand, endothelial cells proved to be sensitive to lack of oxygen. It is evident from the data in Fig. 1 that brief periods of anoxic incubation (1.5 and 3 min, respectively) cause a pronounced decrease of ATP with a corresponding increase in ADP and AMP levels. Simultaneously, remarkable amounts of adenosine are formed and released from the cells into the incubation medium.

It appears from all these observations that the extremely high adenine nucleotide levels are very likely a specific feature of cultured endothelial cells. This view is further supported by determinations of activity values of enzymes involved in degradation and synthesis of adenine nucleotides (Table 2). While 5'-nucleotidase activity in endothelial cells exceeds by far that of myocar-

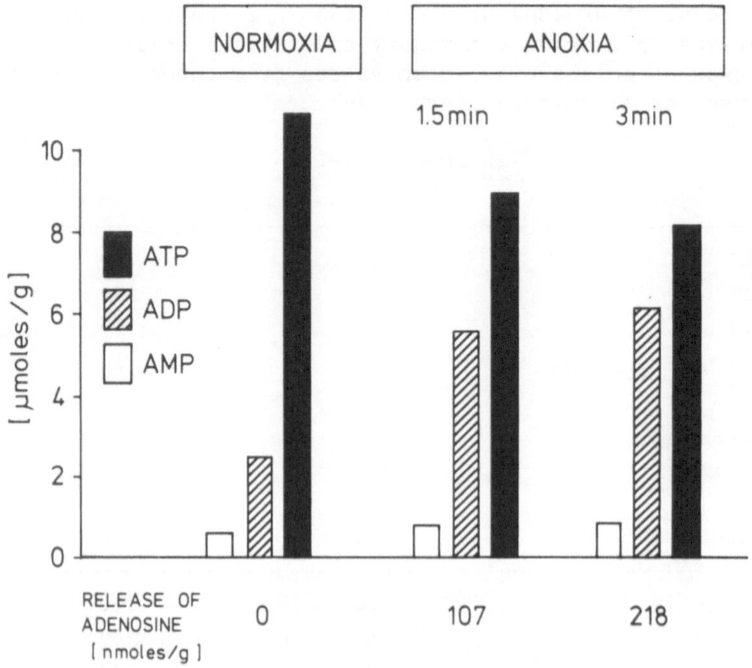

Fig. 1: Influence of anoxia on levels of adenine nucleotides in coronary endothelial cells and on the release of adenosine into the medium (mean values from 3 experiments)

dial tissue, the opposite holds true for adenosine deaminase, the activity of which is much higher in the myocardium. These differences in the pattern of enzyme activities may reasonably explain that endothelial cells contain adenosine in rather high amounts compared with the small quantities of this nucleoside found in the myocardium. As is further evident from the data in Table 2, activities of G-6-PDH and PRPP-synthetase, indirectly involved in the biosynthesis of nucleotides, proved to be much higher in endothelial cells than in cardiac tissue. These findings are in accordance with results from preliminary studies, in which by use of 1-^{14}C-glycine and ^{14}C-labeled purine bases purine nucleotide synthesis in endothelial cells was shown to proceed via salvage and de novo pathways.

SUMMARY

Endothelial cells from coronary vessels of guinea pig hearts were isolated, cultivated and morphologically characterized.- Cells

Table 2: Enzyme activities in cultured coronary endothelial cells and in cardiac tissue from guinea pigs.

	Specific activity [nmoles/min·mg]	
	Endothelial cells	Cardiac tissue
5'-Nucleotidase [E.C. 3.1.3.5]	95	13.6
Alkaline phosphatase [E.C. 3.1.3.1]	14.6	24
AMP deaminase [E.C.3.5.4.6]	1.1	2.2
Adenosine deaminase [E.C. 3.5.4.4]	3.4	28
Glu-6-P-dehydrogenase [E.C. 1.1.1.49]	12.7	4.2
PRPP-synthetase [E.C. 2.7.6.6]	6.58	1.9
APR-transferase [E.C. 2.4.2.7]	0.7	0.2
GPR-transferase [E.C. 2.4.2.8]	0.3	0.1
Adenylate cyclase [E.C. 4.6.1.1]	0.1	0.4
Phosphodiesterase [E.C. 3.1.4.1]	2.1	15

from confluent cultures contained adenine nucleotides and their dephosphorylated degradatives in exceptionally high amounts.- Adenine nucleotide levels were only slightly influenced by the growth state of the cultures and remained stable during incubation for three days in purine-free medium. In contrast, brief incubation of endothelial cells under anoxic conditions resulted in a substantial breakdown of adenine nucleotides associated with an enhanced formation and release of adenosine.- Measurements of specific activities of enzymes involved in adenine nucleotide synthesis and degradation lend additional support to the view that a very active adenine nucleotide metabolism is a typical feature of cultured coronary endothelial cells.

REFERENCES

1) F. Clementi and G. E. Palade, Intestinal capillaries. Permeability to peroxidase and ferritin, J. Cell. Biol. 41:33 (1969)
2) E. A. Jaffe, L. W. Hoyer, and R. L. Nachman, Synthesis of von Willebrand factor by cultured human endothelial cells, Proc. Natl. Acad. Sci. USA 71:1906 (1974)
3) B. V. Howard, E. J. Macarak, D. Gunsen, and N. A. Kefalides, Characterization of the collagen synthesized by endothelial cells in culture, Proc. Natl. Acad. Sci. USA, 73:2361 (1976)
4) T. M. Hollis and L. A. Rosen, Histidine decarboxylase activity of bovine aortic endothelium and intima-media, Proc. Soc. Exp. Biol. Med. 141:978 (1972)
5) M. A. Gimbrone Jr. and R. W. Alexander, Angiotensin II stimulation of prostaglandin production in cultured human vascular endothelium, Science 189:219 (1975)
6) J. Schrader and E. Gerlach, Compartmentation of cardiac adenine nucleotides and formation of adenosine, Pflügers Arch. 367:129 (1976)
7) J. Schrader, S. Nees, and E. Gerlach, Evidence for a cell surface adenosine receptor on coronary myocytes and atrial muscle cells, Pflügers Arch. 369:251 (1977)
8) R. M. Berne, Cardiac nucleotides in hypoxia: a possible role in regulation of coronary blood flow, Am. J. Physiol. 204:317 (1963)
9) E. Gerlach, B. Deuticke and R. H. Dreisbach, Der Nucleotid-Abbau im Herzmuskel bei Sauerstoffmangel und seine mögliche Bedeutung für die Koronardurchblutung, Naturwissenschaften 50:229 (1963)
10) S. Nees, A. L. Gerbes, B. Willershausen-Zönnchen, and E. Gerlach, Isolation, culture and morphologic characterization of endothelial cells from coronary vessels, abstract of the 51st Meeting (Spring Meeting) of the Deutsche Physiologische Gesellschaft, Kiel (1979)

DETERMINANTS OF 5-PHOSPHORIBOSYL-1-PYROPHOSPHATE (PRPP) SYNTHESIS IN HUMAN FIBROBLASTS

Kari O. Raivio, Cheri Lazar, Henry Krumholz and Michael A. Becker

Children's Hospital, University of Helsinki, Finland, and Veterans Administration Hospital, San Diego, California 92161, U.S.A.

The synthesis of PRPP in the intact cell depends on the *in vivo* activity of PRPP synthetase and on the availability of its substrates, ribose-5-phosphate (R5P) and ATP. We have studied the regulation of PRPP synthesis in cultured human fibroblasts, particularly the role of intracellular inorganic phosphate (P_i) as an activator of the enzyme and the supply of R5P for the reaction. The potential pathways of R5P generation include the oxidative and nonoxidative branches of the pentose phosphate shunt (PPS), reutilization of ribose-1-phosphate released in nucleotide catabolism by nucleoside phosphorylase, and phosphorylation of free ribose.

Fibroblasts were cultured in Eagle's MEM using standard techniques. Confluent cultures were trypsinized and the cells were washed and suspended in Krebs-Ringer phosphate buffer containing 5.5 mM glucose (KRPG), and then incubated with shaking at 37°C. The contribution of the PPS to glucose metabolism was calculated on the basis of specific yields of $^{14}CO_2$ from glucose-1- and glucose-6-^{14}C (1). The incorporation of ^{14}C from these same precursors into cellular ATP was measured in the presence of unlabelled adenine and used as an index of carbon flow from glucose via R5P and PRPP into nucleotide. R5P and PRPP were extracted and assayed as previously described (2), and intracellular P_i was measured in cells washed twice with saline and extracted with 10% trichloroacetic acid (3).

In the basal state, the contribution of the PPS to glucose metabolism was calculated to represent 0.8% of glucose utilization. In the presence of 0.1 mM methylene blue (MB), the flux

Table I. EFFECTS OF METHYLENE BLUE (MB) AND 6-AMINONICOTINAMIDE (6-AN) ON PRPP AND RIBOSE 5-P IN FIBROBLASTS

	PRPP		Ribose 5-P	
	nmol/10^6 cells	%	nmol/10^6 cells	%
Control	0.30 ± 0.14 (9)	100	0.25 ± 0.08 (5)	100
MB	0.35 ± 0.18 (9)	117	0.43 ± 0.11 (5)	172
6-AN	0.23 ± 0.13 (9)	77	0.20 ± 0.07 (5)	80

through the oxidative branch increased 10-20-fold. When 6-aminonicotinamide (6-AN), an inhibitor of 6-phosphogluconate dehydrogenase, was added to the culture medium (10 μg/ml) 24 hr before harvesting, the flux through the oxidative PPS decreased to 1/6 of the basal rate. This was accompanied by over 100-fold increase in 6-phosphogluconate concentration in the cells.

The concentrations of PRPP and R5P in fibroblasts, incubated for 30 min in KRPG, are shown in Table I. Neither MB nor 6-AN had a significant effect on PRPP levels, despite the marked variations in the rate of the oxidative PPS. When PRPP generation was assessed on the basis of nucleotide synthesis from 0.1 mM ^{14}C-adenine, the rate under control conditions (1.52 nmol/30 min/10^6 cells) was not significantly altered by MB or 6-AN. R5P levels were clearly elevated by MB and barely affected by 6-AN. It thus appears that PRPP synthesis is not responsive to changes in the rate of the oxidative PPS nor to elevated R5P levels, when incubated in KRPG (16 mM P_i).

The incorporation of label from glucose-1-^{14}C into cellular ATP was clearly higher than from glucose-6-^{14}C under basal conditions of incubation (Fig. 1). When MB was present, the label in ATP from glucose-1-^{14}C decreased, but in cells pretreated with 6-AN the pattern of incorporation was similar to that in the control situation. Taken together with the findings on PRPP and R5P concentrations, the results suggest that the nonoxidative branch of the PPS is capable of supplying all the R5P required for PRPP synthesis, even when the oxidative branch is effectively blocked.

The role of P_i, an important modulator of the activity of purified PRPP synthetase (4), was evaluated in the intact cell by comparing intracellular P_i levels to PRPP concentrations in parallel extracts. Intracellular P_i concentration was dependent on extracellular P_i, although concentrations of 1, 16, and 32 mM in the incubation medium were reflected in much smaller changes intracellularly (6.4, 8.4, and 11.2 nmol per 10^6 cells, respectively). When all the data on parallel P_i and PRPP assays were

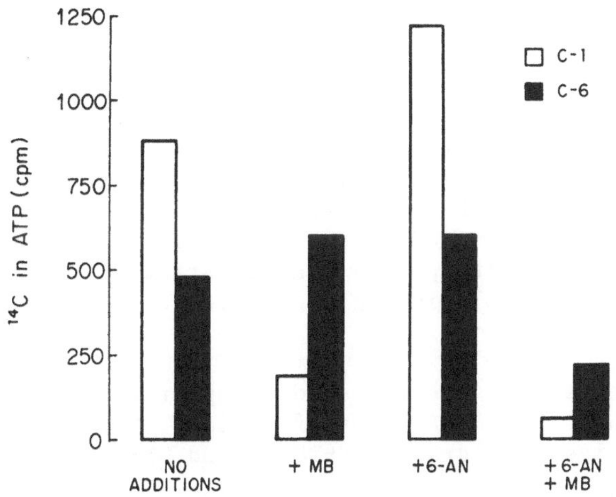

FIGURE 1. Incorporation of radioactivity from glucose-1- and -6-^{14}C into cellular ATP in the presence of 0.1 mM unlabelled adenine.

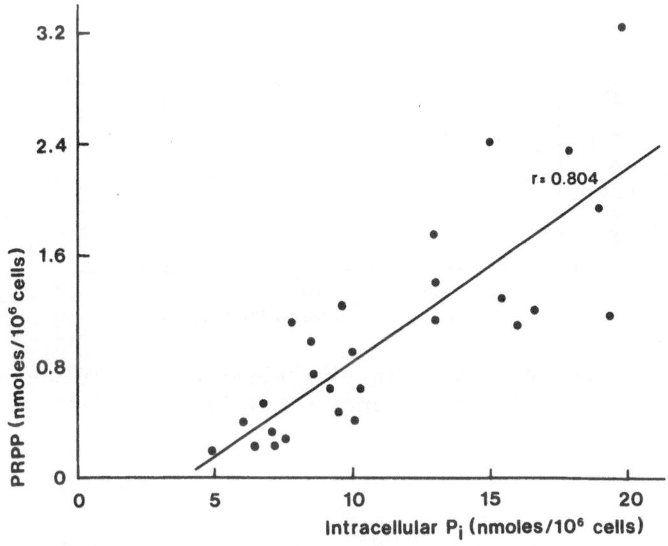

FIGURE 2. Correlation between intracellular phosphate and PRPP concentrations in fibroblasts.

pooled and analyzed, a linear relationship between the two parameters was observed (Fig. 2). This suggests that the activity of PRPP synthetase in the intact cell is regulated by P_i in a similar fashion as the activity of the purified enzyme.

When intracellular P_i is elevated, R5P becomes rate-limiting for PRPP synthesis. This is shown by the finding that elevation of PRPP concentration after MB addition is linearly related to the initial PRPP concentration (Fig. 3), which in turn reflects baseline P_i levels. Purified PRPP synthetase is maximally activated by P_i concentrations that exceed physiological levels by an order of magnitude or more. It appears that substrate limitation does not play a significant role in the regulation of PRPP synthesis in the intact cell.

Acknowledgements: These studies were supported by the Medical Research Council of the Academy of Finland, Veterans Administration Research Service, NIH grants AM-18197 and GM-17702, and the Kroc Foundation.

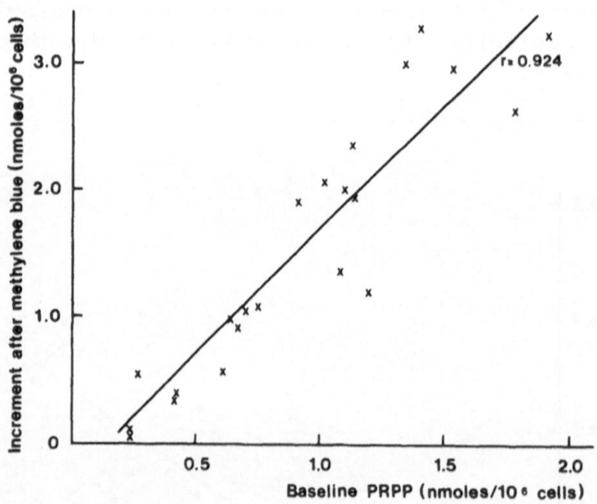

FIGURE 3. PRPP concentration in fibroblasts after incubation with methylene blue as a function of baseline PRPP level.

References:

1. J. Katz and H. G. Wood, J. Biol. Chem. 238:517 (1963).
2. M. A. Becker, Biochim. Biophys. Acta 435:132 (1976).
3. P. S. Chen Jr., T. Y. Turibara and H. Warner, Anal. Chem. 28: 1756 (1956).
4. I. H. Fox and W. N. Kelley, J. Biol. Chem. 246:5739 (1971).

XANTHINE OXIDOREDUCTASE INHIBITION BY NADH AS A REGULATORY FACTOR OF PURINE METABOLISM

Maria M. Jeżewska and Z. W. Kamiński

Institute of Biochemistry and Biophysics
Polish Academy of Sciences
02-532 Warszawa, 36 Rakowiecka St., Poland

INTRODUCTION

Xanthine oxidoreductase occurs in mammalian tissues mostly in a NAD^+-dependent form[1]. In rat liver this dehydrogenase form accounts for over 85 % of total enzyme, the rest being an intermediate dehydrogenase-oxidase form which preferably reacts with NAD^+, but can also use O_2 as an electron acceptor[2]. Both these forms are inhibited by NADH[2], in contrast to the O_2-dependent form[3] arising from them[2]. When catalysed by the O_2-dependent form, two consecutive hydroxylations:

hypoxanthine /Hyp/ ⟶ xanthine /Xan/ ⟶ uric acid /Uri/

proceed, associated with transient accumulation of xanthine[4]. However, the course of this process catalysed by the structurally different[5] NAD^+-dependent form, the activity of which may be influenced by concomitantly formed NADH, has not yet been studied. The cytosolic NAD^+/NADH ratio can be in vivo modified by several factors[6]. It is of interest to examine how its changes could influence the Hyp : Xan : Uri ratio during the Hyp hydroxylation process; namely, changes in Hyp utilization by xanthine oxidoreductase probably may play some role in the regulation of purine-nucleotide biosynthesis by the de novo and the salvage pathways.

In the present study Hyp hydroxylation by the NAD^+-dependent form of the enzyme was examined with respect to the time of reaction, enzymic activity and substrate concentration, under conditions of total reoxidation of NADH or of progressive NADH accumulation.

RESULTS AND DISCUSSION

An enzyme preparation of the stable NAD^+-dependent form, purified 90-fold from 105 000 g supernatant of rat liver homogenate, was used. The course of Hyp hydroxylation was followed with the use of a Cary 118 spectrophotometer with a Repetitive Scan. Absorbance increases were determined at 279 nm /Xan + Uri = Hyp utilization/, at 302 nm /Uri/ and at 340 nm /NADH/ if the accumulation of NADH was not prevented. Molar absorption coefficient 7.58×10^3 $litre \cdot mol^{-1} cm^{-1}$ was used for calculations of the content of Xan and Uri, and a correction for NADH absorbance at 302 nm was made as previously described[2].

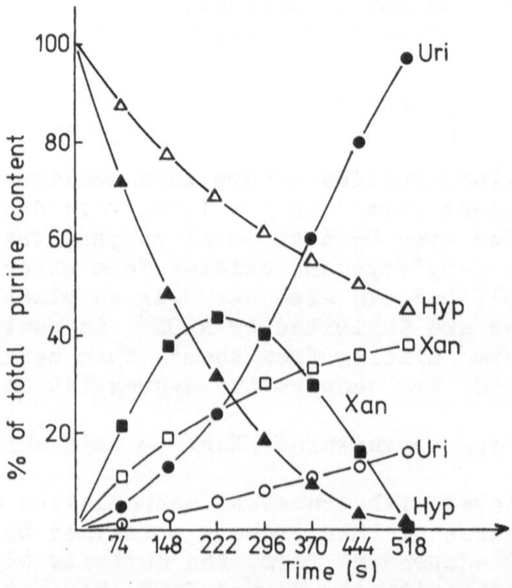

Fig. 1. Time-course of Hyp → Xan → Uri hydroxylation by the NAD^+-dependent form of xanthine oxidoreductase. Incubation mixture: 41 µM Hyp, 350 µM NAD^+, enzymic activity 107 pkat/ml in 50 µM Tris-HCl buffer, pH 8.0, with or without lactate dehydrogenase 40 pkat/ml and 0.5 mM sodium pyruvate.
▲, ■, ● - NADH oxidized, △, □, ○ - NADH accumulating
▲, △ - Hyp, ■, □ - Xan, ●, ○ - Uri.

Time-course of hypoxanthine hydroxylation /Fig. 1/.

When NADH formed was immediately reoxidized, the Hyp disappearence followed a first-order process, Xan initially increased and later dropped, whereas Uri accumulated at a changing rate till total utilization of Hyp. This time-course is identical with that catalysed by the O_2-dependent form of the milk enzyme[4] ; thus, it is not affected by structural changes in the enzyme molecule, associated with the transformation of the NAD^+-dependent to the O_2-dependent form[5] Moreover, the course of Hyp hydroxylation was not changed, when the enzyme was gradually inhibited by accumulating NADH, except that the same stages of this process were attained within a longer time. It seems that under normal physiological conditions xanthine oxidoreductase produces mainly Uri and only small amounts of Xan[7] . It can be seen in Fig. 1 that, whereas Hyp was entirely transformed into Uri in the absence of NADH after 8 min, much more Xan and less Uri was produced, as well as more Hyp was left unreacted during the same time under conditions of the accumulation of NADH.

Changes in enzymic activity and the course of hypoxanthine hydroxylation /Fig. 2/.

Under the experimental conditions used, nearly total transformation of Hyp into Uri, i.e. a state imitating normal physiological state, was achieved with the highest activities of the enzyme and in absence of NADH. If the enzymic activity, still in NADH absence, was reduced by 50 %, the Uri production also decreased with a concomitant rise of the Xan level, but without pronounced changes in Hyp utilization. Only decreases in enzymic activity exceeding 50 % caused simultaneous drops in Xan and Uri production and Hyp utilization. This dependence of the Hyp : Xan : Uri ratio on the enzymic activity resembles closely that obtained for the O_2-dependent form of the milk enzyme[8] . In the case of NADH accumulation, the Xan production remained at a rather constant high level within a broad range of high enzymic activities; instead, Uri formation dropped in proportion to the decrease in Hyp utilization. When low enzymic activities were used, changes in Hyp utilization and in Xan and Uri production proceeded in the same direction as under conditions of total NADH reoxidation.

It is evident that the reduction of the enzymic activity, for instance by allopurinol, could exert various effects on the Hyp : Xan : Uri ratio, depending on the initial level of the enzyme and on the rate of NADH reoxidation.

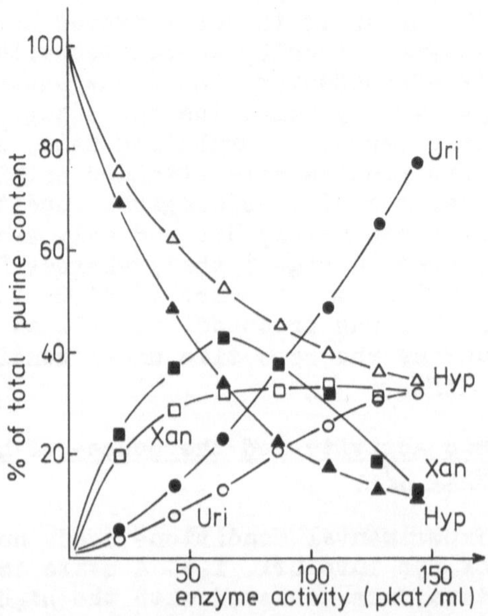

Fig. 2. Effect of the activity of the NAD^+-dependent form of xanthine oxidoreductase on Hyp ⟶ Xan ⟶ Uri hydroxylation.
Experimental conditions and denotations of curves as in Fig. 1. 65 µM Hyp and 175 µM NAD^+ were used.
Reaction time 5 min.

Effect of changes in the level of hypoxanthine on the course of its hydroxylation /Fig. 3/.

For the enzymic activity used, within a certain range of lowest Hyp concentrations, Hyp was hydroxylated entirely to Uri during the reaction time chosen. This range was wider in the absence of NADH and narrower in its presence. Probably, under physiological conditions, when Xan appears in small amounts, the enzyme must act at Hyp concentration slightly above the upper limit of such concentration range. Upon further increase in Hyp concentration, in the absence of NADH, the percentage of Xan in the incubation mixture also rose, and that of Uri initially dropped; subsequently both remained constant and most Hyp was left non-hydroxylated. When NADH accumulated, these changes were parallel, except that the Hyp level was higher and the levels of both other purines - lower. Similar changes in the Hyp : Xan : Uri ratio have been observed for the O_2-dependent form of the milk enzyme[4]; it has been suggested that they may be responsible for the Xan appearence *in vivo* under conditions of enhanced purine catabolism[4].

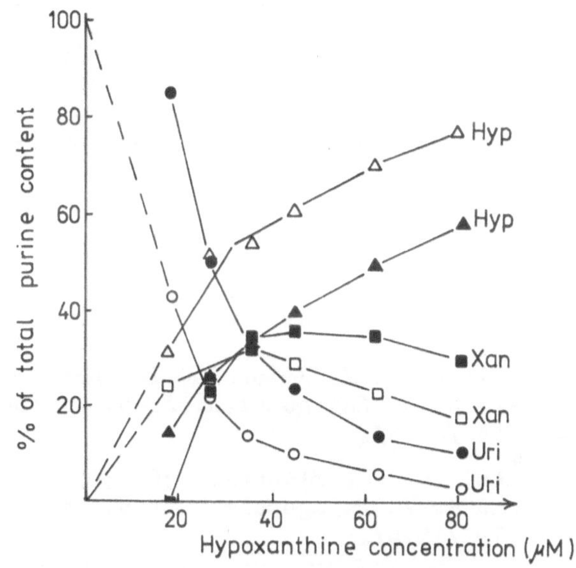

Fig. 3. Effect of substrate concentration on Hyp → Xan → Uri hydroxylation by the NAD⁺-dependent form of xanthine oxidoreductase.
Experimental conditions and denotations of curves as in Fig. 1. 175 µM NAD^+ and enzymic activity 107 pkat/ml were used. Reaction time 5 min.

The present results suggest that if catabolism of purine nucleotide is slow, the supply of Hyp is small, NADH is reoxidized, then Hyp can be nearly entirely converted to Uri. Thus only small amounts of Hyp may be left for the salvage pathway. On the other hand, the reoxidation of 2 NADH formed during transformation of 1 Hyp to 1 Uri can give 6 ATP for the *de novo* biosynthesis of 1 IMP. The more Hyp is supplied, for instance during a transient hypoxia, the less Uri and more Xan are produced. Then, in relation to Hyp utilized, the NADH production decreases and less ATP can be recovered; this recovery is the lower, the slower the reoxidation of NADH during hypoxia. At the same time more Hyp remains non-utilized by the enzyme, increasingly inhibited by accumulating NADH. This Hyp could be transformed into IMP by the salvage pathway, requiring less ATP than the biosynthesis *de novo*. In this manner the action of the NAD^+-dependent form of xanthine oxidoreductase could shift the purine-nucleotide biosynthesis from the *de novo* pathway to the salvage pathway. When the NADH reoxidation is restored during the return to full aerobiosis, accumulated Xan may compete with Hyp for xanthine oxidoreductase and then reduce the utilization of Hyp. Consequently, more Hyp will take part in the salvage pathway, owing to the catalytic properties of xanthine oxidoreductase.

REFERENCES

1. M. G. Batelli, E. Della Corte and F. Stirpe, Biochem. J., 126: 747 - 749 /1972/.
2. Z. W. Kamiński and M. M. Jeżewska, Biochem. J. /1979/ in press.
3. W. R. Waud and K. V. Rajagopalan, Arch. Biochem. Biophys., 172: 354 - 364 /1976/.
4. M. M. Jeżewska, Eur. J. Biochem., 36: 385 - 390 /1973/
5. W. R. Waud and K. V. Rajagopalan, Arch. Biochem. Biophys., 172: 364 - 379 /1976/.
6. K. G. Gumaa, P. McLean and A. L. Greenbaum, Compartamentation in relation to metabolic control in liver, in "Essays in Biochemistry", P. N. Campbell and F. Dickens eds., Academic Press, London, New York /1971/.
7. G. H. Hitchings, Uric acid: chemistry and synthesis, in "Uric Acid", W. N. Kelley and I. M. Weiner eds., Springer-Verlag, Berlin, Heidelberg, New York /1978/.
8. M. M. Jeżewska, Eur. J. Biochem., 46: 361 - 365 /1974/.
9. M. Lalanne and J. Willemot, Int. J. Biochem., 6: 479 - 484 /1975/.

HUMAN PLACENTAL ADENOSINE KINASE: PURIFICATION AND CHARACTERIZATION

Catherine M. Andres, Thomas D. Palella, and Irving H. Fox

Human Purine Research Center
Departments of Internal Medicine and Biological Chemistry, University of Michigan, Ann Arbor, Michigan, 48109

The regulation of adenosine metabolism is an important factor in the determination of the biological properties of this compound[1]. Although the deamination of adenosine has been carefully characterized, only a limited amount of information is available concerning its phosphorylation in mammalian tissue. We have undertaken the purification and characterization of human placental adenosine kinase (E.C.2.7.1.20), which catalyzes the reaction:

Adenosine + ATP ------> AMP + ADP

Adenosine kinase was purified by ion exchange resin batch elution using both DEAE cellulose and CM cellulose and by affinity chromatography on a 5'-AMP-Sepharose 4B column[2,3]. The final purification was 3600-fold with a 24 percent yield. The enzyme was 68 percent pure and had a specific activity of 3.5 μmoles/min/mg.

Physical properties of human placental adenosine kinase were evaluated[2,3]. The subunit molecular weight estimated by SDS polyacrylamide gels is 40,740 daltons; the Stokes radius determined by gel filtration is 26.4 Å with an estimated molecular weight of 37,250. The enzyme thus appears to be a monomer. The $S_{20,W}$ estimated by sucrose gradient ultracentrifugation is 3.50. The isoelectric pH is 5.93. The purified enzyme is indefinitely stable at -70°C, if the enzyme protein has a high concentration of about 1 mg/ml.

The pH optimum is 5.5 under conditions of saturating ATP and adenosine. Studies revealed an absolute requirement for a divalent cation, preferably Mg^{2+}. In comparison to ATP as a phosphate donor, GTP gave 76 percent, deoxy-ATP 42 percent, ITP 38 percent activity.

Deoxyadenosine kinase activity did not copurify with adenosine kinase; in the crude placental supernatant deoxyadenosine kinase activity was 10 percent that of adenosine kinase, and was 0.5 percent in the purified preparation[2,3]. Supernatant deoxyadenosine kinase had a pH optimum of 7.4. No Mg^{++} was required to achieve maximal activity, and high concentrations of Mg^{++} (up to 4.0 mM) did not inhibit the enzyme. It appears most likely that adenosine kinase and deoxyadenosine kinase are distinct enzymes, with adenosine kinase having a small amount of deoxyadenosine phosphorylating activity.

Kinetic studies of adenosine kinase were performed[4,5]. Double reciprocal plots of initial velocity studies were linear and intersecting. The Michaelis constants estimated from replots were 0.4 μM for adenosine and 75 μM for MgATP. Product inhibition studies with AMP and ADP yielded Ki values of 70 μM AMP (competitive with adenosine) and 50-150 μM ADP (non-competitive with ATP). In general, other mono-, di-, and tri-phosphate nucleosides were not inhibitory. Inhibition greater than 50 percent was observed with 100 μM 2'-deoxyadenosine, S-adenosylhomocysteine, deoxymethylthioadenosine, and 6-methylmercaptopurine riboside.

The effects of pH and Mg^{++} were examined simultaneously. Optimum activity was observed at pH 6.5 with Mg:ATP at 2:1, under conditions of saturating ATP; inhibition was noted at higher Mg:ATP ratios at all pH values between 5.5-8.5. A Mg:ATP optimum of 1:1 was noted at pH's other than 6.5.

These data suggest that adenosine kinase may be regulated by: a) Availability of adenosine; b) inhibition of ADP and AMP, but not other nucleotides; and c) relative concentrations of Mg^{++}. In addition, substrate binding may be strongly influenced by pH.

CONCLUSIONS

Human placental adenosine kinase has thus been purified 3600-fold and characterized with respect to molecular weight, substrate specificity, divalent cation requirements, pH optimum, isoelectric pH, and kinetic properties. These data contribute to the information currently available about the regulation of adenosine metabolism, information critical for an understanding of the biological properties of adenosine.

ACKNOWLEDGMENTS

The authors wish to thank Jumana Judeh for the typing of this manuscript. These studies were supported by USPHS grants AM 19674 and 5M01RR42 and a grant from the American Heart Foundation (77-849) and the Michigan Heart Association.

REFERENCES

1. I. H. Fox and W. N. Kelley, The role of adenosine and deoxyadenosine in mammalian cells, Ann. Rev. Biochem. 47:655 (1978).
2. C. M. Andres and I. H. Fox, The phosphorylation of adenosine and deoxyadenosine using purified adenosine kinase, Clin. Res. 26:671A (1978).
3. C. M. Andres and I. H. Fox, Purification and properties of human placental adenosine kinase, (In Preparation).
4. T. D. Palella, C. M. Andres, and I. H. Fox, Regulation of human placental adenosine kinase, Fed. Proc. 38:669 (1979).
5. T. D. Palella, C. M. Andres, and I. H. Fox, Regulation of human placental adenosine kinase, (In Preparation).

LONG-TERM EFFECTS OF RIBOSE ON ADENINE NUCLEOTIDE METABOLISM IN ISOPROTERENOL-STIMULATED HEARTS

H.-G. Zimmer, H. Ibel, G. Steinkopff and H. Koschine

Physiologisches Institut der Universität München
Pettenkoferstr. 12
D-8000 München 2, Germany

INTRODUCTION

It is well established that isoproterenol stimulates cardiac ß-adrenergic receptors[1] and induces a positive inotropic effect which is Ca^{++}-mediated and which eventually leads to a decline of the ATP concentration[2,3]. Isoproterenol also induces an enhancement of the biosynthesis of myocardial adenine nucleotides (AN) in rat hearts[3], which, however, is not of such a magnitude that it has an appreciable effect on the diminution of the ATP concentration. This seems to be mainly due to the fact that cardiac AN biosynthesis is limited by the available pool of 5-phosphoribosyl-1-pyrophosphate (PRPP) which is supplied by the hexose monophosphate shunt[4,5]. Ribose which bypasses the hexose monophosphate shunt has been shown to overcome this limitation thus leading to an elevation of the available PRPP pool and to an increase of the biosynthesis of AN in rat hearts in vivo[4]. On the basis of these findings studies were performed to examine whether ribose may affect the isoproterenol-induced diminution of cardiac ATP concentration when applied as continuous i.v. infusion over a longer period of time.

MATERIAL AND METHODS

Female Sprague-Dawley rats (200-220 g) fed a diet of Altromin were used in all experiments. 1-^{14}C-glycine, 2-^{14}C-glycine and ^{14}C-formate were purchased from Amersham Buchler, Braunschweig. Isoproterenol was obtained from C.H. Boehringer Sohn, Ingelheim. D (-)-Ribose was purchased from Sigma, München. All other chemicals were obtained from MerckAG, Darmstadt and were of analytical reagent grade.

Isoproterenol was dissolved in 0.9% NaCl and administered subcutaneously. Ribose was applied either as single i.v. injection in a dose of 100 mg/kg or as continuous i.v. infusion (200 mg/kg/h) in unanesthetized and unrestrained rats[5]. The concentrations of myocardial ATP were determined after separation by paper chromatography[6]. Rates of the de novo synthesis of cardiac AN were obtained by relating the total radioactivity of myocardial AN to the mean specific activity of the precursor glycine[7].

RESULTS AND DISCUSSION

A first series of experiments was carried out to determine which of the commonly used precursor substances is most suitable for measuring the de novo synthesis of myocardial AN. The data given in Table 1 show that the radioactivity and the rate of biosynthesis of cardiac AN are lower when 1-^{14}C-glycine is used as compared to 2-^{14}C-glycine. By far the highest radioactivity of AN is obtained when ^{14}C-formate is applied as precursor substance. This may be due to the facts that two one-carbon units each from 5,10-methenyl tetrahydrofolate and from 10-formyl tetrahydrofolate contribute to the formation of the purine ring during its de novo synthesis and that the C-2 atom of the purine ring can exchange with ^{14}C-formate[8,9]. Thus the use of ^{14}C-formate as precursor substance does not yield reliable data for the de novo synthesis of AN in this tissue. The increased incorporation of 2-^{14}C-glycine into cardiac AN can be explained by the conversion of 2-^{14}C-glycine to ^{14}C active formate[10,11], which would then result in an additional labeling of the C-2 position of the purine ring.

On the basis of these results only 1-^{14}C-glycine was used as precursor substance for determination of rates of AN biosynthesis in the rat myocardium. As is evident from the data in Table 2, continuous infusion of ribose for 24 hours in normal rats results in an increase of cardiac AN biosynthesis which is of the same order of magnitude as that obtained with a single dose of isoproterenol. The ATP concentration appears to be slightly elevated after 24 hours of ribose infusion. When ribose is constantly applied in isoproterenol-treated animals, cardiac AN biosynthesis turns out to be even more exaggerated. The ATP decline induced by isoproterenol is prevented under these experimental conditions.

Assuming that AN biosynthesis is maintained at the same rate during the entire 24 hour period as measured during the last 60 minutes of the exposure time to isoproterenol and ribose (about 80 nmoles/g/h), then the amount of AN synthesized de novo during this period would be sufficient to balance the isoproterenol-induced loss. It would thus appear that ribose can enhance AN biosynthesis over a longer period of time to such an extent that it can prevent the isoproterenol-induced decrease of the ATP level.

Table 1: Radioactivity of the myocardial tissue extract, mean specific activity of glycine (MSA) as well as radioactivity and rate of biosynthesis of cardiac adenine nucleotides 60 min after i.v. application of 1-^{14}C-glycine, 2-^{14}C-glycine and ^{14}C-formate in rats. In the experiments utilizing ^{14}C-formate as precursor substance, the mean specific activity of formate was not determined, thus the rate of adenine nucleotide biosynthesis could not be calculated.

Precursor substance	n	Radioactivity of tissue extract (DPM/g)	Glycine MSA (DPM/nmole)	Adenine nucleotides: Radioactivity (DPM/g)	Adenine nucleotides: Biosynthesis (nmoles/g/h)
1-^{14}C-glycine	11	64 827	191 073	1 110	5.8
2-^{14}C-glycine	2	103 242	344 435	2 611	7.6
^{14}C-formate	2	124 950	-	67 285	-

Table 2: Effects of isoproterenol (25 mg/kg, s.c.) and ribose on the biosynthesis of adenine nucleotides (AN) and on the ATP concentration in rat hearts in vivo. Ribose was applied either as continuous i.v. infusion (200 mg/kg/h) for 24 hours or as single i.v. injection (100 mg/kg). Mean values ± SEM, number of experiments in parentheses.

	Biosynthesis of cardiac AN (nmoles/g/h)	ATP (µmoles/g)
Control	6.0 ± 0.7 (25)	4.5 ± 0.08 (34)
Ribose infusion for 24 h	18.1 ± 1.3 (4)	4.7 ± 0.08 (4)
Isoproterenol, 24 h	22.5 ± 3.7 (4)	3.2 ± 0.06 (9)
Isoproterenol, 24 h + Ribose infusion	80.1 ± 11.1 (8)	4.6 ± 0.29 (6)
Isoproterenol, 5 h + Ribose injections every hour	103.2 ± 16.2 (6)	4.1 ± 0.13 (14)
Isoproterenol, 5 h + single ribose injection after 4 h	224.0 ± 31.8 (10)	3.4 ± 0.12 (9)

Since ribose has been shown to enhance considerably the available pool of PRPP in the heart[4], it may be assumed that the increase of cardiac AN biosynthesis observed under the influence of isoproterenol and ribose is primarily due to an elevation of the PRPP pool. Additional data also included in Table 2, however, indicate that release of feedback inhibition of PRPP amidotransferase (EC 2.4.2.14), the first and rate-limiting enzyme of de novo purine synthesis, brought about by the ATP decline may also play a role[4,12]. When ribose is i.v. injected five times every hour in isoproterenol-treated rats, ATP concentration reaches an almost normal level and AN biosynthesis turns out to be enhanced to about the same extent as after 24 hours of constant i.v. infusion of ribose. However, when ribose is i.v. injected only once 4 hours after isoproterenol administration, ATP concentration is still markedly reduced, and AN biosynthesis proves to be maximally stimulated. It thus appears that the lower the ATP level, the higher is AN biosynthesis. One may therefore conclude that two mechanisms are involved in bringing about maximal stimulation of AN biosynthesis: elevation of the PRPP pool and release of feedback inhibition of PRPP-amidotransferase activity.

SUMMARY

1. Among the precursor substances tested 1-^{14}C-glycine is the most suitable for measuring rates of AN biosynthesis in rat hearts in vivo.

2. Continuous i.v. infusion of ribose for 24 hours stimulates AN biosynthesis in normal hearts and amplifies the enhancement of this process in hearts of isoproterenol-treated rats. The isoproterenol-induced decline of myocardial ATP is prevented by long-term application of ribose.

3. Elevation of the available PRPP pool mainly mediated by ribose and release of feedback inhibition of PRPP amidotransferase activity by the isoproterenol-induced ATP decline appear to be the optimal conditions required for maximal stimulation of myocardial AN biosynthesis.

This work was supported by the Deutsche Forschungsgemeinschaft (Zi 199/1).

REFERENCES

1. G. A. Robison, R. W. Butcher, I. Øye, H. E. Morgan and E. W. Sutherland, The effect of epinephrine on adenosine 3':5'-

phosphate levels in the isolated perfused rat heart, Mol. Pharmacol. 1:168 (1965)
2. A. Fleckenstein, Specific inhibitors and promoters of Ca^{++} action, in "Calcium and the Heart", P. Harris and L. Opie, eds., pp. 135, Academic Press, London and New York (1971)
3. H.-G. Zimmer and E. Gerlach, Effect of ß-adrenergic stimulation on myocardial adenine nucleotide metabolism, Circulation Res. 35:536 (1974)
4. H.-G. Zimmer and E. Gerlach, Stimulation of myocardial adenine nucleotide biosynthesis by pentoses and pentitols, Pflügers Arch. 376:223 (1978)
5. H.-G. Zimmer and H. Ibel, Effects of isoproterenol and dopamine on the myocardial hexose monophosphate shunt, Experientia 35:510 (1979)
6. E. Gerlach, B. Deuticke, R. H. Dreisbach and C. W. Rosarius, Zum Verhalten von Nucleotiden und ihren dephosphorylierten Abbauprodukten in der Niere bei Ischämie und kurzzeitiger postischämischer Wiederdurchblutung, Pflügers Arch. 278:296 (1963)
7. H.-G. Zimmer, C. Trendelenburg, H. Kammermeier and E. Gerlach, De novo synthesis of myocardial adenine nucleotides in the rat: Acceleration during recovery from oxygen deficiency. Circulation Res. 32:635 (1973)
8. J. M. Buchanan and M. P. Schulman, Biosynthesis of the purines. III. Reactions of formate and inosinic acid and an effect of the citrovorum factor. J. Biol. Chem. 202:241 (1953)
9. J. F. Henderson and G. A. LePage, Purine biosynthesis de novo in mouse tissues and a mouse tumor. J. Biol. Chem. 234:2364 (1959)
10. T. Sato, H. Kochi, N. Sato and G. Kikuchi, Glycine metabolism by rat liver mitochondria. III. The glycine cleavage and the exchange of carboxyl carbon of glycine with bicarbonate. J. Biol. Chem. 65:77 (1969)
11. T. Yoshida and G. Kikuchi, Significance of the glycine cleavage system in glycine and serine catabolism in avian liver. Arch. Biochem. Biophys. 145:658 (1971)
12. J. F. Henderson and M. K. Y. Khoo, In the mechanism of feedback inhibition of purine biosynthesis de novo in Ehrlich ascites tumor cells in vitro. J. Biol. Chem. 240:3104 (1965)

PURINE SALVAGE ENZYMES IN MAN AND *LEISHMANIA DONOVANI*

Thomas A. Krenitsky, George W. Koszalka, Joel V. Tuttle, David L. Adamczyk, Gertrude B. Elion and J. Joseph Marr†
Wellcome Research Labs., Research Triangle Park, NC and
†St. Louis University School of Medicine, St. Louis, MO USA

A comparison of the enzymes of pathogenic protozoa to those of man is of fundamental importance to the search for much needed chemotherapeutic agents. The enzymes involved in purine salvage are of particular interest because most pathogenic protozoa lack the ability to synthesize purines *de novo* and consequently are obligate salvagers of preformed purines.

Leishmaniasis, along with malaria and trypanosomasis, is one of the major insect-borne protozoan diseases of man. This report describes the multiplicity, the levels of activity, and some basic properties of the purine-salvaging enzymes in *Leishmania donovani*. Promastigotes were grown and extracts were prepared as previously described[1,2]. Where possible, a comparison is made with the corresponding enzymes in man.

Phosphoribosyltransferases

Phosphoribosyltransfer activities were present at relatively high levels in extracts of promastigotes (Table I). Purification of these activities revealed that three distinct enzymes were involved: one specific for hypoxanthine and guanine (EC 2.4.2.8), another for adenine (EC 2.4.2.7), and a third for xanthine. (J. V. Tuttle and T. A. Krenitsky, manuscript in preparation). This enzyme pattern is reminiscent of that described for *Lactobacillus casei* but differs markedly from that for *Escherichia coli*[4]. Man and leishmania appear to differ qualitatively with respect to their phosphoribosyltransferases in that no separate xanthine phosphoribosyltransferase is known to exist in man. The levels of the hypoxanthine-guanine and adenine phosphoribosyltransferases in the promastigotes were much higher than those in human tissues[5].

Table I

Phosphoribosyltransfer Activities of Leishmania donovani

	nmol/min/mg protein[a]
Guanine	150
Hypoxanthine	41
Adenine	63
Xanthine	18

[a]Activities were determined as previously described[3] except that guanine was 0.04 mM.

Nucleoside-Phosphorylating Enzymes

Adenosine kinase (EC 2.7.1.20) was present in extracts of promastigotes (Table II) at a level considerably lower than that found in monkey tissues[6]. Activities of deoxycytidine kinase (EC 2.7.1.74) and thymidine kinase (EC 2.7.1.21) were not detected. However, when p-nitrophenylphosphate was used as the phosphate donor instead of ATP, thymidine was phosphorylated (Table II).

Inosine was rapidly hydrolyzed by the promastigote extracts (Table III) and so its ability to be phosphorylated could not be assessed. However, the inosine analogue, 1-ribosylallopurinol, was relatively resistant to cleavage and therefore could be tested as a substrate for phosphorylating enzymes. As with thymidine, phosphorylation of 1-ribosylallopurinol occurred with p-nitrophenylphosphate as the phosphate donor, but not with ATP (Table II). These results indicated that a phosphotransferase rather than a kinase was involved. The product was readily hydrolyzed by 5'-nucleotidase from Crotalus atrox venom but not by 3'-nucleotidase from rye grass, indicating that the enzyme involved was a nucleoside phosphotransferase (EC 2.7.1.77) rather than a nucleotide phosphotransferase. These two types of enzymes are distinguished by their preferred substrates and by the position at which phosphorylation occurs[7]. With nucleoside phosphotransferase, nucleosides are the preferred substrates and the 5'-monophosphate derivatives are the products. With nucleotide phosphotransferase, nucleotides are the preferred substrates and 2'- or 3'-monophosphate derivatives are the products. Phosphotransferase activity has been reported to be present in human tissues[8,9].

Table II

Nucleoside-Phosphorylating Activities of *Leishmania donovani*

	nmol/min/mg protein	
	p-nitrophenyl phosphate	ATP
Adenosine	-	0.49[a]
2'-Deoxycytidine	-	<0.04[b]
2'-Deoxythymidine	0.5[c]	<0.01[b]
1-Ribosylallopurinol	3.1[c]	<0.004[a]

[a]Assay mixtures contained 50 mM NaPipes; 0.1 mM ^{14}C-nucleoside; 1 mM ATP; 0.2 mM $MgCl_2$; and an ATP-regenerating system at pH 6.8 and 30°. Chromatography was performed on Whatman 3MM paper using n-butanol/propionic acid/water (45:23:32 v/v).

[b]Reaction mixtures contained 100 mM Tris, 0.25 mM ^{14}C-nucleoside, 12.5 mM $MgSO_4$, 10 mM ATP, and an ATP-regenerating system at pH 7.4 and 25°. Chromatography was performed on PEI-cellulose plates in water.

[c]Reaction mixtures contained 1.25 mM ^{14}C-1-ribosylallopurinol or 0.46 mM ^{14}C-thymidine, 200 mM sodium acetate and 200 mM p-nitrophenylphosphate at 37° and pH 5.4. Chromatography was performed as in "b".

Nucleoside-Cleaving Enzymes

Some nucleosides were cleaved very rapidly by the promastigote extracts (Table III). Purine 2'-deoxyribonucleosides were cleaved more rapidly than were their corresponding ribonucleosides. Deoxythymidine and deoxycytidine were relatively resistant to cleavage. In contrast, uridine was the most rapidly cleaved ribonucleoside.

Four distinct enzymes capable of catalyzing the cleavage of nucleosides were separated, the details of which will be published elsewhere[2]. Three of these enzymes were nucleoside hydrolases and one was a nucleoside phosphorylase. One of the hydrolases is completely novel in that it has no known counterpart in nature. It is specific for 2'-deoxyribonucleosides of purines and is referred to as purine 2'-deoxyribonucleosidase. In decreasing order of efficiency, it cleaves deoxyinosine, deoxyguanosine, and deoxyadenosine. Another hydrolase is specific for purine ribonucleosides and is similar to purine ribonucleosidase (EC 3.2.2.1). It cleaves guanosine and inosine but not adenosine or xanthosine. The third hydrolase cleaves nucleosides of both purines and pyrimidines but is specific for the ribosyl moiety. In decreasing order of substrate efficiency, it cleaves uridine, xanthosine, cytidine, and inosine. Because it cleaves uridine most effectively, it is referred to as pyrimidine ribonucleosidase (EC 3.2.2.8). The phosphorylase isolated from the promastigotes readily synthesizes nucleoside from adenine and ribose-1-phosphate or 2'-deoxyribose-1-phosphate. This activity has also been recently detected in extracts of *Leishmania tropica*[10].

The only enzyme known to cleave purine nucleosides in human tissues is purine nucleoside phosphorylase, which in contrast to the nucleoside phosphorylase from *L. donovani* does not efficiently utilize adenine or adenosine as a substrate. Nucleoside hydrolases have not been detected in human tissues.

Table III

Nucleoside-Cleaving Activities of *Leishmania donovani*

Base moiety of nucleoside	nmol/min/mg protein[a] 2'-deoxyribonucleoside	ribonucleoside
Hypoxanthine	920	140
Guanine	300	75
Adenine	150	17
Xanthine	-	47
Thymine	< 2	-
Uracil	-	370
Cytosine	< 4	82

[a]Assays were performed as previously described[2].

Deaminating Enzymes

Neither adenosine deaminase (EC 3.5.4.4) or AMP deaminase (EC 3.5.4.6) was detectable (Table IV). However, both adenine and guanine were deaminated. These activities were separated on a Sephacryl S-200 column, indicating the presence of both an adenine deaminase (EC 3.5.4.2) and a guanine deaminase (EC 3.5.4.3) Human tissues contain guanine deaminase; but rather than adenine deaminase, they contain adenosine deaminase and AMP deaminase.

Table IV

Deaminating Activities of Leishmania donovani

	nmol/min/mg protein
Guanine	11[a]
Adenine	25[a]
Adenosine	<0.05[b]
Adenylate	<0.08[a]

[a]Reactions were monitored spectrophotometrically at 25° and pH 7.4 in 100 mM Tris·HCl, 5mM $MgSO_4$ with 0.037 mM guanine ($\Delta\varepsilon$ = 4.23 $mM^{-1}cm^{-1}$ at 246 nm), 0.1 mM adenine ($\Delta\varepsilon$ = 5.7 at 265) or 0.1 mM adenylate ($\Delta\varepsilon$ = 8.45 at 263).

[b]Assayed as previously described[11].

Oxidizing Enzymes

Hypoxanthine was not detectably oxidized (<0.03 nmol/min/mg protein) by the extracts with oxygen or with ferricyanide as the electron acceptor. This indicated the absence of detectable levels of xanthine oxidase (EC 1.2.3.2) or xanthine dehydrogenase (EC 1.2.1.37). Similarly, N^1-methylnicotinamide was not detectably oxidized by the extracts with oxygen as the electron acceptor, indicating that detectable quantities of aldehyde oxidase (EC 1.2.3.1) were not present. Human tissues contain both xanthine oxidase and xanthine dehydrogenase activities, which appear to be catalyzed by different forms of the same enzyme[12]. Human tissues also contain low levels of aldehyde oxidase[13].

The qualitative differences in the enzymes of purine salvage between man and Leishmania donovani can be summarized thusly. Leishmania, but not man, possesses a separate xanthine phosphoribosyltransferase, three nucleoside hydrolases, and an adenine deaminase. The following enzymes found in human tissues were not detectable in Leishmania: deoxycytidine kinase, thymidine kinase, adenosine deaminase, adenylate deaminase, xanthine dehydrogenase/oxidase, and aldehyde oxidase. It remains to be seen if these qualitative enzyme differences can be exploited chemotherapeutically.

REFERENCES

1. Berens, R.L. and Marr, J.J., J. Parasitol. 64, 160 (1978).

2. Koszalka, G.W. and Krenitsky, T.A., J. Biol. Chem. 254, In press (1979).

3. Miller, R.L., Ramsey, G.A., Krenitsky, T.A., and Elion, G.B., Biochemistry 11, 4723-4730 (1972).

4. Krenitsky, T.A., Neil, S.M., and Miller, R.L., J. Biol. Chem. 245, 2605-2611 (1970).

5. Rosenbloom, F.M., Kelley, W.N., Miller, J., Henderson, J.F., and Seegmiller, J.E., J. Am. Med. Assoc. 202, 103-105 (1967).

6. Krenitsky, T.A., Miller, R.L., and Fyfe, J.A., Biochem. Pharmacol. 23, 170-172 (1974).

7. Brunngraber, E.F. and Chargaff, E., Biochemistry 12, 3005-3012 (1973).

8. Brawerman, G. and Chargaff, E., Biochim. Biophys. Acta 15, 549-559 (1954).

9. Brawerman, G. and Chargaff, E., Biochim. Biophys. Acta 16, 524-532 (1955).

10. Konigk, E., Tropenmed. Parasit. 29, 435-438 (1978).

11. Krenitsky, T.A., Tuttle, J.V., Koszalka, G.W., Chen, I.S., Beacham, L.M., Rideout, J.L., and Elion, G.B., J. Biol. Chem. 251, 4055-4061 (1976).

12. Della Corte, E., Gozzetti, G., Novello, F., and Stirpe, F., Biochim. Biophys. Acta 191, 164-166 (1969).

13. Krenitsky, T.A., Tuttle, J.V., Cattau, E.L., and Wang, P., Comp. Biochem. Physiol. 49B, 687-703 (1974).

REGULATION OF PURINE SAVAGE ENZYMES IN *E. COLI*

Roy A. Levine
Milton W. Taylor

Department of Biology
Indiana University
Bloomington, IN 47405

INTRODUCTION

There is no doubt that in eukaryotic cells there is an interplay between the *de novo* purine biosynthetic pathway and purine salvage enzymes, or products of these enzymes. Feedback inhibition of the first specific enzyme of the pathway, PRPP amidotransferase, occurs via AMP or GMP and repression of the pathway is possibly controlled by the product of the salvage enzymes, APRT and HGPRT (Henderson, 1972).

We were interested in examining the interaction of these pathways in *E. coli*, particularly in light of the observations of Hochstadt-Ozer and Stadtman (1971) that the purine phosphoribosyl transferases of *E. coli* are derepressed when *E. coli* is grown in the presence of substrates of these enzymes, and that this derepression can be enhanced 30-40 fold if the purine *de novo* pathway is blocked. We have previously looked for derepression of Chinese hamster cell APRT under similar conditions and did not observe any effect of purine concentration on APRT or HGPRT activity.

MATERIALS AND RESULTS

Overnight cultures of E. coli K12 were grown in the absence of purines, except when a Pur^- strain was used. An exponentially growing culture was diluted into the appropriate media (with supplemental adenine and/or hypoxanthine) and harvested during mid-log phase. Cells were washed twice in 10mM MgS04, sonicated, centrifuged, and dialyzed overnight against 2,000 volumes of tris-EDTA buffer

Table 1. Specific Activities of Phosphoribosyl Transferases in *E. coli* K12 W1485

Addition to Media	HPRT		APRT	
	S.A.	%Control	S.A.	%Control
no purine	26.0	100	17.9	100
10 μg/ml hypoxanthine	19.1	73	16.9	94
100 μg/ml hypoxanthine	20.7	79	15.7	88
10 μg/ml adenine	22.6	87	14.7	82
100 μg/ml adenine	19.1	73	7.6	42

Legend

W1485 was grown in Vogel-Bonner's minimal media supplemented with different concentrations of purines. S.A.=specific activity, nmoles/min/mg protein.

pH 7.6 and assayed for APRT and HPRT specific activity. Similar experiments were done with cells harvested in stationary phase. The specific activity of a culture grown in purine-free media was used as a standard. It is obvious from Table 1 that there was no apparent induction (derepression) of APRT or HPRT at any concentration of purine. On the contrary, there was, if anything, a slight decrease in HPRT and APRT activity in the presence of these bases. The significance of this is not clear. These experiments have been repeated a number of times.

Table 2. Specific Activity of APRT and HPRT Under Conditions of no *De novo* Purine Synthesis

Addition to Media	HPRT		APRT	
	S.A.	%Control	S.A.	%Control
no purine	20.8	100	18.4	100
adenine + hypoxanthine (20 μg/ml each)	18.2	88	17.4	95
purines + thymidine + aminopterin($3x10^{-6}$M)	20.0	96	16.6	90
purines + thymidine + aminopterin($3x10^{-5}$M)	19.7	95	18.4	100

Table 3. APRT and HPRT Activity in Pur⁻mutants of E. coli K12

Strain	Addition to Medium	HPRT		APRT	
		S.A.	%Control	S.A.	%Control
W1485	---------	20.8	100	18.4	100
W1485	20 μg/ml adenine + hypoxanthine	18.2	88	17.5	95
W1485 Pur⁻	20 μg/ml adenine + hypoxanthine	17.5	84	14.6	79
W1485 PurA	40 μg/ml adenine + 20 μg/ml hypoxanthine	23.3	112	23.8	129

Since Hochstadt-Ozer and Stadtman (1971) reported that the highest specific activity of APRT was induced following treatment of the cells with amethopterin to block de novo purine biosynthesis, we attempted to repeat this experiment by blocking de novo purine biosynthesis with the folic acid inhibitor aminopterin, using concentrations of $3x10^{-5}$ M and $3x10^{-6}$ M. As shown in Table 2, even under these conditions, and with the additions of free purine bases there was no induction of APRT or HPRT activity.

A superior way of blocking de novo purine biosynthesis would be the utilization of a mutant strain of E. coli unable to carry out de novo purine biosynthesis. We have utilized a Pur⁻A and a still unidentified Pur⁻ strain of W1485. The strains are totally dependent on an exogenous purine source for growth. Again, as shown in Table 3, there was no effect of exogenous purines on APRT or HPRT activity.

DISCUSSION

We have demonstrated, contrary to other reports that APRT and HPRT are constitutive (non-inducible) genes in E. coli. Similar results have been found in this laboratory with mammalian APRT. To ascertain that we were able to detect enzyme induction under our conditions, we have looked at another related inducible system, adenosine deaminase (Nygaard, 1976). Under our growth conditions we have found a 10.6 fold stimulation of enzyme activity when purine bases were added.

ACKNOWLEDGEMENTS

This work was supported by U.S.P.H.S. grants GM18924 and SO7 RR 7031. We thank Kevin Olson for many of the enzyme assays.

REFERENCES

Henderson, J.F., 1972, Regulations of purine biosynthesis, ACS Monograph, 170, Washington, D.C.

Hochstadt-Ozer, J. and Stadtman, E.R., 1971, J. Biol. Chem., 240: 5294-5303.

Nygaard, P., 1976, Purine Metabolism in Man, Advance in Experimental Medicine and Biology, 76A: 186-195.

PURINE TRANSPORT AND THE CELL CYCLE

M.P. Rivera, M.R. Grau, J. Rigau, A. Goday

Instituto de Farmacología, C.S.I.C.
Jorge Girona Salgado s/n
Barcelona (34). España.

INTRODUCTION

Some of the earliest changes that have been detected following the stimulation of quiescent cells have been changes in the membrane permeability and in the transport of nutrients[1,2]. There is, futhermore, increasing evidence that the growth of some animal cells may be regulated, at least in part, by the availability of essencial nutrients[3].

Since the life cycle is an expression of cell proliferation the study of uptake mechanisms of nutrients into the cell at different stages of the cycle may be one way of understanding the relationship between transport activity and cell growth properties.

The definition of this relationship requires the kinetic determination of initial rates of transport adequately differentiated from intracellular metabolism.

In this study we have established the kinetics of entry of guanine and guanosine into L5178Y cells in different life cycle stages.

METHODS

Murine leukemic lymphoblasts (L5178Y) were harvested from BDF_1 hybrid mice (C57BL/6 DBA_2) 10 days after i.p. tumor inoculation of $2x10^6$ cells.

Transport studies were performed on suspension cultures of L5178Y lymphoblasts incubated in vitro ($2x10^6$ cells/ml) in Fisher medium at 37ºC with increasing concentrations (0.5-100uM) of (8-^{3}H)guanosine (10 Ci/mmol) and (8-^{3}H)guanine (15Ci/mmol).

The incubations were stopped by dilution (10 times) with ice saline (NaCl 0.85%) and centrifugation at 4ºC. The radioactivity incorporated into the whole cells and cold acid soluble (CAS) fraction was determined by liquid scintillation counting. The acid insoluble radioactivity was calculated by differences between whole cell and CAS fraction values.

Cells in different life cycle stages were achieved by chemical synchronization (TdR/colcemid)[4] and volume selection[5].

The purine nucleoside phosphorilase (PNP), with (8-^{3}H) guanosine as substrate, was assayed by a modification of the method of Yamada[6]. The separation of the products of enzyme reaction was carried out by ascending paper chromatography (Whatman 1) with 5% aq. Na_2HPO_4 satd. with isoamyl alcohol, both aq. and nonaq. phases being present in the trough[7].

RESULTS

Uptake of guanosine in L5178Y cells in the exponential growth

The uptake of guanosine was studied over a concentration range from 0.5 to 100 uM. The rate of entry of the nucleoside was linear during the first 2.5 minutes and revealed a kinetics of saturation over a concentration range from 0.5-16 uM (Fig.1).

An incubation time of 60 seconds was chosen to determine the kinetic constants of entry: Vmax= $2.51x10^{-5}$um/10^6cell/min; Km= 3.58 uM.

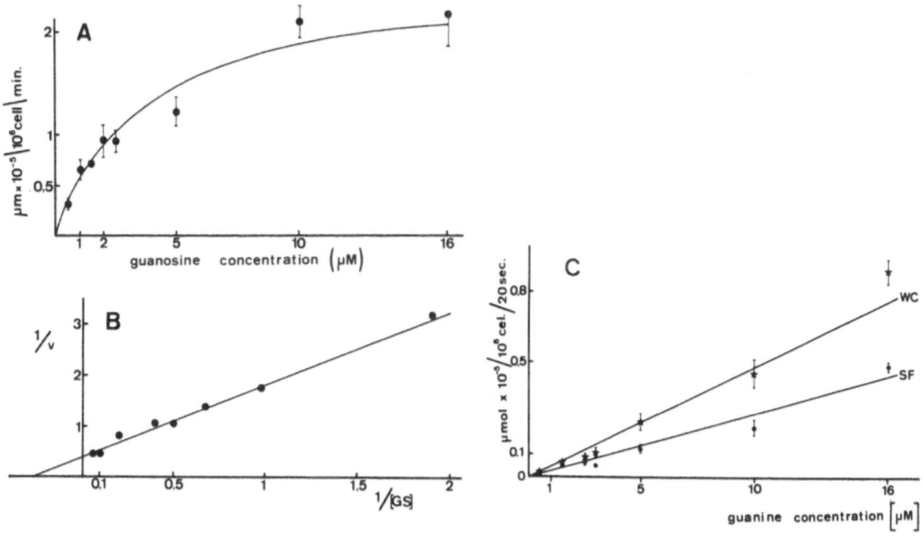

Fig.1: A) Uptake of guanosine by L5178Y cells in exponential growth. B) Reciprocal plot of initial uptake rate of guanosine, Vmax= 2.51×10^{-5}umol/10^6cell/min, Km= 3.58 uM.
C) Entry of guanine into total cell (★) and cold acid soluble fraction (✱).

Practically all the radioactivity taken up by the cell was located in the acid soluble fraction (91%) showing that relatively little, if any, is incorporated in nucleic acids.

The nucleoside incorporated in the acid soluble fraction is metabolized by PNP. The kinetic parameters of this enzyme, determined in cell extracts, were: Vmax= 1.76×10^{-3}umol/10^6cell /min and Km= 23 uM. The Vmax of the enzyme is two orders of magnitude higher than that determined for the process of entry. The PNP shows less affinity for guanosine than for the transport system; in spite of this fact, 95% of the nucleoside is found as nucleotides[8].

All these results indicate that the rate-limiting step of incorporation of guanosine by L5178Y cells is the transport process across the cell membrane.

Uptake of guanine by L5178Y cells in exponential growth

The rate of entry of guanine was linear during 60 seconds. An incubation time of 20 seconds was chosen since during this period the uptake was a linear function for all concentrations tested. The guanine entry was studied over a concentration range from 0.5 to 100 uM (Fig.1). No saturation was found over this concentration range.

The total amount of guanine taken up by whole cell was lower than that of guanosine, when both concentrations in the medium ranged from 0.5 to 10 uM.

The radioactivity incorporated into nucleic acids represents approximately 37% of the total guanine taken up by the cell.

Incorporation of guanosine and guanine in different phases of the cell cycle

L5178Y cells in S phase incorporate guanosine following similar kinetics to that described for cells in exponential growth (Fig.2). The kinetic constants in S phase, Vmax= 2.27x 10^{-5} umol/10^6 cell/min and Km= 6.89 uM, show a slight loss of affinity of the transport system that may be attributed to the method of selection used[9].

In G_1 phase the Vmax is 0.97x10^{-5} umol/10^6 cell/min, which indicates a lower capacity of uptake of the nucleoside for cells in this phase. The Km values were the same for G_1 and S phases.

The PNP activity measured in different stages indicates that the different Vmax values cannot be attributed to the enzyme.

Less than 10% of the total radioactivity taken up by the cell is found as acid insoluble fraction in all the growth stages.

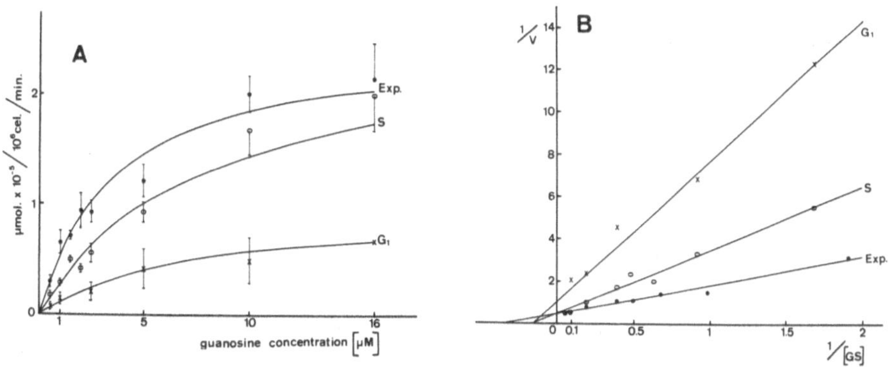

Fig.2: A) Uptake of guanosine by L5178Y cells in: exponential growth (✱), S phase (○) and G_1 phase (✕). B) Double-reciprocal plots of transport rates.

The amount of guanine taken up by the cells in S and G_1 phases is similar to that found in exponential growth. It does not reveal saturation and is distributed equally in acid soluble and insoluble fractions whatever the phase in which the cells are found.

DISCUSSION

The guanosine appears to enter L5178Y cells by a carrier mediated process as is indicated by: (i) the saturability observed over the concentration range studied, (ii) the low level of intracellular nonmetabolized substrate and (iii) the fact that the rate of uptake in the whole cell is lower than the enzyme activity (PNP) measured in cell extracts. This is also an evidence that intracellular metabolism is not rate-limiting step for the incorporation.

We have observed that at 100 uM of guanosine in the medium, the amount incorporated in the cells is greater than that expected by the transport system. This fact suggests the existence, in these cells, of a mechanism of simple diffusion for the nucleoside[10].

The finding that the rate of guanine incorporation increases linearly with the guanine concentration in the medium suggests that the base enters the cells by a simple diffusion mechanism. This idea is also supported by the fact that the entry of guanine does not vary with respect to the phase of the life cycle of the cell.

The higher rate of entry of guanosine in S phase with respect to G_1 phase shows a parallelism between the uptake and DNA synthesis. This difference may be attributed either to a periodical synthesis of the carrier or to a decrease in its activity in G_1 phase. However, we cannot, at this stage, establish the possible causal relationship that may exists between the synthesis of nucleic acids and the carrier activity.

REFERENCES

1- A. Hershko, P. Mamont, R. Shields, G.M. Tomkins; Pleistypic Response, Nature new. Biol., 232:206 (1971).

2- O.H. Petersen; Cell Membrane Permeability Change: An important step in Hormone Action, Experientia 30:1105 (1974).

3- R.W. Holley; Control of growth of mammalian cells in cell cultures, Nature, 258:487 (1975).

4- Y. Doida, S. Okada; Synchronization of L5178Y cells by succesive treatment with excess thymidine and colcemid, Exptl. Cell Res., 48:540 (1967).

5- J. Rigau, M.P. Rivera, M.R. Grau, F.G. Valdecasas; Comportamiento de células leucémicas de ratón (L5178Y) en gradientes de Ficoll, Arch. Farmacol. Toxicol., 4:164 (1978).

6- E.W.Yamada; The Phosphorolysis of Nucleosides by Rabbit Bone Marrow, J. Biol. Chem., 193:497 (1951).

7- C.E. Carter;J. Am. Chem. Soc., 72:1466 (1950).

8- A. Goday, M.R. Grau, I. Jadraque, M.P. Rivera; Purine salvage pathway in leukemic cells, 3th International Symposium on Purine Metabolism in Man, Madrid (1979).

9- M.P. Rivera, J. Rigau. M.R. Grau, F.G. Valdecasas; INfluence of the synchronization method on nucleoside transport, 7th International Congress of Pharmacology, Paris (1978).

10- P.G.W. Plagemann, D.P. Richey; Transport of nucleosides, nucleic acid bases, Cholina and glucose by animal cells in culture, Biochem. Biophys. Acta., 344:263 (1974).

HYPOXANTHINE TRANSPORT IN HUMAN ERYTHROCYTES

Costantino Salerno and Alessandro Giacomello

Istituti di Reumatologia e di Chimica Biologica della Università di Roma e Centro di Biologia Molecolare del C.N.R., Roma, Italia.

De novo synthesis of purines does not appear to take place in mature human erythrocytes because the enzymes for the pathway are absent[1]. Therefore purine bases should be normally have to be supplied exogenously to erythrocytes.

It is well established that hypoxanthine is able to penetrate the human erythrocyte membrane[2]. The uptake of hypoxanthine mediated by its incorporation into IMP is dependent on the activity of hypoxanthine guanine phosphoribosyltransferase (HGPRT) and on intracellular avaibility of phosphoribosylpyrophosphate (PRPP)[3]. PRPP formation in erythrocytes is stimulated by a sufficiently high inorganic phosphate (Pi) level in the suspending medium[4].

In the present paper the conditions governing the uptake and release of hypoxanthine by intact human red blood cells are studied. Human red cells were prepared from blood freshly drawn in heparin and washed with 0.9% NaCl with removal of white cells by aspiration. They were suspended in equal volume of medium containing $5x10^{-2}$M TRIS, pH 7.4, $5x10^{-3}$M glucose, and an appropriate amount of NaCl to give isotonic solution, and incubated at 37°C.

As shown in fig. 1, the rate of uptake of hypoxanthine by erythrocytes was constant over the interval studied, a linear function of Pi concentration up to $1.2x10^{-2}$M, and independent of the hypoxanthine concentration (from $1x10^{-6}$M to $1x10^{-5}$M) in the suspending medium. In the absence of hypoxanthine, PRPP content of erythrocytes increased with increasing Pi concentration[4]. When erythrocytes were preincubated in a medium containing Pi but not hypoxanthine, and transferred to a fresh medium containing ($8-^{14}C$) hypoxanthine but not Pi, the PRPP content of erythrocytes (determined according to ref. 4) decreased and the uptake of the labeled purine was very rapid. The amount of radioactive material incorporated was proportional to the preincubation time at a fixed Pi concentration and to Pi concentration at a fixed preincubation time (fig. 2). Chromatographic analysis according to ref. 5 of the incorporated radioactive material showed that ^{14}C-IMP was the only detectable radioactive compound present in the cells.

After a lag time, a considerable amount of the incorporated radioactive material was released by erythrocytes into the surrounding medium. The radioactive compound released by the erythrocytes was identified as hypoxanthine by paper chromatography . Increasing Pi concentration in the preincubation medium markedly increased the lag time before hypoxanthine release (fig. 3). The lag disappeared and the rate of ^{14}C-hypoxanthine release increased when cold hypoxanthine or guanine was added in the suspending medium.

Studies of the turnover of purine nucleotides in rabbit erythrocytes in vitro[6], led to the conclusion that IMP can be sequentially degraded to inosine and hypoxanthine. Such a mechanism could explain the drop of intracellular ^{14}C-IMP derived from exogenously supplied ^{14}C-hypoxanthine and the increase of ^{14}C-hypoxanthine in the surrounding medium. Since we did not find appreciable amounts of ^{14}C-inosine in either erythrocytes or suspending medium, the rate of

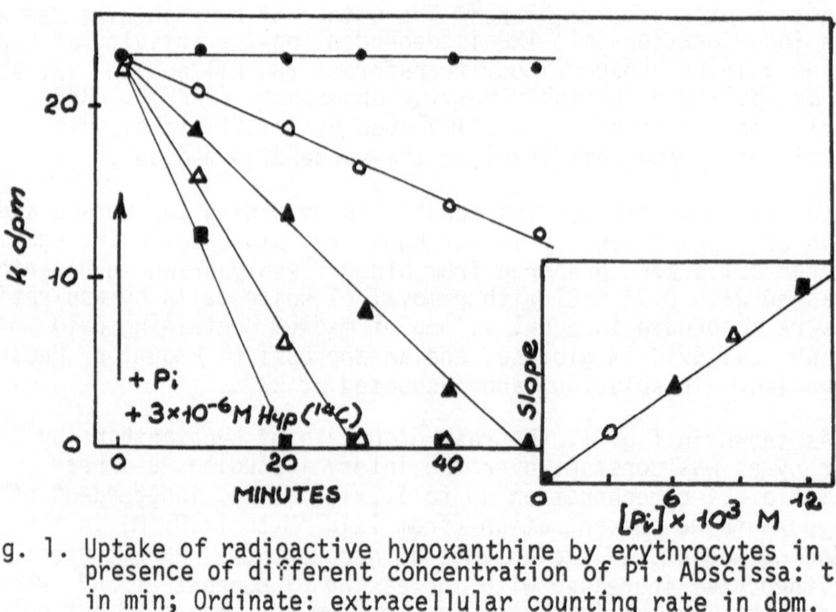

Fig. 1. Uptake of radioactive hypoxanthine by erythrocytes in the presence of different concentration of Pi. Abscissa: time in min; Ordinate: extracellular counting rate in dpm. Inset: rate of uptake of radioactive hypoxanthine as a function of Pi concentration. All other conditions were as described in the text.

phosphorolysis of inosine to hypoxanthine must be much faster than that of hydrolysis of IMP to inosine (rate limiting step in the degradation sequence of IMP). The lag time preceding the release of hypoxanthine could be attributed to reaction between the released hypoxanthine and PRPP to form new IMP. In agreement with this hypothesis we have found that (i) the lag time and cellular PRPP content increased with increasing Pi concentration in the preincubation meyium, (ii) the lag disappeared in the presence of cold purine bases which can react with PRPP, and (iii) no appreciable amount of intermediate metabolites accumulated during the lag time.

In the presence of guanine the rate of ^{14}C-hypoxanthine release increased and was almost equal to that obtained adding cold hypoxanthine to the suspending medium. Guanine concentration in the external medium decreased while the intracellular concentration of GMP increased with an IMP-GMP exchange (data not reported in this paper). Taking into account that HGPRT has been recently reported to catalyze the following reaction[7]

$$\text{IMP + guanine} \rightleftarrows \text{hypoxanthine + GMP}$$

the possibility that this enzyme is involved in the IMP-GMP exchange should be considered.

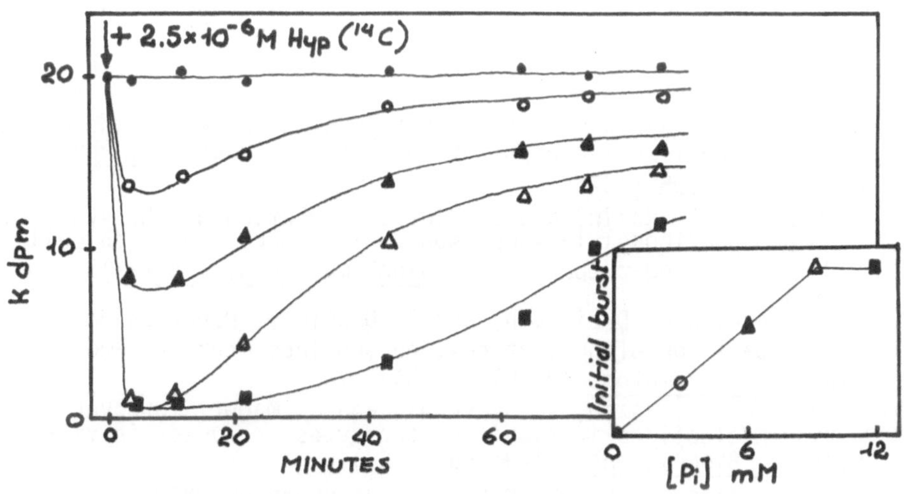

Fig. 2. Uptake and release of radioactive hypoxanthine by erythrocytes preincubated for 40 min in a medium containing Pi but not hyp. and transferred to a fresh medium containing ^{14}C-hyp. but not Pi. Abscissa: time in min; Ordinate: Extracellular counting rate in dpm. All other conditions were as described in the text. Inset: maximal amount of radioactive material incorporated as a function of Pi concentration in the preincubation medium.

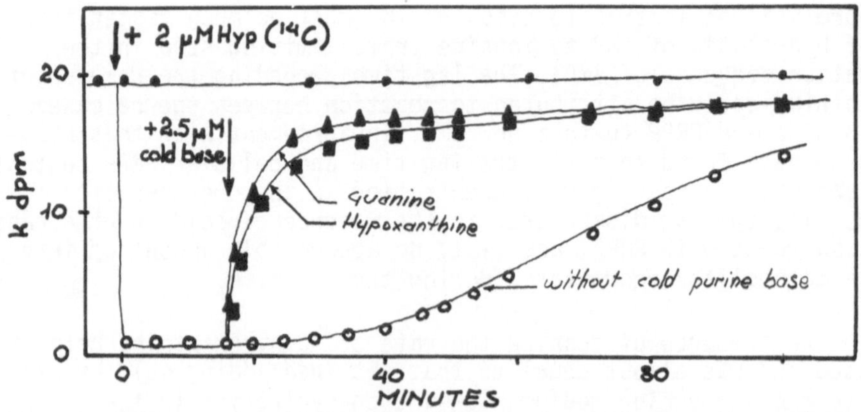

Fig. 3. Effect of cold hypoxanthine and guanine on the release of intracellular radioactive material derived from exogenously supplied ^{14}C-hypoxanthine. All experimental conditions were as described in fig. 2. Pi preincubation concentration was: ● , none; ○ , ▲ , ■ , 9×10^{-3} M.

REFERENCES

1. G.L. Brewer, General red cell metabolism, in "The red blood cell," D. MacN. Surgeror, ed., Academic Press, New York (1974).
2. U.V. Lassen, Hypoxanthine transport in human erythrocytes, Biochim. Biophys. Acta 135: 146 (1967).
3. W. Gutensohn, Hypoxanthine phosphoribosyltransferase and hypoxanthine uptake in human erythrocytes, Hoppe-Seyler's Z. Physiol. Chem. 356: 1105 (1975).
4. A. Hershko, A. Razin, and J. Mager, Regulation of the synthesis of 5-phosphoribosyl-1-pyrophosphate in intact red blood cells and in cell-free preparations, Biochim. Biophys. Acta 184: 64 (1969)
5. E. Gerlach, R.H. Dreisbach, and B. Deuticke, Paper chromatographic separation of nucleotides, nucleosides, purines and pyrimidines, J. Chromatog. 18: 81 (1965).
6. A. Hershko, A. Razin, T. Shoshani, and J. Mager, Turnover of purine nucleotides in rabbit erythrocytes - studies in vitro, Biochim. Biophys. Acta 149: 59 (1967).
7. A. Giacomello and C. Salerno, Role of human hypoxanthine guanine phosphoribosyltransferase in nucleotide interconversion, These Proceedings.

UPTAKE OF ADENOSINE IN HUMAN ERYTHROCYTES

M. Kraupp, P. Chiba and M. M. Müller

Department of Medical Chemistry and

2nd Department of Medicine,

University of Vienna, Austria

INTRODUCTION

The transport of nucleosides across mammalian cell membranes is performed by a transport system with broad specificity for purine and pyrimidine nucleosides. For this uptake process a two component system is discussed; one for low physiological and another for high concentrations of nucleosides (1, 2, 3). In human blood platelets a close association of the high affinity carrier for adenosine with adenosine kinase (AK, E. C. 2.7.1.20) was postulated, indicating phosphorylation being an integral part of adenosine transport as described in bacteria (3, 4). In contrast at high concentrations adenosine transported unchanged into the platelets and is then converted by adenosine deaminase (ADA, E. C. 3.5.4.4) and purine nucleoside phosphorylase (PNP, E. C. 2.4.2.1) into inosine and hypoxanthine.

The present study was undertaken to investigate the uptake of adenosine into human erythrocytes at physiological concentrations and to characterize this process. In addition experiments with and without erythrohydroxynonyladenosine (EHNA), a potent inhibitor of ADA and AK (5), should prove the role of ADA and AK for adenosine uptake.

MATERIALS AND METHODS

Erythrocytes were prepared from heparinized blood by centrifugation, washed 3-times with Krebs Ringer phosphate solution (pH 7.4) and resuspended in Krebs Ringer containing 16.7 mmol/l glucose, resulting in a 4 % (v/v) solution. Incubations were performed in triplicates with 14C-adenosine (59 mCi/mmol) at 10°-$37^{o}C$ for 7 to 120 seconds. In experiments with EHNA the inhibitor was added 30 minutes before the experiment to give a final concentration of 10 umol/l. The intracellular space was determined by incubations with THO and 14C-sucrose. The cells were separated from the incubation medium through a layer of cilicone oil into perchloric acid (10 %, V/V) by centrifugation. Radioactivity of incubation medium and perchloric acid extract was determined separately and the intracellular concentrations od adenosine calculated. Mean values of typical experiments are given in the figures. The experimentsl details of this rapid filtering-centrifugation technique are described elsewhere (6, 7).

RESULTS AND DISCUSSION

Temperature dependence

To measure initial rates of adenosine uptake incubations for 7 seconds were performed. Under these conditions the initial rates were determined as a function of temperature and analyzed by Arrhenius plots. The data (figure 1) show that adenosine uptake exhibit a transition temperature at about $20^{o}C$ indicating alterations of the energy of activation at this temperature by a change of the carrier's conformation. On basis of these experiments at temperatures between $20^{o}C$ and $37^{o}C$ 4.588 kJ/mol whereas at temperatures lower than $20^{o}C$ 15.177 kJ/mol energy of activation is necessary for the transport through the erythrocyte membrane. The addition of EHNA did not alter the temperature dependence of adenosine transport, indicating that ADA and AK are not primarily involved in the transport process. On basis of the mentioned exüeriments all further studies were performed at $18^{o}C$ and at $25^{o}C$.

Time dependence

Figure 2 shows the time dependence of adenosine uptake. The uptake process shows linearity upto 15 seconds of incubation and comes to an end within

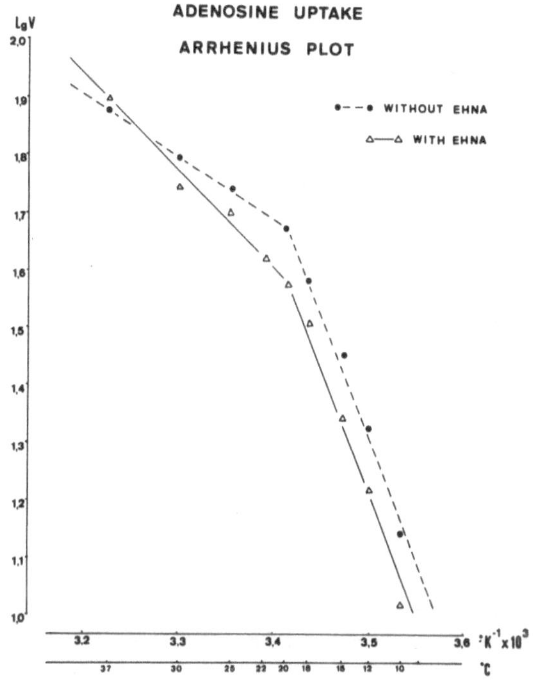

Figure 1. Temperature dependence of adenosine uptake. Incubation time: 7 seconds. Adenosine concentration in the incubation medium: 45 µmol/l.

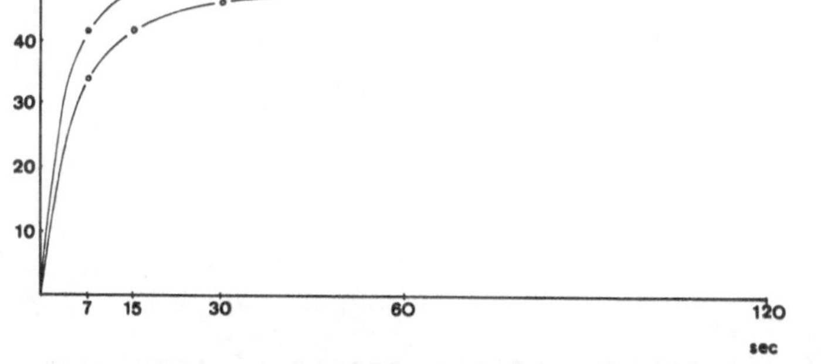

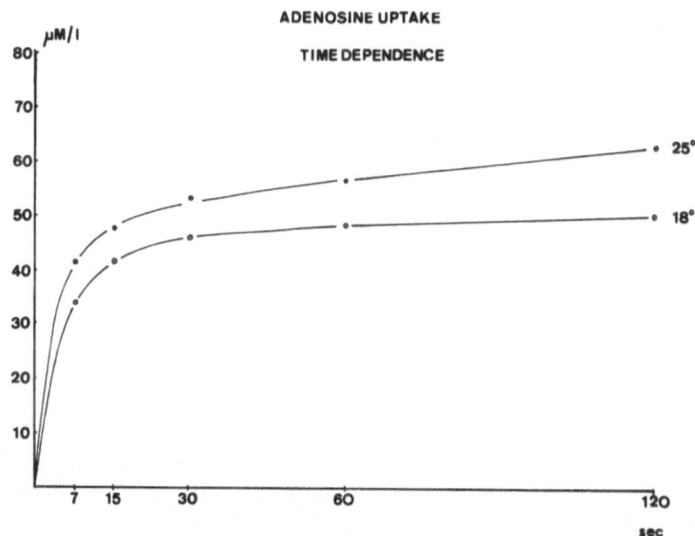

Figure 2. Time dependence of adenosine uptake. Adenosine concentration in the incubation medium: 30 µmol/l.

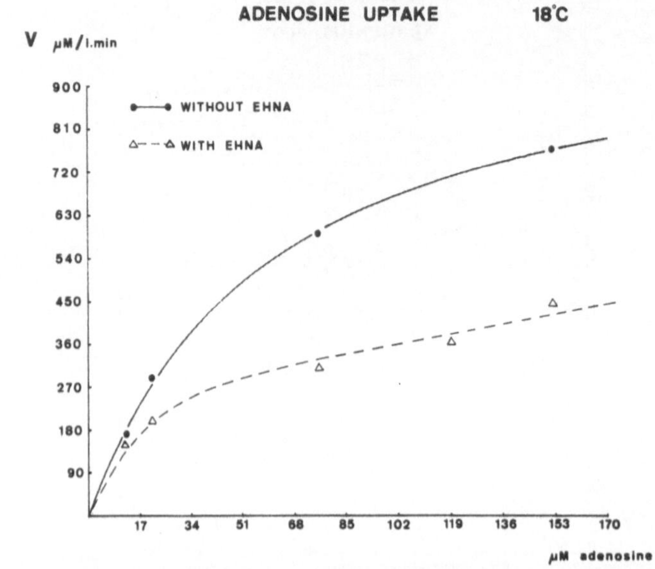

Figure 3. Concentration dependence of adenosine uptake at 18°C. Incubation time: 7 seconds. EHNA concentration in the incubation medium: 10 μmol/l.

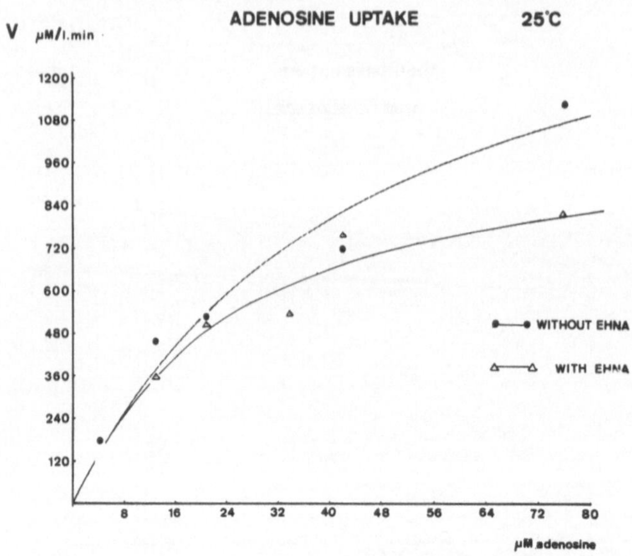

Figure 4. Concentration dependence of adenosine uptake at 25°C. Conditions as in fig. 5.

2 minutes at 18°C. In contrast in experiments at 25°C no saturation could be observed, indicating that intracellular metabolism of adenosine might be active. The uptake was performed against a concentration gradient.

Kinetics

Adenosine uptake at 18°C and 25°C was measured in the presence and absence of EHNA (figures 3, 4) using adenosine concentrations from 10 to 153 µmol/l. Again the preincubation with EHNA resulted in a decreased influx of adenosine but without changing the characteristics of the uptake process. Statistical treatment of the data showed that the transport of adenosine into erythrocytes could be best characterized by a carrier mediated faciliated diffusion (figure 5). The carrier mediated process can be described by a Michaelis-Menten kinetic with apparent K_M-values of 15 µmol/l and 47 µmol/l and apparent V_{max}-values of 755 µmol/l.min. and 1,731 µmol/l.min for incubations at 18°C and 25°C respectively. The fact that the addition of EHNA does not alter the uptake characteristics of adenosine hints that ADA and/or AK are not associated with the adenosine carrier in erythrocyte membranes.

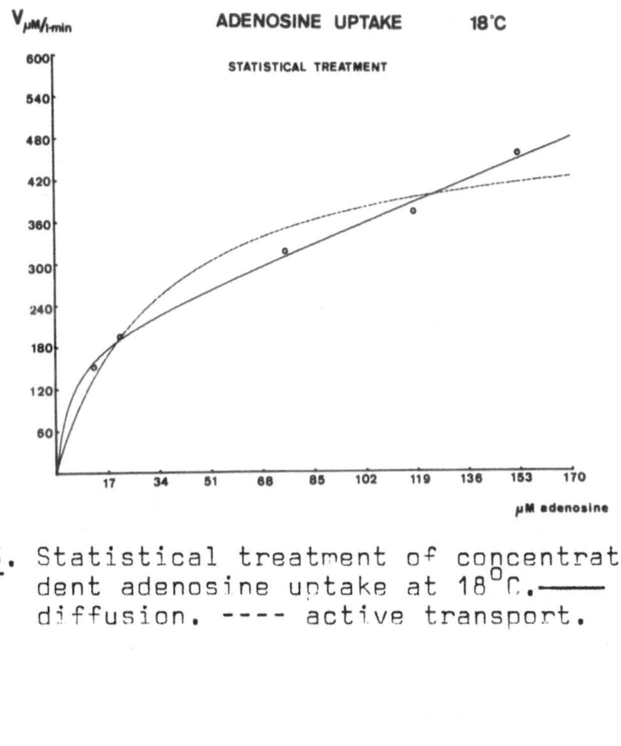

Figure 5. Statistical treatment of concentration dependent adenosine uptake at 18°C. —— faciliated diffusion. ---- active transport.

The data presented suggest that the adenosine uptake into human erythrocytes followes a carrier mediated faciliated diffusion. The adenosine translocator operates very fast and seems to be rate limiting for adenosine metabolism in erythrocytes.

REFERENCES

1. Plagemann P. G. W. (1971)
Biochim. Biophys. Acta 233, 688 - 698.

2. Pearson J. D., Carleton J. S., Hutchings A., Gordon J. L. (1978)
Biochem. J. 170, 265 - 271.

3. Sixma J. J., Lipds J. P. M., Trieschnigg A. M. C., Holmsen H. (1976)
Biochim. Biophys. Acta 443, 33 - 48.

4. Hochstadt-Ozer J. (1972)
J. biol. Chem. 247, 2419 - 2426.

5. Schaeffer H. J., Schwender C. F. (1974)
J. Med. Chem. 17, 6 - 8.

6. Klingenberg M., Pfaff E. (1967)
In Methods of Enzymology, eds. Colowick S. P., Kaplan N. O., Vol. 10, p.680 - 684, Academic Press, New York.

7. Müller M. M., Falkner G. (1977)
Adv. Exp. Med. Biol. 76B, 131 - 138.

EFFECT OF ACTINOMYCIN D ON *IN VIVO* PURINE BIOSYNTHESIS IN HAMSTER CELLS

Milton W. Taylor, Kailash C. Gupta, and Ludmila Zawistowich

Department of Biology
Indiana University
Bloomington, IN 47401

INTRODUCTION

In the course of studying the possible induction of enzymes associated with purine biosynthesis in the Chinese Hamster Cell line, V79, we wished to ascertain whether any newly synthesized short-lived mRNA species, or proteins were required for regulation of the *de novo* pathway. This was done by measuring *in vivo* purine biosynthesis in the presence of an inhibitor of RNA synthesis (Actinomycin D), and inhibitors of protein synthesis (cycloheximide and puromycin). Although we could not, at this stage, find evidence of a need for new mRNA or proteins, a direct effect of both actinomycin D and puromycin on *de novo* purine biosynthesis was noted.

MATERIALS AND METHODS

All experiments were done with V79, a Chinese hamster lung cell line. *De novo* purine biosynthesis was measured by measuring the accumulation of ^{14}C-formate or ^{14}C-glycine labelled formylglycineamide ribotide (FGAR) in the presence of 20 μM azaserine, (Taylor et al. (1978)). Total purine pools, and total *de novo* purine biosynthesis was measured as described by Hershfield and Seegmiller (1977). RNA synthesis and protein synthesis was measured by labelling with ^{3}H-uridine (5 μc/ml) or C^{14} protein hydrolysate (1 μc/ml), respectively.

The effects of inhibitor on FGAR biosynthesis was measured as follows: V79 cells were pregrown in 25 cm^2 flasks in F12 + 5% fetal calf serum until subconfluent monolayers formed. The monolayers were washed three times with phosphate buffered-saline, refed, and resuspended in 2 ml of MEM-"autopow" containing 5% dialyzed fetal calf serum for 2 hours. 20 µm azaserine was then added followed 10 minutes later by 5 µCi C^{14}-glycine or C^{14}-formate and 400 µm glutamine. Cells were incubated in this media and harvested at the required times. FGAR was measured as described previously by high voltage electrophoresis, (Feldman and Taylor 1974).

RESULTS AND DISCUSSION

Initial experiments indicated there was a 70% inhibition of FGAR biosynthesis when cells were incubated in the presence of actinomycin D (1 µg/ml). This suggested that new mRNA, and possibly a newly synthesized, short lived protein were required to maintain in vivo purine biosynthesis. However, when the same measurements were made in the presence of cycloheximide, under conditions in which there was 99% inhibition of protein synthesis, there was no detectable effect of the inhibitor on purine biosynthesis over this time period.

Time course experiments were done comparing the amount of labelled FGAR accumulating in the presence of actinomycin D (1 µg/ml) cycloheximide (10 µg), and puromycin (100 µg/ml) (Fig. 1a, b, c). The effect of actinomycin D was rapid, although the rate of FGAR accumulation continued to be linear with time. Although there was no obvious effect of puromycin on FGAR accumulation up to 60 minutes (Fig. 1c), at later times there was complete inhibition of FGAR accumulation. The kinetics of inhibition in the presence of actinomycin D, and puromycin obviously differ. Under identical conditions there was no effect of cycloheximide (Fig. 1b).

We have examined in more detail the effect of different concentrations of actinomycin D. As can be seen from Table 1, as little as 0.1 µg/ml actinomycin D can lead to 70% inhibition of FGAR biosynthesis. At 5 µg/ml this rises to 85% inhibition. At 0.1 µg/ml there is only 80% inhibition of RNA synthesis (Fig. 2). This level of actinomycin D selectively inhibits the synthesis of ribosomal RNA (Perry & Kelley, 1970). At 5 µg/ml actinomycin D there is 99% inhibition of RNA synthesis.

These data, including the lack of effect of cycloheximide suggest that the effect of actinomycin D is not at the level of mRNA synthesis. Purine biosynthesis appear to be much more sensitive to the inhibitory effects of actinomycin D than mRNA synthesis (5 µg/ml is needed to inhibit mRNA synthesis). This may explain the observation of many that RNA virus synthesis is affected by actinomycin D.

If we eliminate a role for new mRNA synthesis and protein synthesis on FGAR biosynthesis, the following explanation for this effect has to be explored: Degradation of nucleic acid (RNA) in

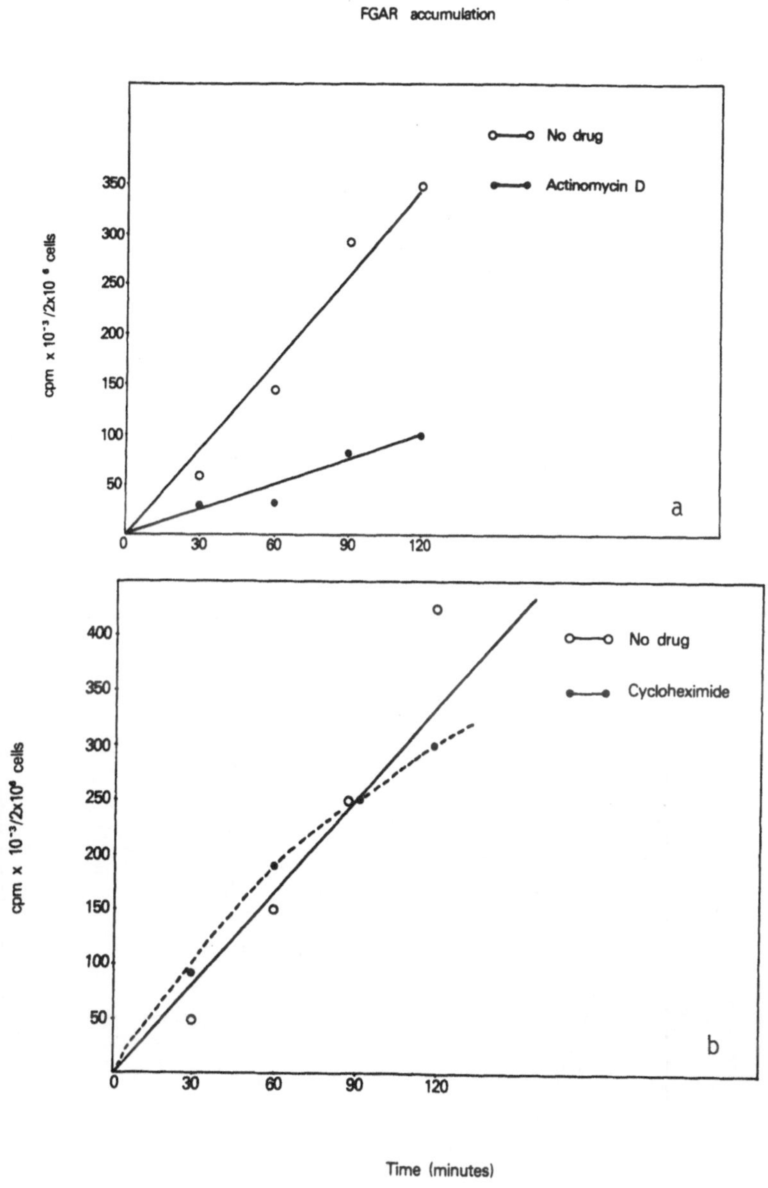

Fig. 1(a),(b). Accumulation of labelled formylglycineamide ribotide (FGAR) with time. (a) 1 μg/ml actinomycin D; (b) 10 μg/ml cycloheximide. 1(c) follows on p. 82.

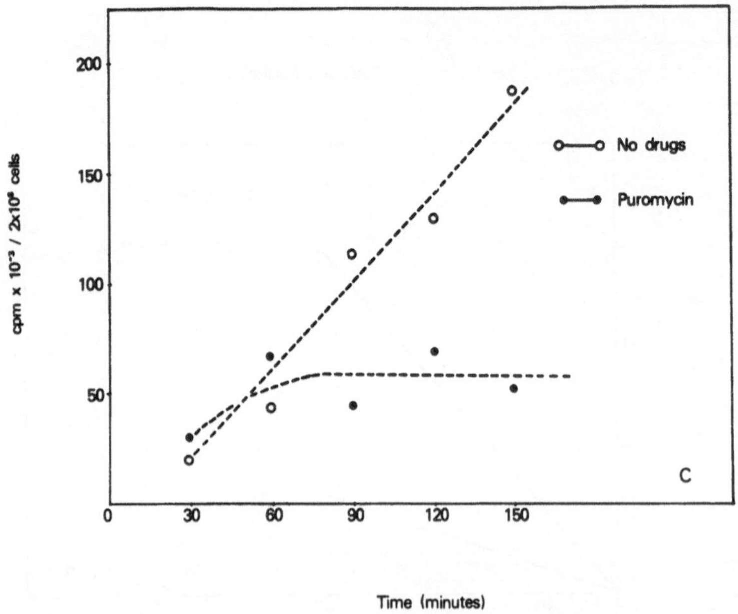

Fig. 1(c). Accumulation of labelled FGAR with time. 100 μg/ml puromycin.

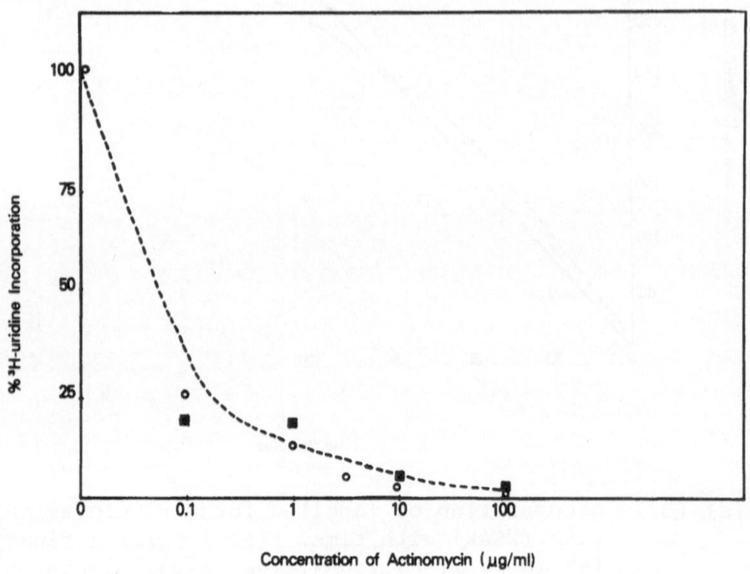

Fig. 2. RNA synthesis in presence of actinomycin D.

Table 1. Percent Inhibition of FGAR Formation After 1 Hour in Presence of Actinomycin D

Act. D Concentration	Cpm/100 μg Protein	%Inhibition
0	6740	-
0.25 μg/ml	2060	70
0.5 μg/ml	2262	66
1 μg/ml	1976	71
5 μg/ml	1056	85

the presence of actinomycin D (or puromycin), leading to feedback inhibition or repression of the biosynthetic pathway.

Alternatively, actinomycin D may also have a direct effect on enzymes involved in _de novo_ purine biosynthesis or substrate. Since puromycin is a structural analogue of adenine, it may be acting by direct feedback inhibition.

If the _de novo_ pathway was undergoing repression in the presence of actinomycin D, this should be manifest in a loss of enzyme activity of the first enzyme of the pathway. We measured the specific activity of PRPP amidotransferase in control and actinomycin D treated cells. The specific activity in the absence of actinomycin D was 55 nmole/mg/hr., and in the presence of actinomycin D was 46.5 nmoles/mg/hr. These results are not considered significantly different. Thus there does not appear to be an effect on the first enzyme of the pathway.

We are currently investigating whether there is any effect on pool sizes, and on _in vitro_ FGAR biosynthesis.

ACKNOWLEDGEMENT. This research was supported by a PHS grant GM 18924.

REFERENCES

Feldman, R.F., and Taylor, M.W., 1974, Purine mutants of mammalian cell lines. I Mutant lacking FGAR amidotransferase activity, _Biochem. Genetics_, 12:393.

Hershfield, M.S., and Seegmiller, J.E., 1977, Regulation of _de novo_ purine synthesis in human lymphoblasts, _Jour. Biol. Chem._, 252:6002.

Perry, R.P., and Kelley, D.E., 1979, Inhibition of RNA synthesis by actinomycin D: Characteristic dose response curve of different RNA species, J. Cell Physiol., 76:127.

Taylor, M.W., Tokito, M.K., Gupta, K.C., and Pipkorn, J., Regulation of purine de novo biosynthesis in Chinese hamster ovary cells: effect of glutamine starvation, Bioch. Acta., 517:1.

PURINE CATABOLISM IN ISOLATED HEPATOCYTES :

INFLUENCE OF COFORMYCIN

Georges Van den Berghe, Françoise Bontemps
and Henri-Géry Hers

Laboratoire de Chimie Physiologique, Université
de Louvain and International Institute of Cellular
and Molecular Pathology, B-1200 Brussels (Belgium)

The nucleoside antibiotic coformycin[1], a potent inhibitor of adenosine deaminase, has also been shown to interact with purified muscle AMP deaminase[2]. In rat liver extracts, maximal inhibition of adenosine deaminase and of AMP deaminase was observed with 10^{-7}M and 5 x 10^{-5}M coformycin respectively. In isolated rat hepatocytes the basal production of allantoin (30-40 nmol/min/g of cells) was not influenced by the addition of 10^{-7}M coformycin, whereas the metabolization of adenosine was strongly inhibited. In hepatocytes in which the adenine nucleotide pool had been prelabelled with ^{14}C adenine, 5 x 10^{-5}M coformycin caused a 85 % inhibition of the basal production of allantoin and a complete suppression of the incorporation of ^{14}C in the end products of purine catabolism.

The addition of fructose (1 mg/ml) to the incubation medium caused a rapid degradation of ATP without an equivalent increase in ADP and AMP. This depletion of the adenine nucleotide pool was accompanied by pronounced transient increases in IMP and in the rate of production of allantoin. In the presence of 5 x 10^{-5}M coformycin, the fructose-induced breakdown of ATP was not modified but no rise in IMP could be detected and the rate of production of allantoin increased only slightly. As a result the depletion of the adenine nucleotide pool proceeded much more slowly. The concentration of AMP, however, increased 14-fold and radioactive adenosine was formed from ^{14}C-labelled adenine nucleotides. In the presence of 10^{-7}M coformycin, the fructose effect was not modified and labelled adenosine was not detected.

These results are in agreement with our hypothesis that the hepatic catabolism of the adenine nucleotides and hence the formation of uric acid and allantoin are controlled by the activity of AMP deaminase[3]. They constitute further evidence that 5'-nucleotidase is inactive on AMP, unless the concentration of this nucleotide rises to unphysiological levels[4].

REFERENCES

1. T. Sawa, Y. Fukagawa, H. Homma, T. Takeuchi and H. Umezawa, Mode of action of coformycin on adenosine deaminase, J. Antibiot., Ser. A. 20:227-231 (1967).
2. R.P. Agarwal and R.E. Parks Jr., Potent inhibition of muscle 5'-AMP deaminase by the nucleoside antibiotics coformycin and deoxycoformycin, Biochem. Pharmacol. 26:663-666 (1977).
3. G. Van den Berghe, M. Bronfman, R. Vanneste and H.G. Hers, The mechanism of adenosine triphosphate depletion in the liver after a load of fructose. A kinetic study of liver adenylate deaminase. Biochem. J. 162:601-609 (1977).
4. G. Van den Berghe, Ch. van Pottelsberghe and H.G. Hers, A kinetic study of the soluble 5'-nucleotidase of rat liver. Biochem. J. 162:611-616 (1977).

Supported by NIH grant AM 9235, the Fonds de la Recherche Scientifique Médicale and the Fonds de Développement Scientifique, Université de Louvain. G. Van den Berghe is Onderzoeksleider of the Belgian Nationaal Fonds voor Wetenschappelijk Onderzoek.

We thank Professor H. Umezawa, Institute of Microbial Chemistry, Tokyo, Japan, for a generous gift of coformycin.

INACTIVATION OF HYPOXANTHINE GUANINE PHOSPHORIBOSYLTRANSFERASE BY GUANOSINE DIALDEHYDE : AN ACTIVE SITE DIRECTED INHIBITOR

L.A. Johnson, R.B. Gordon, B.T. Emmerson

University of Queensland Department of Medicine,
Princess Alexandra Hospital, Ipswich Road,
Brisbane 4102, Australia

Congenital deficiencies of the enzyme HGPRT are invariably associated with metabolic overproduction of purines, which manifest clinically as hyperuricaemia and gout (1). In cases of severe enzyme deficiency (Lesch Nyhan syndrome), these manifestations are accompanied by a bizarre neurological disorder (2). Production of the enzyme deficiency in laboratory animals would be a considerable experimental advantage in studying these disorders. This might be achieved by the use of active-site directed irreversible inhibitors of the enzyme (3). Although a wide range of substrate analogues have been synthesised, most have exhibited poor binding compared to the substrates. A more successful approach has been the modification of the enzyme reaction product GMP. GMP-dialdehyde, the periodate oxidation product of GMP, has been shown to be a specific irreversible inhibitor of HGPRT in cell extracts but was unable to penetrate the intact cell membrane (4, 5). The experiments reported here show that guanosine dialdehyde (GDA), the periodate oxidation product of guanosine, is also a relatively specific but less potent inhibitor of HGPRT and appears to be able to act intracellularly. The specific interaction between GDA and HGPRT is superimposed on non-specific interactions between GDA and proteins which render the inhibitor unsuitable for use in vivo. However, the results suggest that suitable modification of the ribose moiety of guanosine may be a fruitful approach to the development of an in vivo inhibitor.

MATERIALS AND METHODS

GDA was prepared by a modification of published methods (6, 7). The GDA was lyophilised and stored dessicated at -30°C. This material gave a single UV absorbing band on high voltage paper electrophoresis in bisulphite (7) and borate (8) buffer systems.

Table 1. Enzyme inhibitor mixtures were preincubated at 37° in 50 mM NEM pH 7.4. At 10 min and 120 min, aliquots were removed and diluted into enzyme assay mixtures, to give a final protein concentration of 0.1 - 0.3 mg/ml (hemolysate) or 0.01 mg/ml (partially purified enzyme).

THE EFFECTS OF PROTEIN CONCENTRATION AND SUBSTRATES ON THE INACTIVATION OF HGPRT BY GDA

Protein Concentration	Other Additions	% Inactivation by 0.2 mM GDA 10 min	120 min
Erythrocyte hemolysate			
1 mg/ml	none	56	82
30 mg/ml	none	26	44
Partially purified enzyme			
0.5 mg/ml	none	54	90
"	5 mg/ml albumin	32	75
"	0.1 mM G	50	91
"	2 mM PRPP	56	87
"	2 mM PRPP + 5 mM Mg^{++}	<1	<1

Assays using cation-exchange high pressure liquid chromatography revealed the presence of minor peaks of guanosine (0.6%) and guanine (0.2%). ^{3}H-GDA was prepared from (8-^{3}H)-guanosine, which had been purified by high voltage paper electrophoresis. HGPRT was partially purified from human erythrocytes (9). This preparation had a protein concentration of 1 mg/ml and a HGPRT activity of 31 nmoles/min/mg protein, representing a 50-fold purification. In experiments on inactivation, the partially purified enzyme at a concentration of 0.5 mg/ml was preincubated with various concentrations of GDA and then diluted 50-fold into the enzyme assay mixture. HGPRT, APRT and PNP activities were assayed by standard procedures (8).

RESULTS

In preliminary experiments, erythrocyte hemolysate was used as the source of HGPRT. These experiments showed time-dependent inhibition by GDA indicating the occurrence of "irreversible" inactivation. However, high hemolysate protein concentrations also protected the enzyme against inactivation (Table 1). Therefore haemoglobin was removed on CM-Sephadex and the partially purified enzyme used for further studies.

The protective effect of high protein concentrations was also observed with the partially purified enzyme in the presence of bovine

Table 2. Enzyme-inhibitor mixtures contained 0.5 mM inhibitor and 0.05 mg protein/ml of partially purified enzyme in 50 µl 50 mM NEM pH 7.4, 5 mM $MgSO_4$ and were preincubated at 37° for 30 min. The appropriate substrates were then added in a volume of 5 µl and residual enzyme activity assayed. Controls contained no inhibitor. GMA = guanosine monoaldehyde; GTA = guanosine tri-alcohol.

SPECIFICITY OF GDA INHIBITION AND EFFECT OF INHIBITOR MODIFICATIONS

0.5 mM Inhibitor	% Inhibition in 30 min HGPRT	APRT	PNP
GDA	76.3	11.6	3.2
IDA	33.6		
ADA	< 1.0	50.2	
PYR P	97.6	89.2	8.1
GR	14.7		
GMA	13.9		
GTA	< 1.0		
GDA + NH_2OH	19.1		
GDA + $NaHSO_3$	< 1.0		

serum albumin (Table 1). Saturating concentrations of the substrate guanine (G) (or hypoxanthine), had no effect on the inactivation of HGPRT by 0.2 mM GDA and, in the absence of Mg++, neither did 2 mM PRPP. In the presence of 5 mM Mg++ however, 2 mM PRPP completely protected the enzyme from inactivation for at least 2 hours. These results are consistent with proposed ordered reaction mechanisms for HGPRT in which the Mg++ complex of PRPP is the first substrate to add to the enzyme (10, 11).

To determine non-specific binding of GDA to proteins, the hemolsyate and the partially purified enzyme were preincubated with ^{3}H-GDA. Unbound inhibitor was then removed by gel filtration on Sephadex G25 columns. With hemolysate at high protein concentrations (30 mg/ml) considerable binding occurs. However, at lower protein concentrations (0.3 - 0.5 mg/ml) only a small fraction (1-2%) of the ^{3}H-GDA was protein bound. The non-specific binding of GDA to proteins could involve the formation of Schiff bases with lysine amino groups and thioacetals with sulphydryl groups, and probably accounts for the partially protective effect of high protein concentration against inactivation of HGPRT (Table 1).

The relative specificity of GDA inhibition is shown in Table 2. GDA considerably inhibited HGPRT but had little effect on APRT or PNP. Inosine dialdehyde (IDA) was a less effective inhibitor of

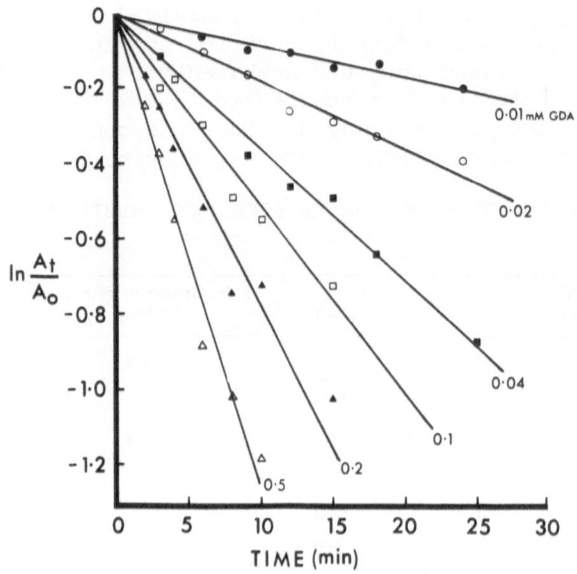

Figure 1. Time dependent "irreversible" inactivation of HGPRT by GDA. Enzyme (0.5 mg protein/ml) was preincubated at 37° with the inhibitor concentrations shown. At intervals 1 µl aliquots were withdrawn and added to 50 µl of enzyme assay mixture. The observed rate constants of inactivation (k_{obs}) were calculated from the slopes for each inhibitor concentration.

HGPRT than GDA. Adenosine dialdehyde (ADA) had no effect on HGPRT but substantially inhibited APRT. Chemical modifications of the aldehyde groups of GDA as outlined in Table 2 either abolished or severely diminished the effectiveness of the inhibitor. These results suggest that the aldehyde groups are important for inactivation, however, the weak inhibition by guanosine (GR) and GMA and the complete lack of inhibition of GTA suggest that a ring structure via internal hemiacetal formation (6) may be important in binding to the enzyme active site.

The inactivation of HGPRT by GDA follows pseudo-first order kinetics as shown in Figure 1. A double reciprocal plot of the apparent rate constants (slopes of the lines) against inhibitor GDA concentration demonstrated a saturation effect and gave estimates of the inhibition constant $K_{i(app)}$ = 0.19 mM, and the limiting rate constant, k_{+2} = 0.18 min^{-1}.

Although the affinity of HGPRT for GDA is considerably lower than its affinity for guanine (K_m = .004 mM) and GMP (K_i = .014) (10) the partially purified enzyme was extensively inactivated by moderate concentrations (0.2 mM) of GDA. The maximum inactivation obtained was 94% with 2 mM GDA after 2 hours at 37°. Much lower concentrations (.01 mM) of GMP-dialdehyde inactivate HGPRT to a similar extent (5). Thus the phosphate group makes a considerable contribution to binding but it also prevents entry into the cell (5).

The ability of the inhibitor GDA to act intracellularly was tested by incubating a 1% suspension of erythrocytes in PBS with 0.2 mM GDA at 37°. The cells were sampled at intervals, washed and lysed and assayed for HGPRT and APRT activities. The APRT activity served as a control for non-specific effects of GDA. The maximum inhibition of HGPRT was 37% in 2 hours. In this time APRT activity declined 7%. However, strong agglutination of the cells was observed.

Thus the relatively poor inhibition of intracellular HGPRT and the cell agglutination make GDA unsatisfactory for in vivo use. These effects are probably due to non-specific interactions resulting from the high chemical reactivity of the aldehyde groups. Substituting less reactive alkylating groups, e.g. halomethylketo, may provide a more effective in vivo inhibitor of HGPRT.

REFERENCES

1. Kelley, W.N., Rosenbloom, F.M., Henderson, J.F. and Seegmiller, J.E. (1967). Proc. Nat. Acad. Sci. (Wash) 57:1735.
2. Lesch, M. and Nyhan, W.L. (1964). Am. J. Med. 36:561-570.
3. Baker, B.R., "Design of Active-Site-Directed Irreversible Enzyme Inhibitors. The Organic Chemistry of the Active Site". John Wiley and Sons, New York, 1967.
4. Gutensohn, W. and Huber, M. (1975). Hoppe-Seylers Z. Physiol. Chem. 56:431.
5. Gutensohn, W. and Jahn, H. (1977). Hoppe-Seylers Z. Physiol. Chem. 358:939.
6. Khym, J.X. and Cohn, W.E. (1960). J. Am. Chem. Soc. 82:6380.
7. Hansske, F. and Cramer, C. (1977). Carbohydr. Res. 54:75.
8. Emmerson, B.T., Thompson, C.J. and Wallace, D.C. (1972). Ann. Intern. Med. 76:285.
9. Krenitsky, T.A., Papaioannou, R. and Elion, G.B. (1969). J. Biol. Chem. 244:1263.
10. Henderson, J.F., Brox, L.W., Kelley, W.N., Rosenbloom, F.M. and Seegmiller, J.E. (1968). J. Biol. Chem. 243:2514.
11. Krenitsky, T.A. and Papaioannou, R. (1969). J. Biol. Chem. 244:1271.

ROLE OF HUMAN HYPOXANTHINE GUANINE PHOSPHORIBOSYLTRANSFERASE IN NUCLEOTIDE INTERCONVERSION

Alessandro Giacomello and Costantino Salerno

Istituti di Reumatologia e di Chimica Biologica della Università di Roma e Centro di Biologia Molecolare del C.N.R., Roma, Italia

It is well established that human Hypoxanthine-Guanine Phosphoribosyltransferase (HGPRT, EC 2.4.2.8) catalyzes the reactions:

$$\text{Hypoxanthine} + \text{PRPP} \xrightleftharpoons{Mg^{2+}} \text{IMP} + PP_i \qquad (1)$$

$$\text{Guanine} + \text{PRPP} \xrightleftharpoons{Mg^{2+}} \text{GMP} + PP_i \qquad (2)$$

Deficiency of HGPRT is associated with purine overproduction and elevated cellular concentration of phosphoribosylpyrophosphate (PRPP).[1,2] The high cellular PRPP content has been generally attributed to a failure to utilize PRPP for nucleotide synthesis by HGPRT [1,2]. However the following observations are against such a conclusion:

a) evidence has been presented of an increased PRPP synthesis in HGPRT deficient cells[3]

b) deficiency of Adenine-Phosphoribosyltransferase (APRT, EC 2.4. 2.7) the closely related enzyme which catalyzes the reaction:

$$\text{Adenine} + \text{PRPP} \xrightleftharpoons{Mg^{2+}} \text{AMP} + PP_i \qquad (3)$$

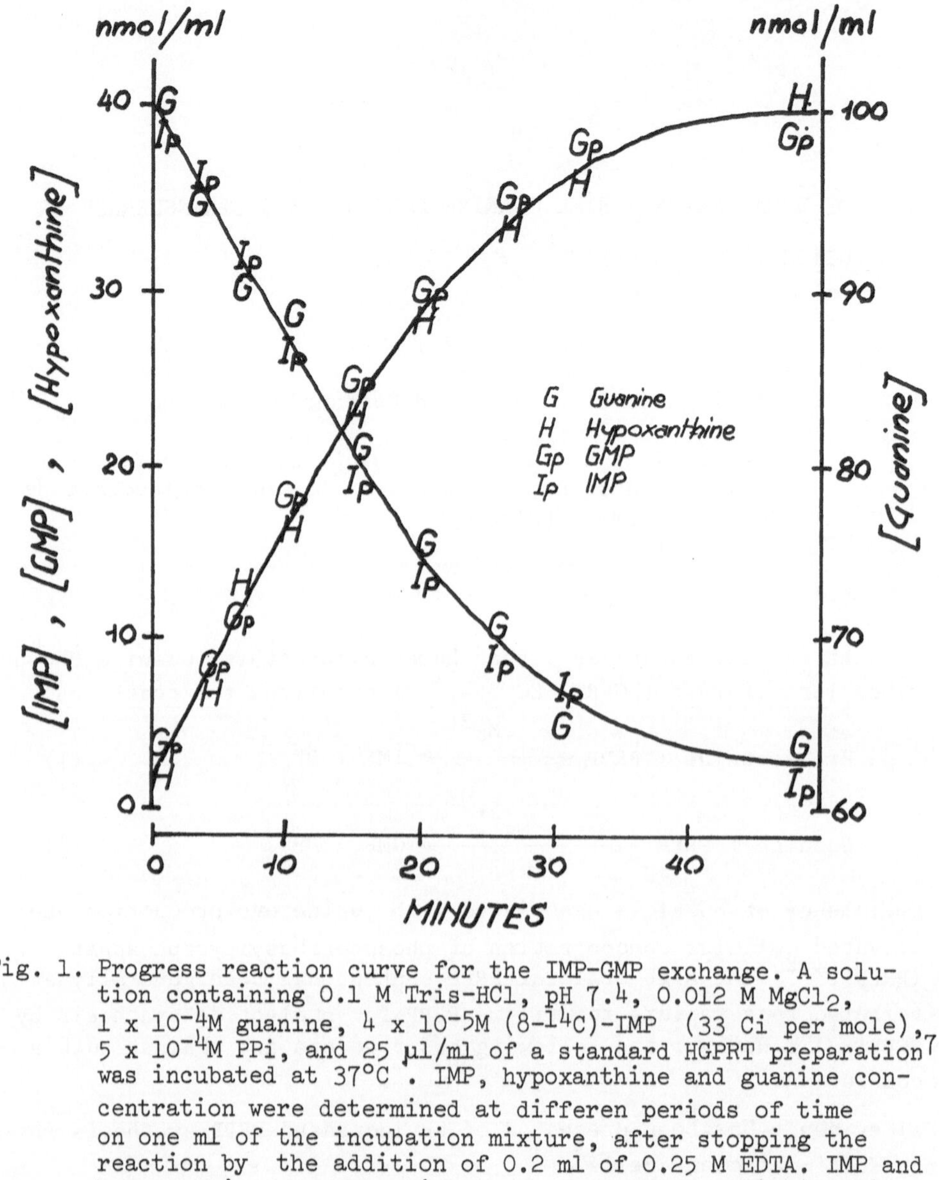

Fig. 1. Progress reaction curve for the IMP-GMP exchange. A solution containing 0.1 M Tris-HCl, pH 7.4, 0.012 M $MgCl_2$, 1×10^{-4}M guanine, 4×10^{-5}M (8-^{14}C)-IMP (33 Ci per mole), 5×10^{-4}M PPi, and 25 µl/ml of a standard HGPRT preparation[7] was incubated at 37°C . IMP, hypoxanthine and guanine concentration were determined at differen periods of time on one ml of the incubation mixture, after stopping the reaction by the addition of 0.2 ml of 0.25 M EDTA. IMP and hypoxanthine concentrations were measured by scintillation spectrometry after separation by paper chromatography. Guanine was measured by the enzymatic spectrophotometric method of Kalckar[9]. GMP concentration in the incubation mixture was determined by the NADH coupled enzyme method of Grassl[10] after protein precipitation according to Hurlbert et al.[11]

is not associated with elevated cellular concentration of PRPP[4], although the Michaelis constant for PRPP reported for this enzyme is considerably lower than that for HGPRT[5], and a high quantity of PRPP should normally be utilized by APRT as suggested by the finding that, in subjects homozygous for deficiency of this enzyme, adenine and its oxidative products account for up 25 per cent of total urinary purine metabolytes[6].

Thus the role of HGPRT in purine metabolism is not yet completely understood.

We are presenting evidence that HGPRT may have a role in nucleotide interconversion.

When a solution at pH 7.4 containing HGPRT, purified from human erythrocytes to apparent electrophoretic homogeneity, IMP guanine, PP_i, and magnesium ions was incubated at 37°C, IMP and guanine were consumed with formation of an equimolar amount of GMP and hypoxanthine (Fig. 1), according to the following reaction

$$\text{IMP} + \text{Guanine} \xrightleftharpoons{Mg^{2+},\ PP_i} \text{GMP} + \text{Hypoxanthine} \qquad (4)$$

Fig. 2A shows the differential absorption spectra recorded at 20 min intervals between the solution employed to demonstrate the IMP-GMP exchange and a solution containing the same substrates without PP_i which is necessary for the reaction to occur. These differential absorption spectra have minima, maxima and isosbestic points of the theoretical spectrum obtained by subtracting at each wavelength the value of the change in extinction coefficient for the formation of IMP from hypoxanthine (Fig. 2C) from that for the formation of GMP from Guanine (Fig. 2B). This result confirms the stoichiometry reported above for the reaction. No appreciable amount of PRPP was formed during the IMP-GMP exchange.

Failure to separate HGPRT and IMP-GMP exchange activities by heat treatment according to Olsen and Milman[7], and by gel filtration on Sephadex G150[12], suggests that reactions 1, 2 and 4 are catalyzed by the same molecular species.

As shown below, the HGPRT catalyzed IMP-GMP exchange might be thought to take place by the coupling of IMP pyrophosphorolysis (reaction 1) and guanine ribotidation by PRPP (reaction 2)

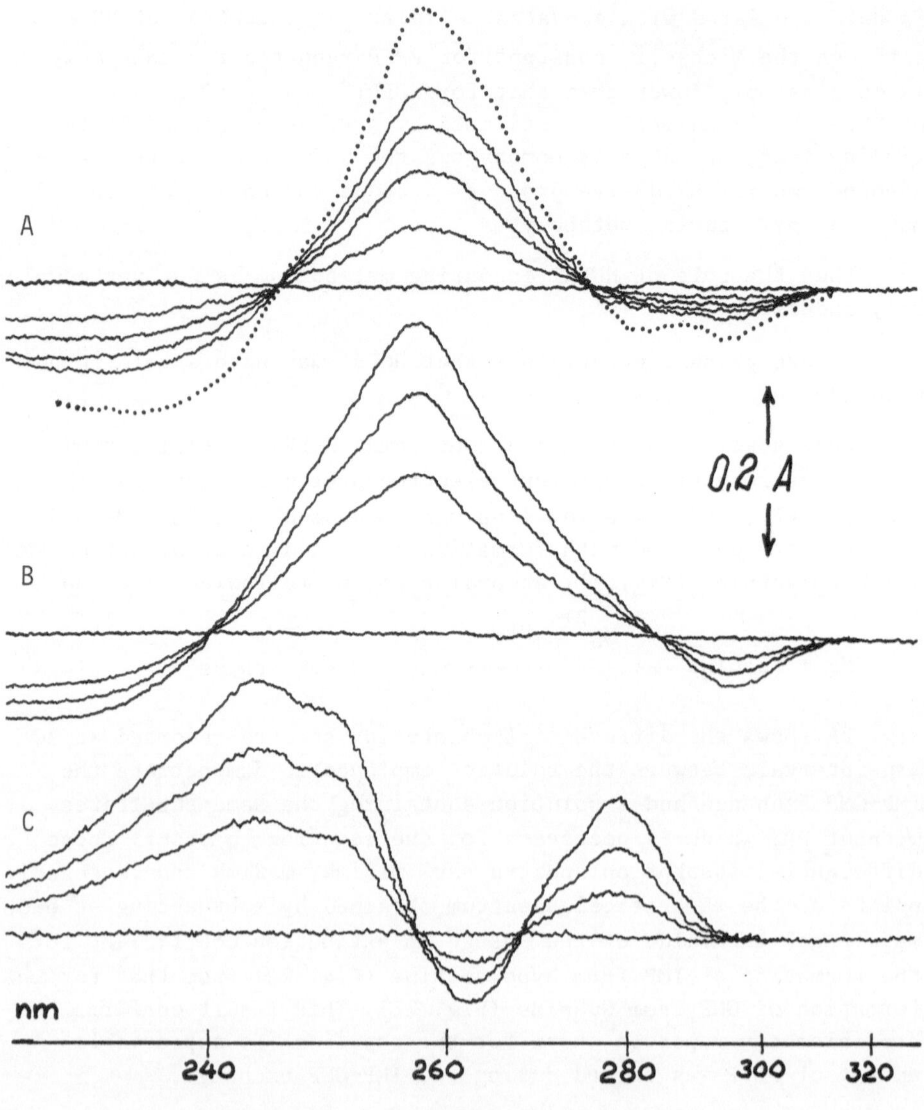

Fig. 2. Differential absorption spectra recorded at 20 min intervals of the hypoxanthine guanine phosphoribosyltransferase catalyzed reactions.

A) IMP + guanine ⟶ hypoxanthine + GMP

The sample cell contained the following reaction mixture: 0.1 M Tris-HCl, pH 7.4, 0.012 M $MgCl_2$, 1 x 10^{-4}M guanine, 5 x 10^{-5}M IMP, 5 x 10^{-4}M PP_i, and 0.02 ml of a hypoxanthine guanine phosphoribosyltransferase preparation in a final volume of 2 ml. The reference cell contained the same compounds without PP_i. The theoretical differential spectrum (dotted line) was calculated as described in the text.

B) Guanine + P-Rib-PP ⟶ GMP + PP_i.

The sample cell contained the following reaction mixture: 0.1 M Tris-HCl, pH 7.4, 0.012 M $MgCl_2$, 1×10^{-4}M guanine, 1×10^{-3}M P-Rib-PP, and 0.02 ml of a hypoxanthine guanine phosphoribosyltransferase preparation in a final volume of 2 ml. The reference cell contained the same compounds without P-Rib-PP.

C) hypoxanthine + P-Rib-PP ⟶ IMP + PP_i.

The sample cell contained the following reaction mixture: 0.1 M Tris-HCl, pH 7.4, 0.012 M $MgCl_2$, 1.7×10^{-4}M hypoxanthine, 1×10^{-3}M P-Rib-PP, and 0.02 ml of a hypoxanthine guanine phosphoribosyltransferase preparation in a final volume of 2 ml. The reference cell contained the same compounds without P-Rib-PP.

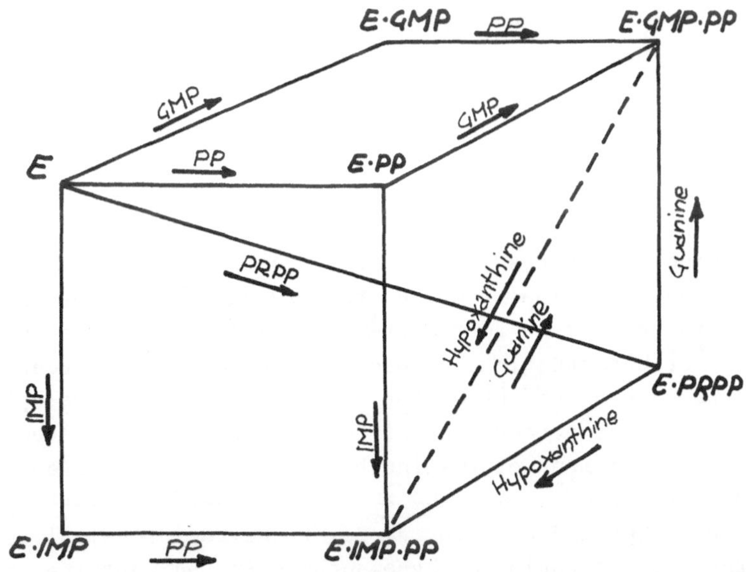

Fig. 3. Model for the HGPRT catalyzed IMP-GMP exchange according to the mechanism proposed in Eq. 5. The broken line represents the Theorell-Chance mechanism for guanine binding and hypoxanthine release postulated to justify the deviation from parallel of the lines obtained in fig. 4.

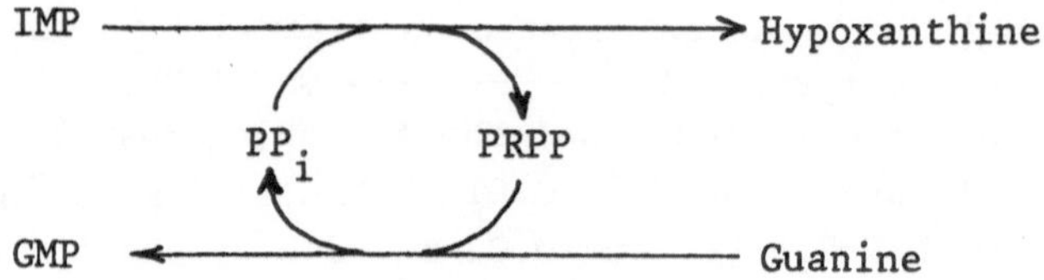

(5)

Steady state kinetics of these two reactions have been studied[5,13,14,15]. Assuming that the IMP-GMP exchange occurs according to Eq. (5), Fig. 3 shows the minimum kinetic model which still fits the experimental data (summarized in table 1). The observation that no appreciable amount of PRPP is formed during theIMP-GMP exchange raises the interesting question whether or not the EPRPP complex is formed during the reaction. At a fixed IMP and PP_i concentration, the initial rate of hypoxanthine production was greater in the presence (reaction 4) than in the absence (reaction 1) of guanine. The maximal increase in the initial rate was achieved at concentrations of guanine lower than 1.5×10^{-6}M. Guanine activation cannot be followed by the spectrophotometric coupled xanthine oxidase assay[14,15] below such concentrations owing to its consumption during the IMP-GMP exchange. Therefore the activation pattern has been studied by comparing double reciprocal plots of the initial velocity against the concentration of IMP or PP_i, at a constant concentration of the second substrate, in the absence and at saturating concentrations of guanine.

According to the reaction mechanism considered above (Fig. 3), one would expect guanine to affect the 1/v axis intercept since it combines with one enzyme form different from that reacting with IMP and PP_i[16]. Furthermore, in the absence of hypoxanthine, guanine is predicted to have no effect on the slope of the reciprocal plots since the points of attachement of IMP and PP_i are separated from the point of attachment of guanine by the irreversible step of hypoxanthine release[16]. As shown in Fig. 4, guanine affects the 1/v axis intercept and also the slope of the reciprocal plots. The steady state concentration of hypoxanthine in the xanthine oxidase coupled assay employed to follow hypoxanthine production is at least 1000 fold lower than that which can justify the deviation from parallel of the lines in fig. 4[15]. Therefore it must be admitted that the points of addition of IMP and PP_i are connected to the point of attachment of guanine by reversible steps only[16].

Although more complex mechanisms cannot be excluded, the simplest assumption to eliminate the irreversible step of hypoxanthine release, which separates the points of addition of the substrates, is that guanine can bind to the enzyme before the release of hypoxanthine. Taking into account that guanine and hypoxanthine bind to the same enzyme form, a Theorell Chance mechanism[16] for guanine binding and hypoxanthine release might be postulated (fig. 3, dotted line).

According to this mechanism the PRPP complex is not formed during the IMP-GMP exchange.

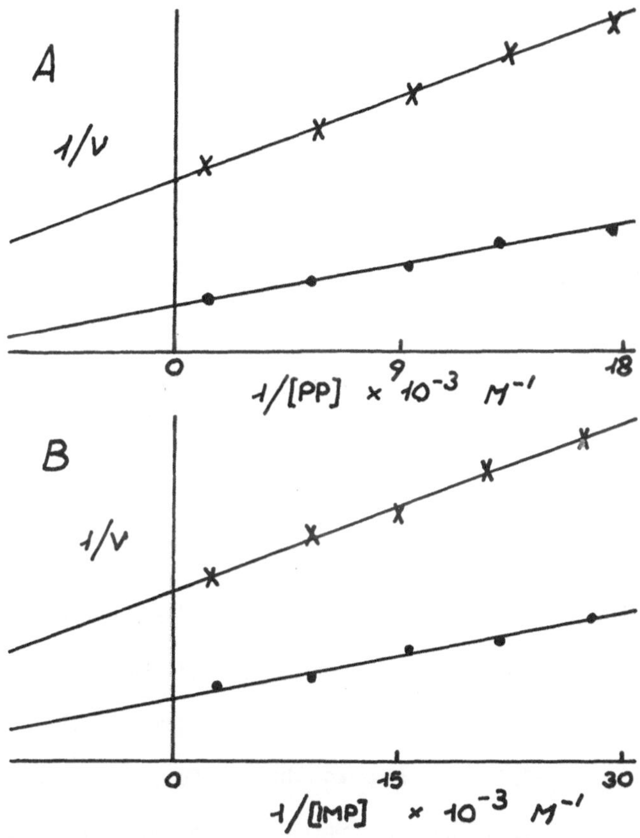

Fig. 4. Guanine activation of IMP pyrophosphorolysis with PP (A) and IMP (B) as the variable substrate.
The reaction was followed at 37°C in the presence of an excess of XOD activity (0.08 IU/ml). Tris-HCl and $MgCl_2$ assay mixture concentrations were 0.1 and 0.012 M respectively.
A) The concentration of IMP was held constant at 1.3 x 10^{-4}M. The concentrations of guanine were: X , none; •, 8.3 x 10^{-5}M.
B) The concentration of PPi was held constant at 1 x 10^{-4}M. The concentrations of guanine were: X none; • , 2.1 x 10^{-5}M.

Table 1. Product and alternate substrate or product inhibition pattern for reactions (1) and (2).

Product and alternate substrate or product inhibitor	Varied substrate	Fixed substrate	Inhibition pattern	Ref.
IMP	PRPP	Hypoxanthine	Competitive	5,13,14
GMP	PRPP	Guanine	Competitive	5,14,15
PPi	PRPP	Hypoxanthine	Competitive	14
PPi	PRPP	Guanine	Competitive	15
IMP	Hypoxanthine	PRPP	Noncompetitive	5,13,14
GMP	Guanine	PRPP	Noncompetitive	5,14,15
PPi	Hypoxanthine	PRPP	Noncompetitive	5,14
PPi	Guanine	PRPP	Noncompetitive	5,15
PRPP	IMP	PPi	Competitive	14
PRPP	PPi	IMP	Competitive	14
Guanine	Hypoxanthine	PRPP	Competitive	5,15
Guanine	PRPP	Hypoxanthine	Noncompetitive	15
GMP	Hypoxanthine	PRPP	Noncompetitive	5,13,15
GMP	PRPP	Hypoxanthine	Competitive	5,13,15
GMP	IMP	PPi	Competitive	15
GMP	PPi	IMP	Noncompetitive	15

REFERENCES

1. F.M. Rosenbloom, J.F. Henderson, I.C. Caldwell, W.N. Kelley, and J.E. Seegmiller, Biochemical bases of accelerated purine biosynthesis de novo in human fibroblasts lacking hypoxanthine guanine phosphoribosyltrasferase, J. Biol. Chem. 243: 1166 (1968).
2. J.B. Wyngaarden and W.N. Kelley, "Gout and hyperuricemia", Grune & Stratton, New York (1976).
3. G.H. Reem, Phosphoribosylpyrophosphate overproduction, a new metabolic abnormality in the Lesch Nyhan syndrome, Science 190: 1098 (1975).
4. I.H. Fox, J.C. Meade, W.N. Kelley, Adenine phosphoribosyltransferase deficiency in man - report of a second family, Am. J. Med. 55: 614 (1973).
5. I.F. Henderson, L.W. Brox, W.N. Kelley, F.M. Rosenbloom, and J.E. Seegmiller, Kinetic studies of hypoxanthine guanine phosphoribosyltrasferase, J. Biol. Chem. 243: 2514 (1968).
6. K.J. Van Acker, H.A. Simmonds, C. Potter, and J. S. Cameron, Complete deficiency of adenine phosphoribosyltransferase - report of a family, N. Engl. J. Med. 297: 127 (1977).
7. A.S. Olsen and G. Milman, Subunit molecular weight of human hypoxanthine guanine phosphoribosyltransferase, J. Biol. Chem. 249: 4038 (1974).
8. E. Gerlach, R.H. Dreisbach, and B. Deuticke, Paper chromatographic separation of nucleotides, nucleosides, purines and pyrimidine, J. Chromatog. 18: 81 (1965).
9. H.M. Kalckar, Differential spectrophotometry of purine compounds by means of specific enzymes, J. Biol. Chem. 167: 428 (1947).
10. M. Grassl, Guanosine-5'-monophosphate, in "Methods of enzymatic analysis", H.U. Bergmeyer ed., Academic Press, New York (1974).
11. R.B. Hurlbert, H. Schmitz, A.F. Brumm, V.R. Potter, Chromatographic separation of acid soluble nucleotides, J. Biol. Chem. 209: 23 (1954).
12. S.H. Hughes, G.M. Wahl, and M.R. Capecchi, Purification and characterization of mouse hypoxanthine guanine phosphoribosyltransferase, J. Biol. Chem. 250: 120 (1975).
13. T.A. Krenitsky and R. Papaioannou, Human hypoxanthine phosphoribosyltransferase - kinetic and chemical modification, J. Biol. Chem. 244: 1271 (1969).
14. A. Giacomello and C. Salerno, Human hypoxanthine guanine phosphoribosyltrasferase - steady state kinetics of the forwerd and reverse reactions, J. Biol. Chem. 253: 6038 (1978).
15. C. Salerno and A. Giacomello, Human hypoxanthine guanine phosphoribosyltransferase - IMP-GMP exchange: stoichiometry and steady state kinetics of the reaction, in press.
16. I.H. Segel, "Enzyme kinetics - behaviour and analysis of rapid equilibrium and steady state enzyme systems", J. Wiley & Sons New York (1975).

PURIFICATION AND CHARACTERIZATION OF MAMMALIAN ADENINE PHOSPHORIBOSYLTRANSFERASES

Milton W. Taylor
Howard V. Hershey
Department of Biology
Indiana University
Bloomington, IN 47405 U.S.A.

INTRODUCTION

In mammals, the enzyme adenine phosphoribosyl transferase (APRT; AMP: pyrophosphate phosphoribosyl transferase; E.C. 2·42·7) is the only known mechanism by which dietary adenine can be utilized[1]. In humans with reduced or absent APRT activity, renal stones containing substantial amounts of 2,8-dioxyadenine can occur[2]. However, other APRT deficient individuals do not suffer urolithiasis and appear to be asymptomatic. Thus, the role of APRT, and, indeed, the role of adenine, in the metabolism of mammals is unclear.

Even though APRTs from human and rodent sources have very similar mechanisms of action[3,4], the physical and antigenic properties of these enzymes have been reported to be quite different. Human APRT has been reported to be predominately a trimer of identical 12,000 dalton subunits resulting in an active 34,000 dalton enzyme[5]. Minor amounts of activity were also observed at 69,000 daltons and 29,000 daltons which was interpreted as the hexamer and dimer form respectively. However, these biophysical data conflict with data from workers examining the migration of electrophoretic variants. These genetic data indicate that APRT is predominantly a dimer[6]. Although there were minor differences in salt concentrations between these two groups of experiments[5,6], it is unlikely that this accounts for the discrepancy.

Rat liver APRT has a subunit of 20,000 daltons[7,8], and has been reported to be active as the monomer. No evidence was reported for higher multimeric structures of the active enzyme.

As further evidence of the divergence of human and rodent APRTs, antibodies prepared against either partially purified mouse or human enzymes have been reported to be incapable of binding the other enzyme activity[9].

Thus, APRTs from rodents and man have been reported to exhibit substantial structural differences - differences in both subunit size and in the multimeric nature of the native enzyme. Moreover, there also appears to be substantial differences in antigenic properties between mouse and human APRTs. Such dramatic differences in an enzyme which retains a common mechanism of action would be a remarkable evolutionary event or might indicate the presence of several different APRT enzymes. Alternatively, a closer examination of the size, structure, and antigenic properties of human and rodent enzymes might be in order - indeed, the known discrepancy in the multimer structure of human APRT indicates that a re-examination would be of value. The recent development of a rapid purification scheme using affinity chromatography facilitates this study[8].

METHODS AND RESULTS

APRTs from rat liver or Syrian hamster liver were purified by affinity chromatography on AMP-agarose columns essentially as previously described for rat liver[8] with the following exception: the 100,000 x g supernatant of liver was made up to 25% saturation with ammonium sulfate. After centrifugation, the supernatant (which contains the APRT activity) is diluted with one volume of Buffer A (50 mM Tris, pH 7.4; 10 mM $MgSO_4$; 30 mM KCl; 0.02% sodium azide). This 0.5 M ammonium sulfate supernatant is then passed through an AMP-agarose column (P-L Biochemicals). APRT activity binds to the column.

Saline-washed human erythrocytes were lysed by hypotonic shock and dialyzed against 5 mM Tris, pH 7.4. DEAE-cellulose was added at 10 g per 100 mls packed cells. After stirring for 2 hr the DEAE-cellulose was filtered on a Whatman No. 1 filter and washed to remove hemoglobin. The DEAE-cellulose cake was then suspended and stirred in 5 mM Tris, pH 7.4; 100 mM KCl, filtered and the APRT containing filtrate passed through the AMP-agarose column.

After binding the APRT to the AMP-agarose, the column is washed with Buffer A + 0.5 M KCl, followed by Buffer A. APRT is selectively eluted off with Buffer A + 0.5 mM P-ribose-PP. This results in APRT with about 85% purity. Subsequent use of a small DEAE-agarose column can result in 97-99% purity with little loss of enzymatic activity. Table I gives results for the purification of Syrian hamster liver APRT.

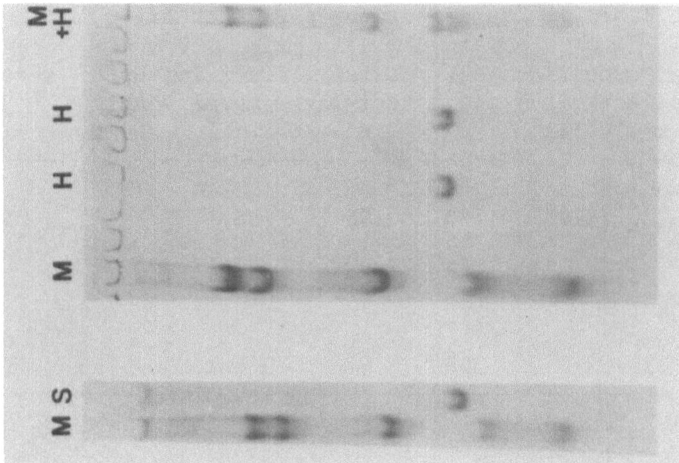

Fig. 1. SDS-PAGE of AMP-agarose purified APRTs. M, markers: heavy chain of IgG, ovalbuin, light chain of IgG, myoglobin, cytochrome c. H, human APRT. S, Syrian hamster liver APRT.

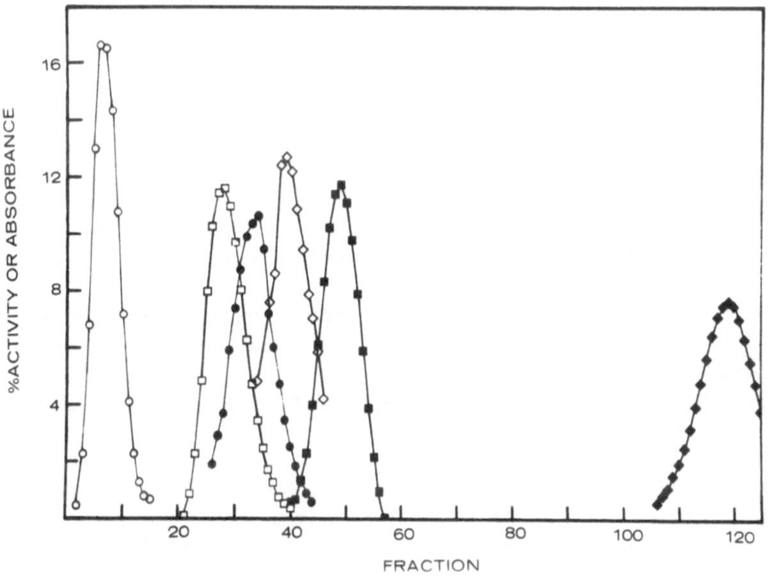

Fig.2. G-100 gel filtration on 83 cm x 1.6cmi.d. column of APRT activity. One ml fractions were collected and absorbance or enzyme activity measured. Open circles: dextran blue 2000, A_{610}. Open squares: human hemoglobin, A_{405}. Closed circles: APRT enzyme activity. Open diamonds: DNase I enzyme activity. Closed squares: myoglobin, A_{405}. Closed diamonds: bromophenol blue, A_{590}.

Table 1. Purification of Hamster liver APRT.

Purification Step	Specific Activity U*/ml	Vol. mls	Total Activity U	Protein mg/ml	Spec. Act. U/mgϕ	Recovery %	Purification -Fold
crude homogenate	$6.7x10^{-2}$	1585	106	58.7	$1.1x10^{-3}$	100	1
100,000xg supt	$7.4x10^{-2}$	1080	79.9	39.1	$1.9x10^{-3}$	75.4	1.7
AMP-agarose purified APRT	$11.6x10^{-2}$	352	41	$20.5x10^{-3}$	5.66	39	5,000

*U - μg AMP formed/min

When the AMP-agarose purified APRT activity from all three mammalian sources is subjected to sodium dodecyl sulfate polyacrylamide gel electrophoresis (SDS-PAGE)[10,11], the major protein band migrates, relative to markers, at about 20,000 daltons (Figure 1). This is in agreement with the previously reported data for the rat liver enzyme[7], but is in considerable disagreement with the reported value of 12,000 daltons for the human erythrocyte enzyme[5].

When the 100,000 x g supernatants of Syrian hamster or rat liver or crude extracts of human erythrocytes are run on a G-100 (Pharmacia) gel filtration column with internal markers of human hemoglobin, pancreatic DNase I, and myoglobin, the major APRT activity of all three extracts migrates as expected for a protein of 40,000 daltons. Figure 2 shows the results for Syrian hamster liver APRT. Rat liver and human erythrocyte APRT give essentially identical results.

Although these results have validity only under these salt conditions (Buffer A), they do show that, under conditions similar to those used for assaying the enzyme, there is little difference in the size of native human and rodent APRTs. Moreover, since the SDS-PAGE indicates a monomer of 20,000 daltons, these data indicate that APRT is a dimer in Buffer A.

The size of native APRT as determined here is in basic agreement with the previous finding for human APRT[5], although that value was somewhat lower (34,000 daltons). Moreover, these data are in agreement with the previous finding that native human APRT is a dimer[6] under certain conditions. Native rat liver APRT is reported to be primarily a monomer of 20,000 daltons with no multimeric structure[7]. However, these data argue that rodent APRTs preferentially form dimers under certain conditions.

Isoelectric focusing and non-denaturing gel electrophoresis of the AMP-agarose purified APRTs from Syrian hamster and human enzymes

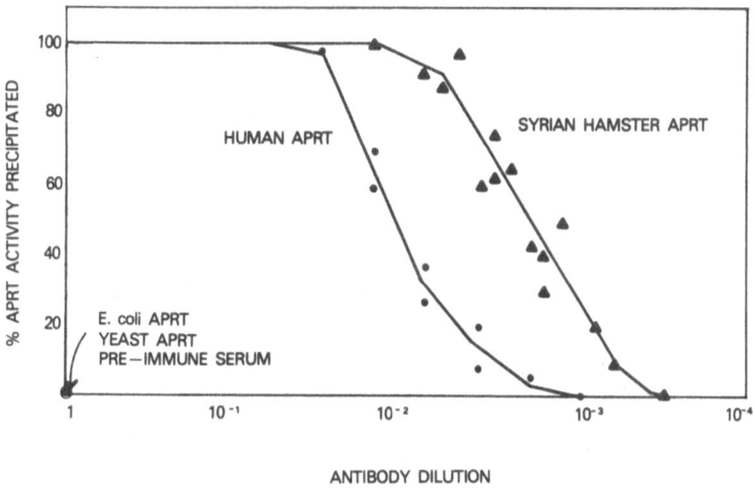

Fig. 3. To tubes were added 5μl of 10mg/ml albumin, 20 ul of APRT containing extract, and 10μl of anti-Syrian hamster APRT antibody. After 2 hr at 4 C, 5μl of fixed S. Aureus (Pansorbin, Calbiochem) was added. After 20 min, the sample was centrifuged and the supernatant assayed for APRT activity.

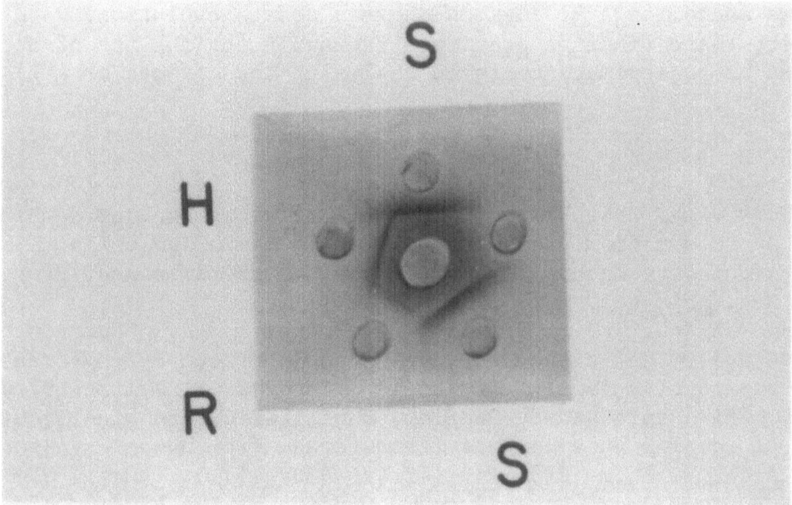

Fig. 4. Ochterlony diffusion of APRTs from various sources. The center well contains antibody against AMP-agarose purified Syrian hamster APRT. The outer wells contain AMP-agarose purified APRTs from Syrian hamster liver (S), human erythrocytes (H), and rat liver (R).

provide confirmatory evidence that the two enzymes differ and that the 20,000 dalton subunit seen in SDS-PAGE is, indeed, the protein which exhibits APRT activity in non-denaturing gels.

In agreement with these biopnysical data, antiboay prepared against AMP-agarose purified Syrian hamster liver APRT strongly crossreacts with human APRT (Figure 3). Assuming similar enzymatic specific activities and correcting for different initial levels of APRT activity, the anti-Syrian hamster APRT sera is capable of reacting about 50% as well against the human erythrocyte enzyme. This is in disagreement with the data from mouse and human APRTs[9].

Ouchterlony diffusion of AMP-agarose purified APRTs against anti-Syrian hamster APRT demonstrate the antigenic similarity of all these APRTs, although the presence of spurs demonstrates that the APRTs are not identical (Fig. 4).

DISCUSSION

The evidence presented in this paper indicate that rodent and human APRTs are similar in subunit size, antigenicity, and, under similar salt conditions, multimeric structure. All three enzymes contain 20,000 dalton subunits and in Buffer A, preferentially form dimers. These data, although considerably more prosaic than the evidence indicating substantial size differences for the APRT peptide and differences in multimeric structure, do provide evidence that the mammalian APRT enzymes share a common evolutionary past. Moreover, these findings provide a rationale for the use of non-primates or non-primate cells in analyzing the effects of APRT or mutants of APRT.

REFERENCES

1. A.W. Murray, D.C. Elliot, and M.R. Atkinson, Nucleotide biosynthesis from preformed purines in mammalian cells: regulatory mechanisms and biological significance, Prog. Nuc. Acid Res. Mol. Biol. 10:87 (1970).
2. K.J. Van Acker, J.A. Simmonds, C. Potter and J.S. Cameron, Complete deficiency of adenine phosphoribosyltransferase report of a family, New Engl. J. Med. 297:127-132 (1977).
3. S.K. Srivastava and E. Bentler, Purification and kinetic studies of adenine phosphoribosyltransferase from human erythrocytes, Arch. Biochem. Biophys. 142:426-434 (1971).
4. R.E.A. Gadd and J.F. Henderson, Studies of the binding of phosphoribosylpyrophosphate to adenine phosphoribosyltransferase, J. Biol. Chem. 245:2979-2984 (1970).
5. C.B. Thomas, W.J. Arnold, and W.N. Kelley, Human adenine phosphoribosyltransferase purification, subunit structure, and substrate specificity, J. Biol. Chem. 248:2529-2535, (1973).

6. W. Mowbray, B. Watson, H. Harris, A search for electrophoretic variants of human adenine phosphoribosyltransferase, Ann. Human Genetics 36:153 (1972).
7. J.G. Kenimer, L.G. Young and D.P. Groth, Purification and proprties of rat liver adenine phosphoribosyltransferase, Biochim. Biophys. Acta 384:87-101 (1975).
8. H.V. Hershey and M.W. Taylor, Purification of adenine phosphoribosyltransferase by affinity chromatography, Prep. Biochem. 8(6):453 (1978).
9. K.R. Held, B. Kahan, and R. DeMars, Adenine phosphoribosyltransferase and hypoxanthine-guanine phosphoribosyltransferase immunoprecipitation reactions in human-mouse and human-hamster cell hybrids, Humangenetik 30:23-34 (1975).
10. K. Weber and Osborn, The reliability of molecular weight determinations by dodecyl sulfate-polyacrylamide gel electrophoresis, J. Biol. Chem. 244:4406-4412, (1969).
11. P. Maisudaira, and D.R. Burgess, SDS microslab linear gradient polyacrylamide gel electrophoresis, Anal. Biochem. 87:386-396, (1978).

THE EFFECT OF PHOSPHORIBOSYLPYROPHOSPHATE ON STABILITY AND CONFIGURATION OF HYPOXANTHINEGUANINEPHOSPHORIBOSYLTRANSFERASE AND ADENINEPHOSPHORIBOSYLTRANSFERASE FROM HUMAN ERYTHROCYTES

W. Gröbner, N. Zöllner

Medical Policlinic, University of Munich

Pettenkoferstraße 8a, Munich

5-Phosphoribosylpyrophosphate (PRPP) is a substrate of hypoxanthineguaninephosphoribosyltransferase (HGPRT) and adeninephosphoribosyltransferase (APRT). It stabilizes these purine phosphoribosyltransferases against thermal inactivation in vitro. Using sucrose density gradient ultracentrifugation we investigated whether this stabilisation is associated with a change in the configuration of these enzymes.

METHODS

Venous blood samples were collected in heparinized tubes and centrifuged at 6oo x g for 1o minutes at room temperature. The buffy coat and plasma were removed and the erythrocytes washed twice with cold o,9% saline. The washed erythrocytes were lysed by freeze thawing and dialysed for 2 hours at 4° against o,o1 M Tris-Cl buffer pH 7,4.

Sucrose gradient ultracentrifugation was performed with a Spinco SW 41 rotor in a Beckman Model L 5o. Isokinetic gradients (1o - 28,2%) of 11,8 ml were prepared according to the method of McCarty et al. (1968). The sucrose was dissolved in o,o1 M potassium phosphate buffer pH 7,4 and other coumpounds were added as indicated in the text. Hemolysate was diluted 1:1o with o,o1 M potassium phosphate buffer pH 7,4, centrifuged for 2o minutes at 1o,3oo x g in order to remove cell membranes, and 2oo µl of the supernatant was applied to the top of the gradient. After centrifugation for 4o hours at 4o,ooo rpm at 4° 235 µl fractions were collected with a Gilson microfractionator. $S_{2o,w}$ values were calculated on the basis of the linear relationship of the sedimentation coefficient to the distance migrated in the isokinetic gradient. Hemoglobin, which was used as an internal standard, was detected by its absorbancy at 43o nm with a spectrophotometer.

The sensitivity of HGPRT and APRT to thermal inactivation was studied by measuring the enzyme activity in the standard assay following heating of dialysed hemolysate at 8o°C and 57°C respectively.

Activity of HGPRT and APRT was assayed as described by Arnold and Kelley (1971). Protein was determined by the method of Lowry et al. (1951).

RESULTS

1 mM MgPRPP stabilises HGPRT from erythrocytes at 8o°C.

Sucrose gradient ultracentrifugation of hemolysate results in the demonstration of one peak of HGPRT activity with a $S_{2o,w}$ of 3,9 ± o,12 ($\bar{x}$ ± s, n = 11; Fig. 1). When hemolysate was incubated at 37°C for 15 minutes with o,3 or 1 mM MgPRPP and then run in a gradient that contained the same concentration of this compound there was a change in the enzyme profile. Under these conditions all of the enzyme has a $S_{2o,w}$ of 5,73 (n = 3; Fig. 1). At +4°C the large form is more stable than the small form (Table 1).

Table 1. Stability of the small and large form of HGPRT at +4°C.

	Enzyme activity (% of initial activity)	
Molecular species	control	7 days
Small form	1oo	57
Large form	1oo	81

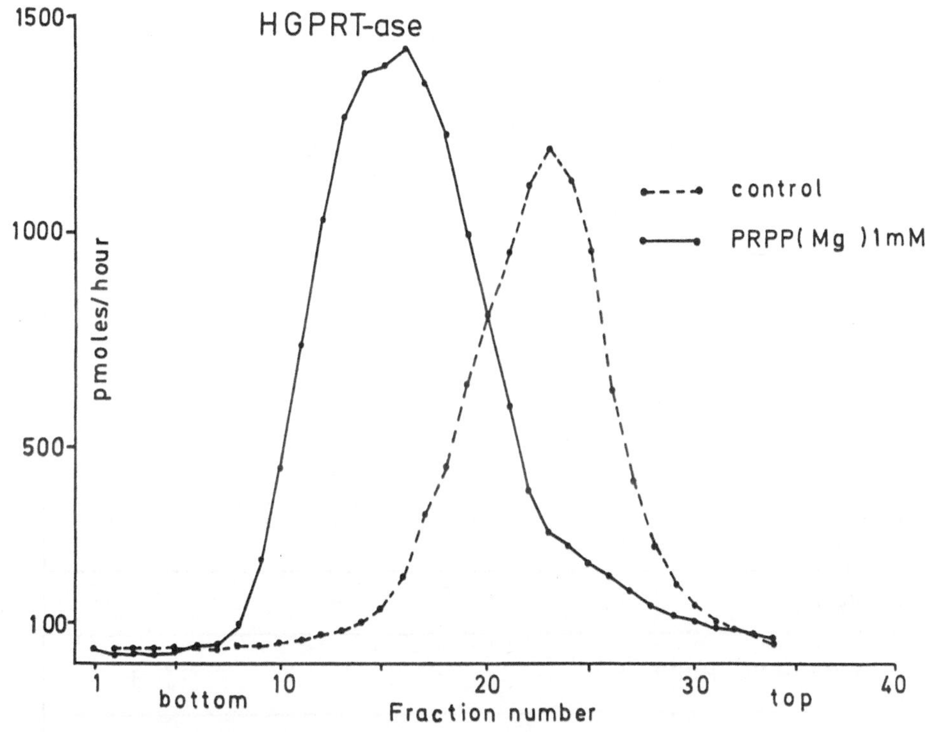

Fig. 1. Sucrose density gradient ultracentrifugation of HGPRT from hemolysate in the absence and presence of 1 mM MgPRPP.

MgPRPP at concentration below o,3 mM, NaPRPP (1 mM), ribose-5-phosphate (1 mM) with or without MgCl (5 mM), Na-pyrophosphate (1 mM) with or without $MgCl_2$ (1 mM) or different concentrations of $MgCl_2$ (1, 5, 1o mM) had no effect on stability and sedimentation coefficient of the enzyme (Table 2).

Table 2. The influence of MgPRPP, NaPRPP, $MgCl_2$, ribose-5-phosphate and Na-pyrophosphate on the sedimentation coefficient of HGPRT from erythrocytes.

	concentration mM	small form $\bar{x} \pm s$ (n=11) S_{20W} 3.9 ± 0.12	large form $\bar{x}$ (n=3) S_{20W} 5.73
–	–	+	
Mg-PRPP	1		+
	0.3		+
	0.03	+	
Na-PRPP	1	+	
$MgCl_2$	1	+	
$MgCl_2$	5	+	
$MgCl_2$ + Na-PRPP	1		+
Ribose-5-Phosphate + $MgCl_2$	1 5	+	
$MgCl_2$ + Na-Pyrophosphate	1 1	+	

APRT has a $S_{2o,w}$ of 3,1. 1mM MgPRPP stabilizes the enzyme without changing its configuration.

DISCUSSION

The results show that stabilisation of unpurified HGPRT from erythrocytes by MgPRPP is accompanied by a change in the molecular size of the enzyme. The sedimentation coefficient in the presence of o,3 or 1 mM MgPRPP is 5,73 and corresponds to the $S_{2o,w}$ of 5,78 observed by Holden and Kelley (1978) for the purified enzyme. Olsen and Milman (1977) found a $S_{2o,w}$ of 5,9 for the purified HGPRT from erythrocytes but prior to ultracentrifugation the enzyme preparation was dialysed for 16 hours against a buffer which contained 6 mM $MgCl_2$ and o,1 mM PRPP. Therefore an effect of MgPRPP on the configuration of the purified enzyme must be assumed. Using sucrose gradient ultracentrifugation Natsumeda et al. (1977) observed a $S_{2o,w}$ of 5,1 for rat liver HGPRT. In the presence of 1 mM MgPRPP and 5 mM $MgCl_2$ a $S_{2o,w}$ of 5,4 was observed. The value of the enzyme with MgPRPP was slightly but significantly larger than control.

References

Arnold, W. N., Kelley, W. N., 1971, J. Biol. Chem., 246:7398.

Holden, J. A., Kelley, W. N., 1978, J. Biol. Chem., 253:4459.

Lowry, O. H., Rosebrough, N. J., Farr, A. L., Randall, R. J., 1951, J. Biol. Chem., 193:265.

McCarty, K. S., Staffort, D., Brown, O., 1968, Anal. Biochem., 24:314.

Natsumeda, Y., Yoshino, M., Tsushima, K., 1977, Biochim. Biophys. Acta, 483:63.

Olsen, A. S., Milman, G., 1977, Biochem., 16:25o1.

ACKNOWLEDGEMENT

This work was supported by the Deutsche Forschungsgemeinschaft.

CHEMICAL MODIFICATION OF HYPOXANTHINE-PHOSPHORIBOSYLTRANSFERASE AND ITS PROTECTION BY SUBSTRATES AND PRODUCTS

Wolf Gutensohn and Heidi Jahn

Institut für Anthropologie und Humangenetik der Universität, Arbeitsgruppe Biochemische Humangenetik, Schillerstr 42, D 8000 München 2, Fed.Rep.Germany

INTRODUCTION

The importance of hypoxanthine-phosphoribosyltransferase (HPRT) to human biochemical genetics need not be stressed in this context. Although in some cases it has not been clear whether deficiencies of this enzyme are due to a structural mutation this seems to be well established in others. Hence a detailed knowledge of the structure and especially of essential amino acid residues seems desirable. Induced by a similar study on quinolinate-phosphoribosyltransferase[1] and parallel to our experiments on specific irreversible inhibition chemical modification of HPRT was investigated.

METHODS

Enzymes. HPRT from human erythrocytes was either partially purified by a batch removal of hemoglobin on DEAE-cellulose or highly enriched by affinity chromatography. Purification of the enzyme from rat brain was carried through the heat step procedure. Enzyme activity was tested in a radiochemical assay and in typical preparations was 0.01 and 4.9 U/mg protein for the partially and highly purified human enzymes respectively and 0.02 U/mg protein for the partially purified rat brain HPRT.

Chemical modofication. The modifying reagents were usually dissolved in 0.05 M sodium phosphate, pH 7.4, and preincubated with the enzyme samples under conditions specified in the tables and figures. For the photooxidation a 0.04% methylene blue solution in 0.05 M phosphate buffer of the appropriate pH was saturated with air. It was then mixed with an equal volume of enzyme solution and irradiated by a 150 W tungsten lamp from 20 cm distance. Further details are given in Fig. 2. Controls were obtained by putting an identical mixture in the dark or by omitting methylene blue.

RESULTS

The basic phenomena of inhibition of HPRT-activity by chemically modifying reagents are summarized in Tab. 1. The decrease of enzyme activity in untreated control samples preincubated for the same time was negligable. Iodoacetate, 2-hydroxy-5-nitrobenzylbromide and acetylimidazole did not affect enzyme activity. Studies on the protection of the enzyme against chemical modification by substrates and products

Table 1.
Chemical modification of HPRT. Enzyme samples were partially purified human erythrocyte enzyme, 2 mg protein in 100 µl.

Modifying reagent	(mM)	% Inhibition of HPRT-activity after preincubation for 30 min	5 h	24 h
N-Ethylmaleimide	0.01	16	3	4
	0.1	33	42	55
	1.0	35	60	68
2,2´-Dinitro-5,5´-dithiodibenzoate	0.01	2	-	1
	0.1	21	-	34
	1.0	44	-	52
p-Chloromercuri-benzoate	0.005	3	0	0
	0.05	2	0	0
	0.5	91	92	92
p-Diazobenzene-sulfonic acid	0.01	0	3	7
	0.1	0	12	0
	1.0	26	44	46
Diethylpyrocarbonate at pH 6.0	0.8	-	-	0
	2.0	-	-	11
	4.0	-	-	39
1,2-Naphthoquinone-4-sulfonic acid	0.1	9	6	5
	1.0	26	33	53
	2.5	-	58	62
	5.0	-	77	83
2,4,6-Trinitrobenzenesulfonic acid	0.1	-	2	0
	1.0	-	9	8
	2.5	-	37	57
	5.0	-	52	85
Glyoxal	3.5	0	-	0
	35	4	-	19
	350	78	-	96

are compiled in Tab. 2. Very similar data, not shown in the table, were obtained with N-ethylmaleimide and 2,2'-dinitro-5,5'-dithiodibenzoate as modifying reagents. For two typical examples the kinetics of inactivation of protected and unprotected samples is demonstrated in Fig. 1. Fig. 2 shows the kinetics of photoinactivation in the presence of methylene blue and the inset demonstrates that the initial velocity

Table 2.
Protection of HPRT by substrates against modification. The enzyme preparation used and the conditions of preincubation and assay were as in Tab. 1, except that substrates were added as indicated.

Modifying reagent	Additions during preincubation				% Inhibition of HPRT-activity after preincubation for	
	PRPP (mM)	Hypox. (mM)	$MgCl_2$ (mM)	GMP (mM)	2 h	24 h
PCMB	-	-	-	-	93	94
0.5 mM	-	-	5.0	-	30	87
	-	1.0	-	-	96	
	-	1.0	5.0	-	91	95
	2.5	-	-	-	95	
	2.5	1.0	-	-	92	
	2.5	-	5.0	-	0	0
	0.05	-	1.0	-	18	
	-	-	-	0.01	62	
Glyoxal	-	-	-	-	81	
175 mM	-	-	1.0	-	82	
	-	1.0	5.0	-	80	
	0.05	-	-	-	83	
	2.5	1.0	-	-	83	
	2.5	-	5.0	-	5	
	0.05	-	1.0	-	53	
	-	-	-	1.0	37	
	-	-	5.0	1.0	42	
Naphtho-	-	-	-	-	66	88
quinone-	-	-	5.0	-		70
sulfonic	-	1.0	-	-		93
acid	-	1.0	5.0	-		86
5.0 mM	2.5	-	-	-		88
	2.5	1.0	-	-		87
	2.5	-	5.0	-		0
	0.25	-	1.0	-	31	
	0.05	-	1.0	-	52	

of this inactivation is pH-dependent. In some cases the inactivation of the enzyme can be reversed. Enzyme inhibited by 2,2′-dinitro-5,5′-dithiodibenzoate is reactivated by treatment with 1 mM dithiothreitol and the effect of PCMB is reversed by 1 mM dithiothreitol or 1 mM mercaptoethanol. The experiment in Tab 3 is based on this reversibility. Photooxidation in the presence of methylene blue may alter sulfhydryl groups in the enzyme protein. However, as shown in the table, enzyme with sulfhydryl groups blocked by previous treatment with PCMB nevertheless is irreversibly inhibited by subsequent photooxidation.

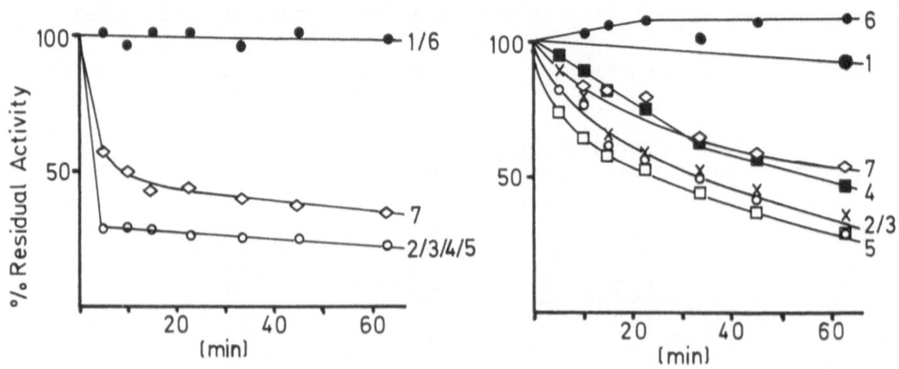

Fig. 1 Kinetics of inactivation of HPRT by 0.5 mM PCMB (left) and 5 mM 2,4,6-trinitrobenzenesulfonic acid (right). Partially purified rat brain enzyme (2 mg protein in 0.5 ml) was incubated at 37° with the following additions: 1. Untreated control 2. Inhibitor alone 3. Inh. + 2.5 mM PRPP 4. Inh. + 5 mM $MgCl_2$ 5. Inh. + 5 mM PP_i 6. Inh. + 2.5 mM PRPP/5 mM $MgCl_2$ 7. Inh. + 5 mM PP_i/5 mM $MgCl_2$.

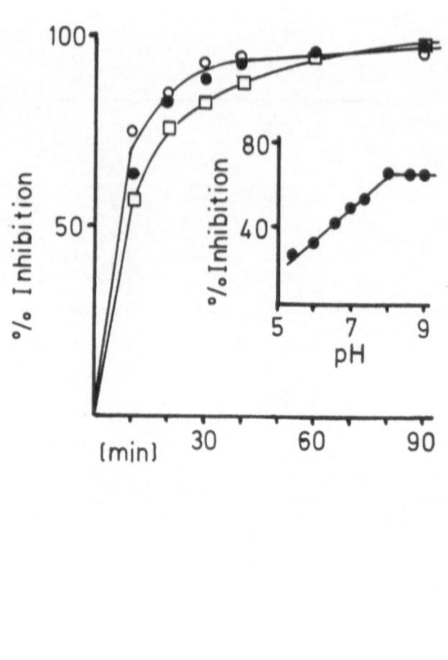

Fig. 2
Photooxidation of HPRT. Homogeneous human enzyme (8 µg protein in 20 µl). Preincubation for 30 min at 22° alone (●), with 1 mM hypox./5 mM $MgCl_2$ (o) or 2.5 mM PRPP/5 mM $MgCl_2$ (□), with BSA 2 mg/ml for stabilisation. Photooxidation at pH 8.5 with methylene blue as described under "Methods". Inset: Partially purified rat brain HPRT (0.5 mg protein in 50 µl), methylene blue in phosphate buffer of given pH. Photooxidation for 15 min.

Tab. 3 Combination of modifying reagents. Samples of partially purified rat brain HPRT (2.5 mg protein in 0.5 ml) were treated as indicated.

Additions		Residual activity, % of control	
1st Incubation period (1 h)	2nd Incubation period (2 h, light)	After inactivation	After dialysis against 1 mM ME
0.5 mM PCMB	Buffer	10.9	57.0
Buffer	0.02% Methylene blue	23.3	24.2
0.5 mM PCMB	0.02% Methylene blue	7.5	12.7

DISCUSSION

The data on inhibition of HPRT by modifying reagents may be criticised for two reasons: 1. Some of the reagents used are known to be not specific enough for only one type of aminoacid residue and 2. even if so, inhibition of enzyme activity might be brought about by induction of conformational changes or steric hindrance rather than alteration of the active center itself.

The specificity of the sulfhydryl reagents used is undisputed. Failure of iodoacetate, in contrast to the other sulfhydryl reagents, to inhibit the enzyme has also been observed with purinenucleoside-phosphorylase[2]. Diethylpyrocarbonate and p-diazobenzenesulfonic acid have been described as histidine-specific reagents. Photooxidation in the presence of methylene blue may alter a number of aromatic aminoacids as well as cysteine and methionine. However, a preferential oxidation of histidine residues has been demonstrated for several proteins. The type of pH-dependence of this oxidation (Inset of Fig. 2) has also been discussed for histidine[2] and an experiment of the type shown in Tab. 3 would exclude cysteine. 1,2-Naphthoquinone-4-sulfonic acid and 2,4,6-trinitrobenzenesulfonic acid are specific for amino groups. However, a distinction between the α-amino group of the N-terminus and the ε-amino group of lysine residues cannot be drawn. On the other hand a blocked N-terminus for human HPRT has been reported[3]. Glyoxal has been used as a reagent for arginine. All the groups and aminoacid residues mentioned sofar would be present in HPRT as shown by two independent analyses of the human enzyme[3,4].

As to the second point of possible criticism the protection experiments shall be discussed. As shown in Tab. 2 and Fig. 1 the results for three types of modifying reagents are very similar. In every case the enzyme is best protected by its substrate PRPP/Mg^{++} even when considerable concentrations of modifying reagent are used as for example with glyoxal. Neither PRPP nor Mg^{++} alone showed a comparable degree of protection. Some protective effect can also be seen with the products PPi/Mg^{++} and GMP which competitively bind to the same site as PRPP/Mg^{++}. Of special importance here is the observation of an intermediate degree of protection obtained with concentrations of substate and product in the range of their K_M and K_i

respectively. All these data would favor the proposal that the modifyable residues are involved in the binding of substrate and products and therefore are an integral part of the active center.

The inactivation of HPRT by glyoxal and the protection of the enzyme by PRPP/Mg^{++} and GMP is very interesting since this is a reaction analogous to the much more specific active-site-directed inhibition by periodate-oxidized GMP or IMP which we had shown earlier. In both cases two closely adjacent aldehyde groups seem to form the reactive principle. This would suggest the guanidino group of an arginine residue in close vicinity to the binding site. An arginine residue modifyable by dialdehyde- or diketo-compounds has been reported for the active center of a number of dehydrogenases (see discussion in loc. cit. 5).

There is a striking similarity between the data presented here and those of Taguchi and Iwai[1] obtained for quinolinate-phosphoribosyltransferase. The major difference lies in the protection experiments. In the case of the latter enzyme the second substrate - the acceptor molecule for the phosphoribosyl moiety - quinolinic acid is very effective in protecting the enzyme against all the modifying reagents. In our case the analogous molecule hypoxanthine is totally ineffective. It may be that these similarities reveal a more general principle of the architecture of phosphoribosyltransferases. But further similar studies with this group of enzymes must be awaited.

In the meantime this technique of chemical modification has been successfully applied to distinguish betweeen normal HPRT and a structurally mutated enzyme from a patient with gout[6].

Acknowledgment: We thank the Deutsche Forschungsgemeinschaft for financial support.

REFERENCES

1. H. Taguchi, and K. Iwai, Chemical modification of the crystalline quinolinate phosphoribosyltransferase from hog liver, Biochim.Biophys.Acta 422:29 - 37 (1976).
2. A. S. Lewis, and M. D. Glantz, Monomeric purine nucleoside phosphorylase from rabbit liver. Purification and characterization, J.Biol.Chem. 251:407 - 413 (1976).
3. A. S. Olsen, and G. Milman, Human hypoxanthine-phosphoribosyltransferase. Purification and properties, Biochemistry 16:2501 - 2505 (1977).
4. H. Muensch, and A. Yoshida, Purification and characterization of human hypoxanthine/guanine phosphoribosyltransferase, Eur.J.Biochem. 76:107 - 112 (1977).
5. J. Berghäuser, Modifizierung von Argininresten in Pyruvat-Kinase, Hoppe Seyler´s Z.Physiol.Chem. 358:1565 - 1572 (1977).
6. W. Gutensohn, and H. Jahn, Partial deficiency of hypoxanthine-phosphoribosyltransferase: evidence for a structural mutation in a patient with gout, Eur.J.Clin.Invest. 9:43 - 47 (1979).

STRUCTURAL STUDIES OF HUMAN ADENINE PHOSPHORIBOSYLTRANSFERASE PURIFIED BY AFFINITY CHROMATOGRAPHY

Joseph A. Holden, Gary S. Meredith, and William N. Kelley
University of Michigan Medical Center
Department of Internal Medicine
Ann Arbor, Michigan 48109

Adenine phosphoribosyltransferase (APRT) catalyzes the condensation of 5-phosphoribosyl-1-pyrophosphate (PP-ribose-P) with adenine to yield adenosine 5'-monophosphate and PP_i. Interest in the human enzyme has been stimulated by the findings of elevated adenine phosphoribosyltransferase activity in erythrocytes from patients with the Lesch-Nyhan syndrome[1] and by the inherited deficiency of the enzyme described in a number of families.[2,3,4,5] In order to better understand the nature of these alterations of adenine phosphoribosyltransferase activity at the molecular level, it is necessary to define the nature of the normal enzyme. The normal enzyme has previously been purified 33,000-fold from human erythrocytes.[6] The procedure, however, is lengthy and laborious. In this report we describe a more efficient purification procedure for adenine phosphoribosyltransferase. Some of the characteristics of the highly purified enzyme, not previously reported, are described.

Four units of packed red cells were washed twice with 0.15 M NaCl and then frozen at -20°C. The lysate was then diluted with 2 volumes of cold distilled water. The hemolysate was brought to pH 5.8 and the precipitated stroma removed by centrifugation at 4000 xg for 20 min. Hemoglobin was removed by CM-Sephadex batch filtration and the enzyme precipitated by the addition of ammonium sulfate to 60% saturation. After dialysis, the enzyme preparation was applied to a GMP-affinity column. GMP-Sepharose was synthesized by the method of Hughes et al.[7] These initial purification steps are exactly the same as those used for the purification of the related enzyme, hypoxanthine-guanine phosphoribosyltransferase (HGPRT).[8] This affords us the opportunity of simultaneously purifying both enzymes from hemolysate.

After the enzyme preparation was applied to the GMP column, the column was washed extensively with 50 mM Tris-HCl pH 7.4, 10mM $MgCl_2$, 25mM KCl, 1mM dithiothreitol (Buffer A). Adenine phosphoribosyltransferase was eluted by washing the column with Buffer A containing 5mM AMP. Fractions with adenine phosphoribosyltransferase activity were pooled and concentrated in an Amicon ultrafiltration apparatus equipped with a UM-10 membrane. The enzyme obtained from the GMP column was chromatographed on a Sephadex G-75 column (1.1 x 100 cm) equilibrated with 50 mM Tris-HCl pH 7.4, 100 mM NaCl, 10 mM $MgCl_2$, and 1 mM dithiothreitol (Buffer B). The enzyme obtained from this purification (0.4 mg/ml) can be stored for at least 4 weeks at -70° without loss of activity.

The purification of adenine phosphoribosyltransferase is shown in Table I. The enzyme appears to be binding to the iminobispropylamine spacer arm rather than the GMP since columns containing only the spacer arm will retain APRT but not HGPRT (unpublished data). The enzyme obtained from the GMP column procedure is about 30% pure as judged by sodium dodecyl sulfate gels. Further chromatography on a Sephadex G-75 column removes contaminating proteins and yields a preparation with a specific activity 55,000 times greater than in hemolysate. The overall yield is 15% and the enzyme is estimated by sodium dodecyl sulfate gels to be at least 97% pure.

Polyacrylamide gel electrophoresis of the purified enzyme at pH 9.5, pH 8.0 and pH 3.8 yields a single Coomassie Blue-staining band. Electrophoresis of APRT at pH 9.5 on 15%, 12.5%, 10%, and 7.5% polyacrylamide gels also demonstrates only a single Coomassie Blue-staining band. The isoelectric point of the purified enzyme is 4.55.

The subunit molecular weight of the enzyme determined by SDS gel electrophoresis is 18,400 and that determined by gel filtration in 6 M guanidine hydrochloride is 17,300.

The amino acid composition of the enzyme is shown in Table II. Cysteine was determined as S-sulfocysteine after reduction and reaction with tetrathionate.[9] The enzyme contains 33% hydrophobic amino acid residues (leucine, valine, isoleucine, phenylalanine, methionine), which is somewhat greater than the average value of 23.9% for most proteins.[10] The minimum molecular weight of the enzyme calculated from these data is 20,800. The data also yield a partial specific volume of 0.75 cm^3/g.[11]

In order to determine the native molecular weight of the enzyme, we have performed sedimentation equilibrium centrifugation. Centrifugation at 14,000 rpm and at 22,000 rpm yields a linear plot of log c versus r^2 with better than 95% recovery of sample. The

TABLE II
AMINO ACID COMPOSITION OF
ADENINE PHOSPHORIBOSYLTRANSFERASE

	Hydrolysis Time (Hours) 25	52	72		
		nanomoles		Averaged nanomoles	Residues/[a] Subunit
S-sulfocysteine	34.7	33.7	33.7	34.0	2.8
Asparate	127.5	127.7	142.5	132.6	10.9
Threonne	73.1	71.6	72.7	72.5[b]	6.0
Serine	126.1	117.3	115.7	134.0[b]	11.0
Glutamate	246.4	258.4	267.5	257.4	21.2
Proline	143.0	143.4	151.4	145.9	12.0
Glycine	221.0	225.3	241.3	229.2	18.9
Alanine	201.3	203.5	209.0	204.6	16.9
Valine	135.2	152.2	173.4	173.4[c]	14.3
Methionine	12.0	12.1	12.2	12.1	1.0
Isoleucine	82.2	84.9	88.9	88.9[c]	7.3
Leucine	363.9	372.1	390.3	390.3[c]	32.2
Tyrosine	56.1	51.2	57.7	55.0	4.5
Phenylalanine	108.3	105.5	109.5	107.8	8.9
Histidine	-	-	-	27.2[d]	2.2
Lysine	-	-	-	110.5[d]	9.1
Arginine	158.9	161.5	164.5	161.6	13.3
Total					192.5

a) Calculations were based on an average integral value derived from the average nanomoles of proline, glycine, phenylalanine, and alanine.
b) Extrapolated to zero hydrolysis time.
c) Values obtained at 72 hour hydrolysis time.
d) Histidine and lysine were not well separated from each other during these determinations. They were therefore determined separately by a 48 hour hydrolysis and do not represent averaged values.

molecular weight calculated from the sedimentaion data is 38,200. A native molecular weight of 38,200 in conjunction with a subunit molecular weight of about 18,000 suggests that the enzyme is a dimer in the native state. In order to test this hypothesis, we have cross-linked the purified enzyme with dimethylsuberimidate.[12] Two protein species are observed when the reaction mixture is analyzed by SDS gels (Figure 1). We believe that these two protein species correspond to the monomer and the dimer. This is further evidence that the protein exists in its native state as a dimer.

The structure of the enzyme has been analyzed by peptide mapping. Purified APRT (1.0 ml, 1.0 mg) from Step 6 was dialyzed in

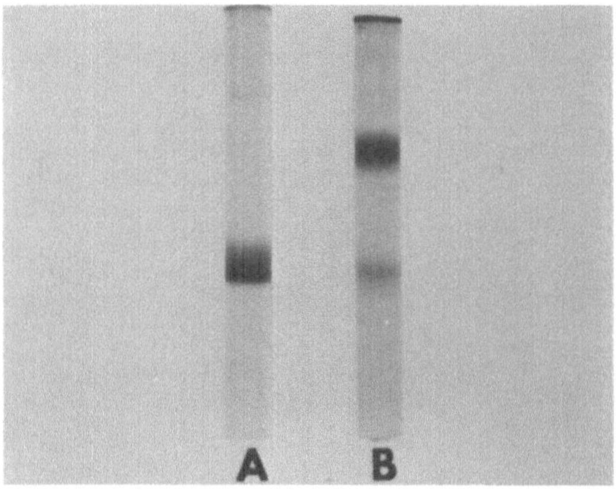

FIGURE I

Sodium dodecyl sulfate gels of cross-linked adenine phosphoribosyltransferase. Purified enzyme (Step 6) was cross-linked with dimethylsuberimidate. Uncross-linked enzyme is shown in A and cross-linked enzyme is shown in B. The anode is at the bottom of the gels and migration is from top to bottom.

1000 volumes of distilled water with three changes over a 36 hour period. The enzyme precipitates under these conditions. The content of the dialysis bag was then transferred to a test tube and lyophilized. The lyophilized enzyme was oxidized by performic acid as described by Hirs.[13] After the oxidation, the enzyme was digested by TPCK-treated trypsin and the sample lyophilized. The tryptic peptides were dissolved in 50 μl of distilled water. To separate the peptides, 5 μl of the tryptic digest was spotted on a cellulose thin-layer chromatogram. Electrophoresis was performed in pyridine-acetic acid-water (10:3:300) at 1000 volts for about one hour. After drying at room temperature, the plate was subjected to chromatography in butanol-pyridine-acetic acid-water (50:33:1:40) at right angles to the direction of electrophoresis. The plate was dried and sprayed with ninhydrin-collidine (150 mg ninhyrin, 150 ml of 95% ethanol, 45 ml of acetic acid, 3 ml collidine). The peptides were revealed by heating the plate at 90° for about 5 minutes.

The result obtained is shown in Figure 2. We observe about 20 peptides from tryptic digests of APRT. This suggests that the subunits are quite similar, if not identical, since there are 44 lysine and arginine residues per molecule.

All of this data suggest that APRT is a dimer of identical subunits with a native molecular weight of 38,000. The native molecular weight is not in agreement with the previously reported

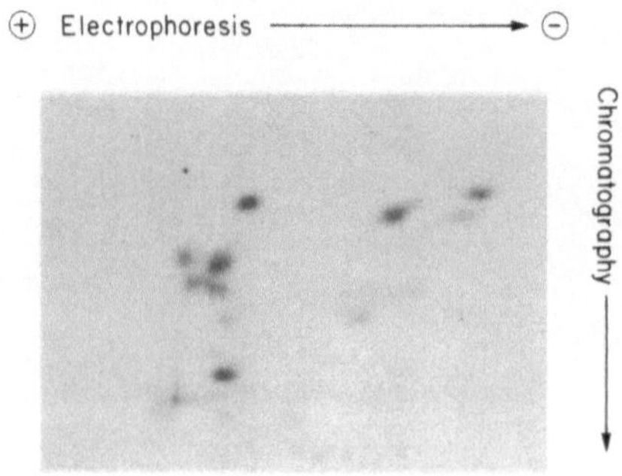

FIGURE II

Tryptic peptide map of adenine phosphoribosyltransferase. Peptides were revealed by spraying with ninhydrin-collidine.

native molecular weight of 34,000.[6] However, this latter value was based on gel filtration and sedimentaion velocity data and assumed a partial specific volume of 0.725 cc^3/gm. Based on the actual $\bar{v}$ of 0.75 cc^3/gm which we calculated from the amino acid composition the previous data yield a native molecular weight of 38,000 for APRT rather than 34,000.

However, the subunit molecular weight is also not in agreement with the value of 11,000 previously reported.[6] This led us to reexamine the former purification procedure. The final specific activity of the enzyme preparation used in that study was 5.08 units/mg with a final recovery of 2.7% as compared to the enzyme obtained from the present purification which has a specific activity of 22 units/mg with a 15% recovery. The differences in the specific activity appear to reflect differences in recovery of enzyme activity. It was not clear why the two different methods of purification would yield protein species of apparently different subunit molecular weight. We, therefore, purified the enzyme again by the former method. The highly purified enzyme obtained from purification on this occasion yielded a single protein band on sodium dodecyl sulfate gel electrophoresis corresponding to a molecular weight of 18,000.

REFERENCES

1. Seegmiller, J.E., Rosenbloom, F.M., and Kelley, W.N. Science 155:1682-1684 (1967).

2. Kelley, W.N., Levy, R.I., Rosenbloom, F.M., Henderson, J.F., and Seegmiller, J.E. J. Clin. Inves. 47:2281-2289 (1968).
3. Fox, I.H., Meade, J.C., and Kelley, W.N. Amer. J. Med. 55: 614-620 (1973).
4. Debray, H., Cartier, P., Temstet, A., and Cendron J. Pediat. Res. 10:762-766 (1976).
5. Van Acker, K.J., Simmonds, H.A., Potter, C., and Cameron, J.S. New Eng. J. Med. 297:127-132 (1977).
6. Thomas, C.B., Arnold, W.J. and Kelley, W.N. J. Biol. Chem. 248:2529-2535 (1973).
7. Hughes, S.H., Wahl, G.M., and Capecchi, M.R. J. Biol. Chem. 250:120-126 (1975).
8. Holden, J.A., and Kelley, W.N. J. Biol. Chem. 253:4459-4463 (1978).
9. Simpson, R.J., Neuberger, M.R., and Liu. T.-Y. J. Biol. Chem. 251:1936-1940 (1976).
10. Hunt, L.T., and Dayhoff, M.O. Composition of Proteins in Atlas of Protein Sequence and Structure p. D-355 (1972).
11. Cohn, E.J., and Edsall, J.T. Proteins, Amino Acids and Peptides p. 370. Reinhold Publishing Corp. New York (1943).
12. Davies, G.E. and Stark, G.R. Proc. Natl. Acad. Sci. 66:651-656 (1970).
13. Hirs, C.H.W. Performic Acid Oxidation in Methods in Enzymology Vol. XI, p. 197-199 (1967).

PHOSPHORIBOSYLPYROPHOSPHATE (PRPP) SYNTHETASE MUTANT IN *Salmonella typhimurium*

B. Jochimsen[§], B. Garber and J.S. Gots

Institute of Biological Chemistry B, University of Copenhagen, Denmark and Department of Microbiology, School of Medicine, University of Pennsylvania, Philadelphia, USA.

PRPP is an essential precursor in the biosynthesis of purine, pyrimidine and pyridine nucleotides and the amino acids histidine and tryptophan. PRPP is also a prerequisite for the utilization of preformed purine and pyrimidine bases and analogs.

Variations in the PRPP pool has been shown to occur in response to carbon source, addition of bases and nucleosides, as well as in starvation experiments.[1,2,3] PRPP synthetase (ATP:ribose-5P pyrophosphotransferase, EC 2.7.6.1) is inhibited by nucleotides, ADP being the most potent inhibitor,[4] and the enzyme level is under repressive control of uridine nucleotides.[5]

In order to study the effects of variations of the PRPP pool, not imposed by addition of bases exogenously which simultaneous results in a raise in nucleotide pools, we sought for a mutant in PRPP synthetase.

The isolation of a mutant in *Salmonella typhimurium* LT7 with a defect in PRPP synthetase arose from a selection protocol designed to give regulatory mutants in GMP reductase: Using a purine requiring strain with additional defects in guanine(xanthine) phosphoribosyltransferase and purine nucleoside phosphorylase,

GP74 : purE66,(proAB47-gpt), pup

which is unable to grow on guanosine (GR) as sole purine source,mutants were selected for their ability to grow

§ Present address: Department of Molecular Biology, University of Aarhus, Denmark.

on guanosine.

Among such mutants, mutants blocked in purine de novo enzymes prior to purE (see figure 1) were a frequent class. Mutants with such de novo blocks were easily detected, being unable to accumulate aminoimidazole ribotide (AIR) during purine starvation. Other studies have indicated AIR as an inhibitor of GMP reductase, rendering strains accumulating this compound unable to grow on guanosine, for which GMP reductase is a prerequisite (fig.1).

One mutant escaped classification as a de novo mutant in that it was still able to form AIR in overnight starvation experiments, however careful examination revealed a significant reduction in AIR formation in mutant GP122 (see figure 2). No defects in the classical purine de novo pathway could be detected, except for the original purE defect. This left us with the alternative that PRPP synthetase might be regarded as an extension of the purine de novo pathway (ie. including the substrates).

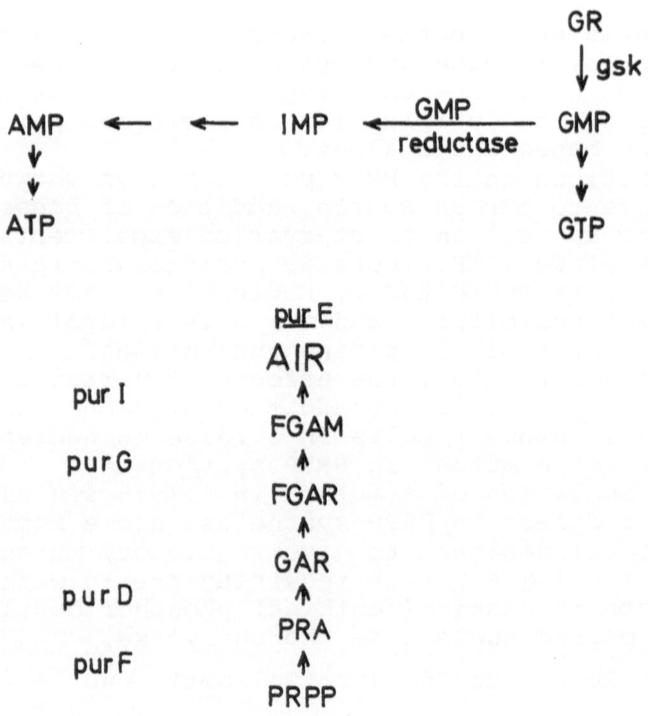

Fig.1 Pathway for the utilization of guanosine, gsk: guanosine kinase, and the de novo pathway leading to the formation of AIR. purX denotes gene symbols for the respective enzymes[6].

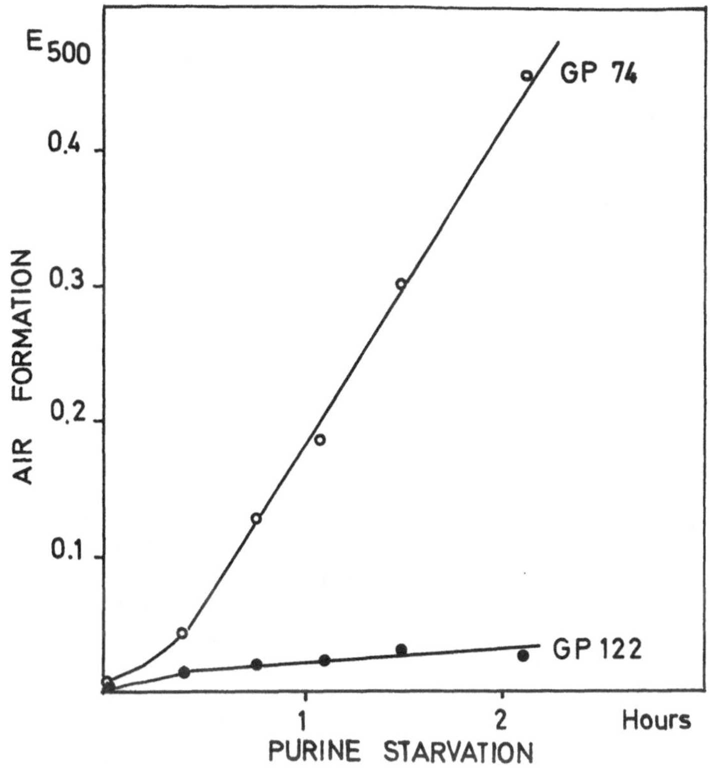

Fig.2 Formation of aminoimidazole ribotide in parent GP74 and in mutant GP122, after shift from media containing hypoxanthine as purine source to purine free media.

In order to establish GP122 as a mutant in PRPP synthetase, enzyme levels were measured and showed a reduction to 25% of that of the parent:

PRPP synthetase activity (nmoles/min/mg)	
GP74	71,3
GP122	16,5

However unequal growth of the two strains forced us to use other methods to show that the mutation resides in PRPP synthetase. This was done by testing the heat stability of the enzyme. Table 1 demonstrates that the enzyme from GP122 is more labile to heat and thereby that the mutation must be located in the structural gene for PRPP synthetase. No alteration was found in K_m towards ribose-5P, (ATP and inhibition pattern by ADP has not been tested).

Heat inactivation		20 min at 56°C(a)/50°C(b)	
	before	after	residual
GP74	57,5	24,1[a]	42%
GP122	8.6	1,3[a]	15%
GP74/F	39,5	28,4[b]	72%
GP122/F	19,3	1,6[b]	8%

Table 1 PRPP synthetase activity (nmoles/min/mg) in dialyzed crude extracts assayed at 37°C before and after heat treatment. Assay measured ribose-5P dependent conversion of ^{32}P ATP into PRPP. /F represents strains carrying an E.coli episome F'128, tn:10, probably carrying E.coli structural gene for PRPP synthetase.

The reduced level of PRPP synthetase activity in GP 122 had a dramatic effect on the pool of PRPP. As seen in table 2 the pool is reduced to less than 10% of that of the parent, under similar growth conditions.

This lowered pool of PRPP stimulated us to study the effect on growth, when the mutant was supplied with PRPP consuming and PRPP sparing compounds. From table 3 it is seen that at least PRPP sparing compounds stimulate growth and that reduction of the phosphate concentration in the media has a drastic effect on growth. The last observation is in accordance with experiments showing phosphate to have a stimulating effect on PRPP synthetase. Addition of tryptophan alone had no effect on growthrate, however when grown without trytophan GP122 showed an enhanced excretion of anthranilate, indicating anthranilate phosphoribosyltransferase reaction to be rate limiting for tryptophan biosynthesis. Two of the pyrimidine de novo enzymes, namely aspartate transcarbamylase and orotate phosphoribosyltransferase has been shown to be derepressed 6 and 3 times respectively.

POOLS / nmoles pr mg dry weight

	purine	gen.	ATP	GTP	PRPP	CTP	UTP
GP74	none	50	5.5	2,6	3,3	1,8	1,8
GP122	none	260	1,9	0,8	<0,1	nd	nd
GP74	GR	52	5,3	3,2	5,1	2,1	2,4
GP122	GR	76	3,9	2,7	<0,5	1,6	2,1

Table 2 Strains used in this study are pur^{+} derivatives of GP74 and GP122. Nucleotides were extracted from exponential growing cells, labeled with ^{32}P, in 1/3 M formic acid, and the nucleotides were separated on PEI plates after chromatography in 0,85 M K-PO_4 pH 3,4.

GP122 shows enhanced resistance to fluorouracil (0,5-2 μg/ml). This could be due either to the lowered PRPP pool or enhanced pyrimidine *de novo* synthesis caused by elevated levels of pyrimidine enzymes, if not a combination hereof.

GP122 GROWTH PROPERTIES/ generation time (min)		
purine supplement: hypoxanthine		
additional supplements		
none	101[a]	320[b]
uracil	103	
tryptophan+histidine	82	
adenine+histidine	77	
guanosine	60	

Table 3 growth monitored in minimal salt media containing a) 50 mM phosphate and b) 1 mM phosphate

The mutation causing the altered PRPP synthetase has been located at 7 min on the new *Salmonella* map.[7] P22 transduction gave a cotransduction frequency of 84% between *prpP-1* (GP122 PRPP synthetase mutation)and the *proAB47* deletion (including *gpt*, guanine phosphoribosyltransferase).An *E.coli* episome F'128 (pro-lac) has been shown to complement the *prpP-1* mutation, but only partly, which could be due to either subunit mixing or low expression of the *E.coli* PRPP synthetase gene in *Salmonella* (see table 1).

A valuable perspective of GP122 could be the possibility to select regulatory mutants in PRPP synthetase, containing high levels of the enzyme and/or high pools of PRPP. Temperature sensitive mutants of PRPP synthetase should also be possible. The selection protocol to be used, can use the fact that lowering phosphate concentration results in impaired growth. Media containing 0,3 mM phosphate + 50 μg/ml hypoxanthine has been used in obtaining revertants in GP122, now able to grow in this media. Such revertants loose the ability to grow on guanosine as sole purine source, the original selection principle for the PRPP synthetase mutant.

PRPP synthetase mutants in *Escherichia coli* has been isolated and characterized by B. Hove-Jensen and P. Nygaard from our laboratory (manuscript in preparation).

REFERENCES

1. Bagnara,A.S. and L.R. Finch, Relationships between Intracellular Contents of Nucleotides and 5-Phosphoribosyl 1-Pyrophosphate in Escherichia coli, Eur.J. Biochem.36:422-427 (1973).
2. Bagnara,A.S. and L.R. Finch, The Effects of Bases and Nucleosides on the Intracellular Contents of Nucleotides and 5-Phosphoribosyl 1-Pyrophosphate in Escherichia coli, Eur.J.Biochem. 41: 421-430 (1974).
3. Sadler,W.C. and R.L. Switzer, Regulation of Salmonella Phosphoribosylpyrophosphate Synthetase Activity in Vivo, J.Biol.Chem. 252: 8504-8511 (1977)
4. Switzer,R.L. and D.C. Sogin, Regulation and Mechanism of Phosphoribosylpyrophosphate Synthetase, J.Biol. Chem. 248: 1063-1073 (1973)
5. Olszowy,J. and R.L. Switzer, Specific Repression of Phosphoribosylpyrophosphate Synthetase by Uridine Compounds in Salmonella typhimurium, J.Bacteriol. 110: 450-451 (1972).
6. Gots,J.S.,C.E.Benson,B.Jochimsen and K.R. Koduri, Microbial models and regulatory elements in the control of purine metabolism, in Ciba Foundation Symposium 48(new series) Elsevier/Excerpta Medica North-Holland 1977.
7. Sanderson, K.E. and P.E. Hartman, Linkage Map of Salmonella typhimurium, Microbiol. Rev. 42: 471-519 (1978).

METHYLMERCAPTOPURINE RIBONUCLEOSIDE TOXICITY IN HUMAN FIBROBLASTS: INHIBITION OF PHOSPHORIBOSYLPYROPHOSPHATE SYNTHETASE AS WELL AS AMIDOPHOSPHORIBOSYLTRANSFERASE

Richard C.K. Yen and Michael A. Becker

Veterans Administration Medical Center
San Diego, California 92161 USA and University of California, San Diego, La Jolla, California

INTRODUCTION

Abundant evidence has been presented to indicate that phosphoribosylpyrophosphate (PRPP) is the rate-limiting substrate for purine synthesis de novo (for review, see ref.1). Cultured cells derived from patients with purine overproduction have increased intracellular PRPP concentrations resulting from either increased PRPP synthesis or diminished PRPP utilization in alternative pathways (2-5). PRPP synthesis is catalyzed by PRPP synthetase (E.C. 2.7.6.1). Several distinct superactive forms of this enzyme have been described in association with increased PRPP generation and increased rates of purine synthesis de novo (3-6). Among these mutant forms of PRPP synthetase are enzymes with: increased maximal velocity of the reaction but normal sensitivity to purine nucleotide inhibition of enzyme activity (7); normal maximal velocity but altered regulatory properties (3,6); both altered catalytic and regulatory properties (8).

The adenosine analogue, methylmercaptopurine ribonucleoside (MMPR), has been utilized to select for cultured cells with increased rates of PRPP formation (9). The rationale underlying this approach is that MMPR blocks purine synthesis de novo at the amidophosphoribosyltransferase (E.C. 2.4.2.14) reaction (10) and that cells blocked in this pathway can survive only by utilizing the PRPP-dependent purine base salvage reactions at increased rates. In rat hepatoma (HTC) cells, a mutant cell line with a PRPP synthetase resistant to nucleotide feedback inhibition has been isolated in a medium containing MMPR, adenine and uridine (9). In addition, medium with MMPR, hypoxanthine and uridine has been successfully employed to permit preferential growth of human fibroblasts bearing a nucleotide-resistant PRPP synthetase (11). If inhibition of amidophosphoribosyltransferase is the sole site at which MMPR toxicity is exerted, any cell with a sufficiently superactive PRPP synthetase should survive in medium containing MMPR and supplemented with a purine base and uridine. Since, however, in both of above cases, the form of PRPP synthetase selected in MMPR-containing medium was feedback-resistant, we have inquired as to whether resistance to feedback inhibition is essential to survival of cells in such medium. We have found that four fibroblast strains (from individuals in three different families) with superactive but feedback-responsive PRPP synthetases failed to survive in supplemented MMPR-containing medium. In contrast, a cell strain (SM) bearing a PRPP synthetase with altered regulatory as well as catalytic properties (8) survived treatment with medium containing MMPR regardless of whether hypoxanthine (or adenine) and uridine were added. These findings suggested that toxicity of MMPR might be mediated by inhibition of PRPP synthesis as well as by inhibition of the amidophosphoribosyltransferase reaction. The following experiments were,therefore, designed to study PRPP metabolism in cultured fibroblasts exposed to MMPR and to assess the effect of MMPR and its metabolites on PRPP synthetase in vitro.

MATERIALS AND METHODS

Fibroblast strains were initiated and propagated in monolayer as previously described (2) from skin biopsy specimens obtained from two normal subjects and five patients (in four families) with superactive PRPP synthetases (4,5,8). Eagle's minimum essential medium (MEM) supplemented with 10% fetal calf serum, 2 mM glutamine, 100 μg/ml streptomycin and 100 units/ml penicillin (G^+ medium) was used routinely for continuous propagation of cells. For overnight incubation prior to the studies described below, G^+ medium and two additional media were employed. These were: G^- medium which was identical to G^+ medium except for deletion of glutamine; and M^+ medium, which was G^+ medium containing 0.2 mM MMPR. Supplementation of M^+ medium with 0.2 - 0.4 mM hypoxanthine (or 0.2 - 0.4 mM adenine) and 0.5 mM uridine was not routinely carried out except as specifically indicated below. As previously demonstrated (12), overnight growth in G^- medium results in reversible inhibition of the early steps of purine synthesis de novo.

The rate of the early reactions of purine synthesis de novo were estimated by measurement of [^{14}C] formate incorporation into formylglycinamide ribotide (FGAR) as previously described (2), except that experiments were carried out in glutamine-free MEM supplemented with 10 mM glutamine (with or without 0.2 mM MMPR) where appropriate. PRPP concentration and generation (5) and activity of PRPP synthetase (5) were measured as previously described.

Assays for Pi concentration (13), adenosine kinase activity (in the presence of an adenosine deaminase inhibitor (EHNA) (14) and production of [^{14}C] MMPR (15) were carried out as previously reported. MMPR-mononucleotide was a gift from Dr. D.J. Nelson of Burroughs-Wellcome Research Laboratories.

RESULTS

Growth curves for a normal fibroblast strain and two strains with superactive PRPP synthetases in G^+ and M^+ media are shown in Figure 1. Although strain HB, derived from a patient with a superactive PRPP synthetase with normal regulatory characteristics, has previously been shown to demonstrate increased PRPP concentration and generation during growth in G^+ medium (16), this strain and the normal fibroblast strain failed to grow in M^+ medium (with or without an added purine base and uridine). Three additional fibroblast strains with excessive PRPP synthetase activities due to increased maximal reaction velocities associated with normal regulatory properties showed similar growth curves in supplemented and unsupplemented M^+ medium. In contrast, strain SM, with a PRPP synthetase altered in regulatory and catalytic functions (8) underwent a doubling in 3 days in M^+ medium (supplemented and unsupplemented) and thereafter achieved a growth rate similar to that of strain SM in G^+ medium. While cells with excessive PRPP generation but normal feedback responsiveness failed to grow in M^+ medium, cells with combined alterations in regulatory and catalytic properties of the enzyme grew. The regulatory defect thus appeared to modulate the growth suppressive effects of M^+ medium,suggesting MMPR toxicity in cells with superactive PRPP synthetases required a feedback-sensitive enzyme. This observation implied that MMPR growth inhibition was mediated, at least in part,by an effect on PRPP synthetase.

To exclude the possibility that survival of SM cells in M^+ medium reflected resistance of amidophosphoribosyltransferase to MMPR metabolites, rates of incorporation of [^{14}C] formate into [^{14}C] FGAR (in the presence of azaserine) were measured in cells grown for 24 h in G^+, G^- and M^+ media. Table 1 shows that rates of purine synthesis *de novo* in normal and SM cells grown in G^- medium and assayed in glutamine-free MEM were diminished in comparison to cells of the same strains grown in G^+ medium and assayed in glutamine-containing MEM. Baseline rates of purine synthesis *de novo* in glutamine-deprived cells could, however, be restored by addition of glutamine to the assay mixture. In contrast, addition of glutamine failed to reverse the inhibition of purine synthesis in MMPR-treated cells. Since both strains incubated with MMPR had basal rates of purine synthesis *de novo* comparable to the respective cell strains deprived of glutamine, growth of strain SM in M^+ medium did not appear to be due to resistance of amidophosphoribosyltransferase in this strain to inhibition by MMPR or its metabolites.

Table 1

RATES OF PURINE SYNTHESIS DE NOVO IN SM AND NORMAL FIBROBLASTS

Culture Medium	Glutamine in Assay	[^{14}C] Formate Incorporation into FGAR	
	mM	cpm/h/10^6 cells ($\times 10^{-3}$)	
		normal	SM
G^+	0	20.7	36.4
	10	68.0	181.9
G^-	0	21.4	29.5
	10	87.8	153.8
M^+	0	23.1	34.4
	10	24.4	33.2

Similar magnitude of inhibition of purine synthesis de novo resulting from incubation with MMPR or from glutamine deprivation should be accompanied by comparable increases in intracellular PRPP concentrations in cells grown in M^+ and G^- media. Less PRPP accumulation after growth in M^+ medium than G^- medium would imply an additional effect of MMPR on PRPP synthesis. We observed that while glutamine-deprived normal, HB, and SM fibroblasts had higher

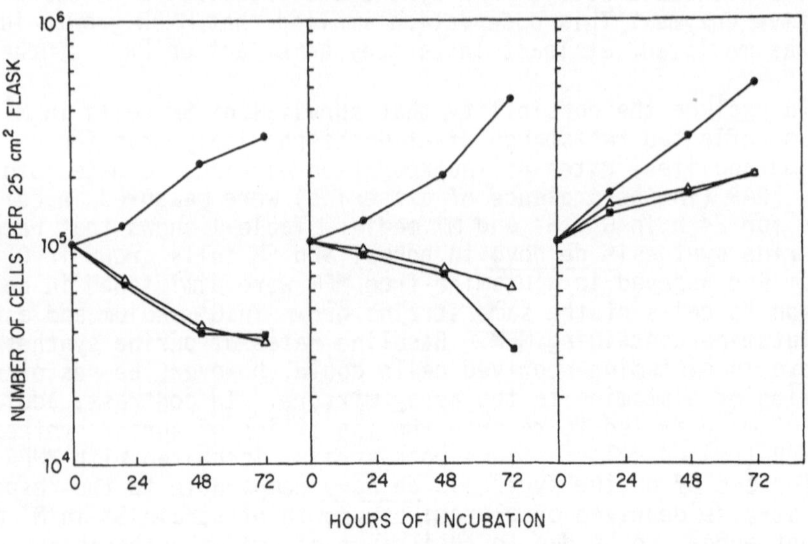

Figure 1. Growth of normal (left), HB (center), and SM (right) fibroblasts in G^+ (●), M^+ (△), and supplemented M^+ (■) media.

concentrations of PRPP(48,71,570 pmoles/10^6 cells, respectively) than cells of these strains incubated in G^+ medium (24,30,147) pmoles/10^6 cells, respectively), normal and HB fibroblasts incubated with MMPR had only 60% and 55%, respectively, of the PRPP concentrations found in normal and HB cells grown in G^- medium. In contrast, concentrations of PRPP were nearly identical in SM cells whether they were incubated in M^+ or G^- media.

Direct measurement of the rates of PRPP generation in intact cells showed that normal fibroblasts incubated for 24 h in M^+ medium incorporated [^{14}C] adenine at an average rate of 18 pmoles/min/10^6 cells, compared with 80 pmoles/min/10^6 cells observed in normal fibroblasts incubated in G^- medium. In contrast, SM fibroblasts incubated in M^+ and G^- medium had similar rates of [^{14}C] adenine incorporation (229 and 246 pmoles/min/10^6 cells, respectively). These studies showed that in normal fibroblasts PRPP synthesis is inhibited during incubation in M^+ medium, an effect to which SM cells appear resistant.

Short-term incubation of erythrocytes with purine nucleosides, including MMPR (at concentrations greater than the 0.2 mM used nere), has been shown to result in diminished PRPP synthesis, in part due to Pi depletion (17). The possibility that the apparent resistance of SM PRPP synthesis to inhibition by M^+ medium resulted entirely from either a higher concentration of intracellular Pi during incubation in M^+ medium or from the greater responsiveness of SM PRPP synthetase to Pi activation (the kinetics of which are hyperbolic for SM enzyme and sigmoidal for normal and HB enzymes) (8) was tested. Normal and SM fibroblasts were found to have nearly identical intracellular Pi concentrations after growth for 24 hours in M^+ medium (2.4 - 2.5 mM). Normal and SM fibroblasts in G^+ medium had Pi concentrations of 2.3 mM and 2.1 mM, respectively. Thus, incubation with 0.2 MMPR did not result in intracellular Pi depletion. In addition, supplementation of M^+ medium with Pi in concentrations up to 10 mM during growth studies increased intracellular Pi concentration of normal and HB cells more than 2-fold, but did not result in growth of these cells in M^+ medium.

Preferential growth of SM cells in M^+ medium was not explained by differences in MMPR metabolism. First, as shown above, normal and SM cells showed similar degrees of inhibition of purine synthesis <u>de novo</u> during incubation in M^+ medium, suggesting that MMPR uptake and phosphorylation was comparable in these cell strains. Second, adenosine kinase activities in normal, HB and SM cells were similar (187.6, 186.0 and 175.3 nmols/h/mg protein, respectively). Third, intracellular concentrations of [^{14}C] MMPR-mononucleotide in normal and SM firbroblasts incubated with [^{14}C] MMPR for 24 hours were similar (1.25 and 1.01 mM, respectively). Thus, studies of MMPR effects on PRPP metabolism and of MMPR metabolism in normal, HB and SM fibroblasts supported the hypothesis that MMPR metabolites act directly on PRPP synthetase to diminish activity of this enzyme.

Inhibition of PRPP synthetase activities in Sephadex G-25-treated normal, HB, and SM fibroblast extracts by MMPR-mononucleotide, the major metabolite of MMPR, was carried out at 1 mM Pi and subsaturating concentrations of substrates (50 μM MgATP, 50 μM ribose-5-phosphate). Addition of up to 1 mM MMPR-mononucleotide resulted in greater than 80% inhibition of normal and HB fibroblast PRPP synthetases; in contrast, inhibition of SM PRPP synthetase was only 30%. In additional studies to delinate the mechanism of MMPR-mononucleotide inhibition of PRPP synthetase, this compound was demonstrated to inhibit purified human erythrocyte PRPP synthetase by a mechanism which is competitive with MgATP and uncompetitive with ribose-5-phosphate.

DISCUSSION

Medium containing MMPR has been utilized in rat hepatoma (HTC) cells to select for a mutant cell line with increased rates of PRPP synthesis due to a feedback-resistant PRPP synthetase (9). The present studies, however, demonstrate that not all mutant cells that synthesize PRPP at an excessive rate and have superactive PRPP synthetases can survive in this selective medium. The data suggest that resistance to MMPR-mononucleotide inhibition of PRPP synthetase is essential to survival of PRPP synthetase mutants during growth in MMPR. While strain SM contains a PRPP synthetase with both abnormal regulatory and catalytic functions (8), the latter alteration does not appear to account for cell survival in M^+ medium. Strain HB and three other fibroblast strains that have similarily increased maximal reaction velocities failed to grow in medium containing MMPR. In addition, the mutant rat hepatoma cell line which survived selection in MMPR did not show increased maximal specific catalytic activity (9).

In addition to confirming that incubation of fibroblasts with MMPR results in inhibition of purine synthesis de novo, these studies showed an inhibitory effect of incubation with MMPR on PRPP synthesis in intact fibroblasts. Inhibition of PRPP generation by MMPR-mononucleotide in fibroblasts with superactive, feedback-sensitive PRPP synthetases provides a plausible explanation for the failure of such strains to survive in M^+ medium even when supplemented with purine bases and uridine.

In their study on inhibition of PRPP synthesis by nucleosides, Planet and Fox (17) demonstrated that MMPR-monophosphate inhibits erythrocyte PRPP synthetase activity in vitro. We observed that normal and HB fibroblast PRPP synthetases and purified erythrocyte PRPP synthetase were inhibited by MMPR-mononucleotide, but that the enzyme derived from strain SM was relatively resistant to MMPR-mononucleotide inhibition. The concentration of MMPR-mononucleotide required for 50% inhibition of normal PRPP synthetase *in vitro* (0.35 mM) is four orders of magnitude greater than the concentration required to achieve a 50% inhibition in purine synthesis *de novo* in lymphoblast cells (18). The latter pathway thus may be more sensitive than PRPP synthesis to inhibition by MMPR-mononucleotide. This suggests the possibility of selecting for cells bearing superactive PRPP synthetases other than feedback-resistant forms in media containing MMPR at reduced concentrations.

REFERENCES

1. M.A. Becker, in Uric acid, Handbook of Experimental Pharmacology Vol. 51 (W.N. Kelley and I.M. Weiner, editors) Springer-Verlag Berlin 1978, p155.
2. F.M. Rosenbloom, J. F. Henderson, I.C. Caldwell, W.N. Kelley, and J.E. Seegmiller, *J.Biol.Chem.* 243:1166 (1968).
3. E. Zoref, A. de Vries, and O. Sperling, *J.Clin.Invest.* 56: 1093 (1975).
4. M.A. Becker, J.J. Meyer, A.W. Wood, and J.E. Seegmiller, *Science* 179:1123 (1973).
5. M.A. Becker, *J.Clin.Invest.* 57:308 (1976).
6. O. Sperling, P. Boer, S. Persky-Brosh, E. Kanarek and A. de Vries, *Rev.Eur.Etud.Clin.Biol.* 17:703 (1972).
7. M.A. Becker, P.J. Kostel, and L.J. Meyer, *J.Biol.Chem.* 250: 6822 (1975).
8. M.A. Becker, K.O. Raivio, B. Bakay, W.B. Adams, and W.L. Nyhan, These proceedings.
9. C.D. Green, and D.W. Martin, Jr. *Proc.Natl.Acad.Sci.* 70: 3698 (1973).
10. J.F. Henderson, and M.K.Y. Khoo, *J.Biol.Chem.* 240:3104 (1965).
11. E. Zoref, A. de Vries, O. Sperling, *Hum.Hered.* 27:73 (1977).
12. K.O. Raivio, and J.E. Seegmiller, *Biochim.Biophys.Acta* 299: 283 (1973).
13. P.S. Chen, T.Y. Toribara, and H. Warner, *Anal.Chem.* 28: 1756 (1956).
14. M.S. Hershfield, F.F. Snyder, and J.E. Seegmiller, *Science* 197:1284 (1977).
15. I.C. Caldwell, J.F. Henderson, and A.R.P. Paterson, *Can.J. Biochem.* 44:229 (1966).
16. M.A. Becker, L.J. Meyer, and J.E. Seegmiller, *Am.J.Med.* 55: 232 (1973).
17. G. Planet, and I.H. Fox, *J.Biol.Chem.* 251:5839 (1976).
18. M.S. Hershfield, and J.E. Seegmiller, *Adv.Exp.Med.Biol.* 76A:19 (1976).

ACKNOWLEDGMENTS

This work was supported by Veterans Administration Medical Research Service and NIH Grant AM 18197.

ADENOSINE KINASE: REGULATION BY SUBSTRATES, MAGNESIUM AND pH

Richard L. Miller and David L. Adamczyk

Wellcome Research Laboratories

Research Triangle Park, N.C. 27709

Adenosine kinase phosphorylates many pharmacologically active nucleosides [1,2]. Several studies have been carried out on the partially purified enzyme from a number of species (see Refs. in 3 and Ref. 4). A striking observation from these reports is the large number of "optimal conditions" reported for the assay of the enzyme from these different sources. In the past, adenosine kinase has been assayed at adenosine concentrations as high as 1000 times that found in vivo. Under these conditions, the enzyme from both Sarcoma 180 cells[5] and rabbit liver[3] has been reported to be inhibited by adenosine. This inhibition and its relationship to the observed optimal assay conditions for the highly purified rabbit liver enzyme[3] is the subject of this report.

In a recent study, Arch and Newsholme[4] observed that adenosine inhibition of adenosine kinase in crude extracts from a variety of sources was pH dependent. This was also observed with the purified rabbit liver enzyme (Fig. 1A). Adenosine inhibition became apparent only at concentrations of adenosine >100 μM when the enzyme was assayed at pH 5.3. In contrast, at pH 6.8, concentrations of adenosine >2 μM were inhibitory. As a consequence, when assayed at 100 μM adenosine, the enzyme appeared to have a sharp optimum at pH 5.2 (Fig. 1B). This finding is in agreement with many previous studies reporting a pH optimum in the range of pH 5.2 - 5.8. As with the enzyme from other sources, the rabbit liver enzyme, when assayed at a fixed magnesium concentration, exhibited a relatively sharp optimum with regard to the ATP concentration (Fig. 1C). The same pattern with respect to magnesium concentration was observed when the enzyme was assayed at a fixed ATP concentration (Fig. 1D). From studies of this type, the

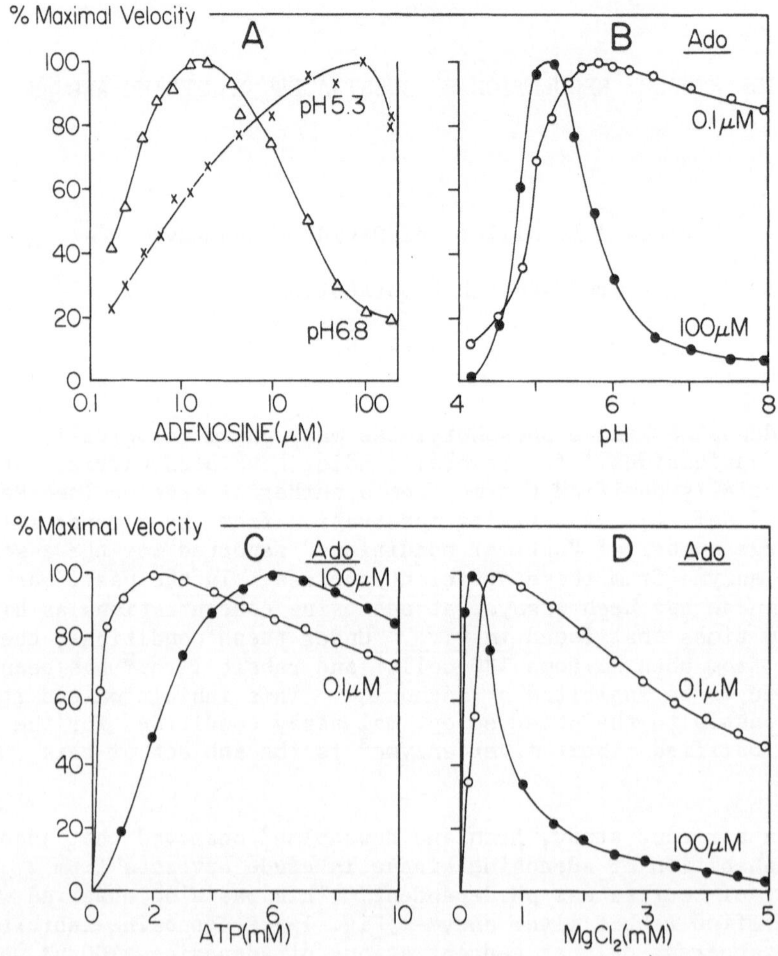

Fig. 1. Effect of reaction conditions on the initial velocity. Assays[3] using 50 mM sodium buffers were as follows: (A) acetate buffer, pH 5.3 or Pipes, pH 6.8; 1 mM ATP and 1 mM $MgCl_2$; (B) overlapping buffers of acetate, cacodylate, Pipes, Hepes; 1mM ATP and 1 mM $MgCl_2$; (C) Pipes, pH 6.8 and 1 mM $MgCl_2$; (D) Pipes, pH 6.8 and 1 mM ATP.

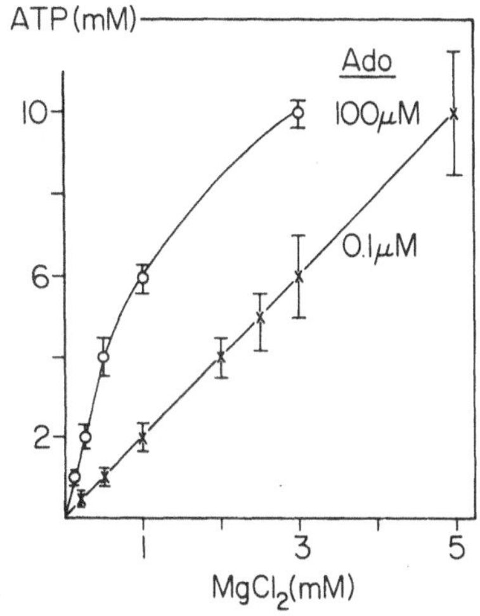

Fig. 2. Effect of adenosine concentration on the optimal ATP/Mg ratio. Assays[3] were conducted in 50 mM sodium Pipes, pH 6.8.

optimal ATP/Mg ratio was found to be a non-linear function of the concentration of both ATP and magnesium (Fig. 2).

When the enzyme was assayed at 0.1 μM adenosine, a concentration well below that which was shown to be inhibitory at pH 6.8 (Fig. 1A), it exhibited a broad pH optimum between 5.3 and 8 (Fig. 1B). In contrast to the findings at 100 μM adenosine, the optimal ATP/Mg ratio appeared to be approximately 2 (Figs. 1C and 1D) and was independent of the concentration of either ATP or magnesium (Fig. 2).

The decline in the activity which was observed upon increasing the concentration of ATP at a fixed magnesium concentration (Fig. 1C) or increasing the concentration of magnesium at a fixed ATP concentration (Fig. 1D) suggests that both free ATP and free magnesium can act as inhibitors of the enzyme reaction.

Further studies demonstrated that the concentration of adenosine required to cause inhibition is influenced by the concentration of both ATP and magnesium. As the magnesium concentration was increased, substrate inhibition by adenosine became apparent at lower adenosine concentrations (Fig. 3A). In contrast, with increasing ATP concentrations, higher adenosine concentrations were required for inhibition to be observed (Fig. 3B). Upon analysis, the data in Fig. 3 were found to fit the standard model for substrate inhibition, $v=(VA)/[K_m + A + (A^2/K_i)]$, thus suggesting that at higher adenosine concentrations, two

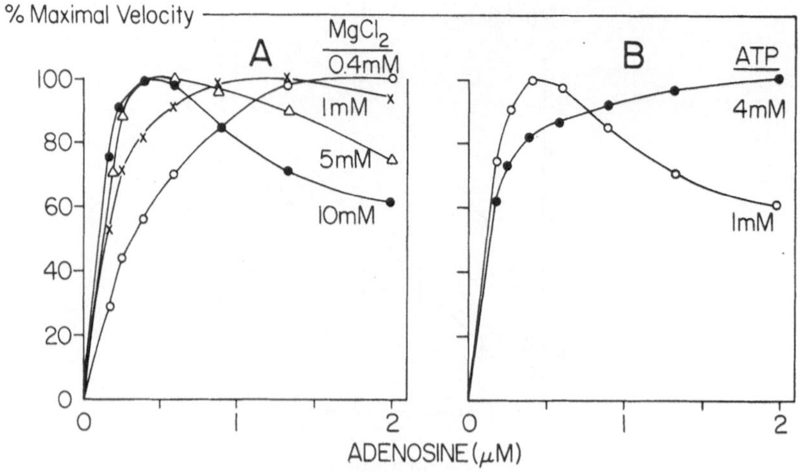

Fig. 3. Effect of the concentration of $MgCl_2$ and ATP on the inhibition by adenosine. Assays[3] in 50 mM sodium Pipes, pH 6.8 contained (A) 1 mM ATP; (B) 10 mM $MgCl_2$.

molecules of adenosine are capable of binding to the enzyme. The K_i for adenosine calculated under a variety of conditions (ie. varying pH and different ATP and magnesium concentrations) was approximately 4 μM.

In conclusion it appears that the activity of rabbit liver adenosine kinase can be controlled by a variety of interrelated factors which include:

A) substrate inhibition by adenosine,

B) inhibition by free ATP and free magnesium, and

C) by pH which affects both the enzyme and the equilibrium involved with the formation of the ATP-Mg complex.

References

1. H. P. Schnebli, D. L. Hill, and L. L. Bennett, Jr., Purification and Properties of Adenosine Kinase from Human Tumor Cells of Type H. Ep. No. 2, J. Biol. Chem. 242:1997 (1967).

2. R. L. Miller, D. L. Adamczyk, W. H. Miller, G. W. Koszalka, J. L. Rideout, L. M. Beacham, III, E. Y. Chao, J. J.

Haggerty, T. A. Krenitsky, and G. B. Elion, Adenosine Kinase from Rabbit Liver. II. Substrate and Inhibitor Specificity, J. Biol. Chem. 254:2346 (1979).

3. R. L. Miller, D. L. Adamczyk, and W. H. Miller, Adenosine Kinase from Rabbit Liver. I. Purification by Affinity Chromatography and Properties, J. Biol. Chem. 254:2339 (1979).

4. J. R. S. Arch and E. A. Newsholme, Activities and Some Properties of 5'-Nucleotidase, Adenosine Kinase and Adenosine Deaminase in Tissues from Vertebrates and Invertebrates in Relation to the Control of the Concentration and the Physiological Role of Adenosine, Biochem. J. 174:965 (1978).

5. A. Y. Divekar and M. T. Hakala, Adenosine Kinase from Sarcoma 180 Cells. N^6-Substituted Adenosines as Substrates and Inhibitors, Mol. Pharmacol. 7:663(1971).

ADENOSINE AND DEOXYADENOSINE KINASE FROM RAT LIVER

Nobuaki Ogasawara, Yasukazu Yamada and Haruko Goto

Department of Biochemistry, Institute for Developmental Research, Aichi Prefectural Colony, Kasugai, Aichi-480-03, Japan

INTRODUCTION

Recent studies demonstrated that dATP concentration was markedly increased in erythrocytes from patient with adenosine deaminase deficiency[1,2]. Deoxyadenosine is toxic to T lymphoblast in the presence of an inhibitor of adenosine deaminase. By the addition of deoxyadenosine, dATP accumulated in the cell and the level of dATP well corelated with the cytotoxicity[4,5].

We are interested in whether adenosine and deoxyadenosine are phosphorylated by separate enzyme activities or by the same enzyme activity. Previously Streeter et al[6]. copurified ribavirin, adenosine and deoxyadenosine kinase activity from rat liver, but concluded that adenosine and deoxyadenosine kinase are separate enzyme activities and that phosphorylation of ribavirin is associated with the latter activity.

MATERIALS METHODS

ATP, phosphoenolpyruvate, NADH, pyruvate kinase and lactate dehydrogenase were purchased from Boehringer, Mannheim. Nucleosides were obtained from Sigma Chemical Co. and Boehringer. Sephadex and AMP-Sepharose 4B were obtained from Pharmacia.

Nucleoside kinase was assayed by enzymatically coupling the formation of ADP to NADH oxidation. NADH oxidation was measured by following decrease in absorbance at 340 nm minus 400 nm on a 356 Hitachi two wavelength double beam spectrophotometer at 30°C. Unless otherwise specified, the assay mixtures contained 64 mM Tris-

HCl, pH 7.5, 3.8 mM EDTA, 180 mM KCl, 5 mM $MgCl_2$, 24 mM ammonium sulfate, 1 mM ATP, 0.5 mM phosphoenolpyruvate, 0.1 mM NADH, 5 units of pyruvate kinase, 13.8 units of lactate dehydrogenase, 25 μM adenosine or 1 mM deoxyadenosine and appropriate amount of kinase in a final volume of 1 ml. Reactions were initiated by the addition of nucleoside and 1 unit of enzyme activity is defined as the amount catalyzing the phosphorylation of 1 nmole of nucleoside in 1 min.

RESULTS AND DISCUSSION

Purification of Enzyme

Fresh rat liver(40 g) was homogenized in 120 ml of buffer A which consisted of 10 mM Tris-HCl, pH 7.5, and 0.15 M KCl. The homogenate was centrifuged at 20,000 x g for 30 min and recentrifuged at 105,000 x g for 90 min. The supernatant was saved and solid ammonium sulfate was added to 40 % of saturation. After centrifugation, solid ammonium sulfate was added to the supernatant to 80 % of saturation. The solution was centrifuged and the pellet was dissolved in buffer B; 10 mM Tris-HCl, pH 7.5; 10 mM $MgCl_2$; 1 mM dithiothreitol.

The solution was dialyzed against buffer B overnight, and applied to a column of DE-52 ($3.2\ cm^2$ x 20 cm) equilibrated in buffer B. After washing with 600 ml of buffer B, the kinase activity was eluted with buffer B containing 0.1 M KCl. The fractions con-

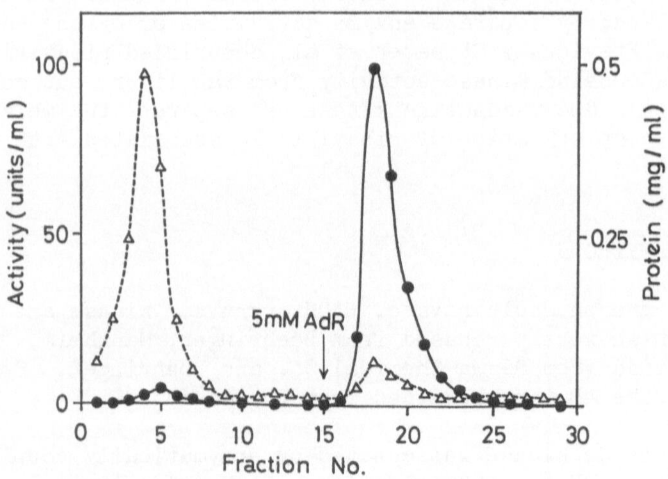

Fig. 1. AMP-Sepharose affinity chromatography. Kinase was adsorbed to AMP-Sepharese and eluted with 5 mM deoxyadenosine. (●), deoxyadenosine kinase activity. (Δ), protein.

Table 1. Purification of Adenosine and Deoxyadenosine Kinase

Fraction	Total Volume	Total Activity[b]	Total Protein	Specific Activity[b]	AR/AdR[a]
	(ml)	(units)	(mg)		
AS (45-80%)	30.0	5490	777.0	7.1	0.575
DE-52	4.2	2163	58.8	36.7	0.631
Sephadex G-100	24.0	2148	14.4	147.2	0.704
DE-52	13.2	2085	7.13	292.4	0.683
AMP-Sepharose	8.4	1476	0.87	1688.8	0.725

[a]Adenosine kinase activity/deoxyadenosine kinase activity.

[b]Deoxyadenosine as substrate.

taining kinase activity were pooled and concentrated by the addition of solid ammonium sulfate to 90 % of saturation. After centrifugation, the pellet was dissolved in 4-5 ml of buffer B.

The solution was then applied to a column of Sephadex G-100 (5.2 cm^2 x 45 cm) equilibrated in buffer A. The column was eluted with the same buffer. The fractions containing kinase activity were pooled and dialyzed against buffer B.

The G-100 fraction was again applied to a column of DE-52 (0.75 cm^2 x 10 cm) equilibrated in buffer B. The column was eluted with a linear gradient consisting of 75 ml of buffer B and 75 ml of buffer B containing 0.1 M KCl. The fractions containing kinase activity were pooled and dialyzed against buffer C; 10 mM potassium phosphate, pH 7.0; 1 mM dithiothreitol.

The 2nd DE-52 fraction was applied to a column of AMP-Sepharose 4B(0.75 cm^2 x 10 cm) equilibrated in buffer C. After washing with buffer C, the column was eluted with buffer C containing 5 mM deoxyadenosine (Fig. 1). The fractions containing kinase activity were pooled and dialyzed against buffer B.

The stepwise purification of deoxyadenosine kinase is summarized in Table 1. The specific activity of the final preparation is about 1700 and represents a 240 fold purification over the original supernatant and a recovery of 27 % of the initial activity.

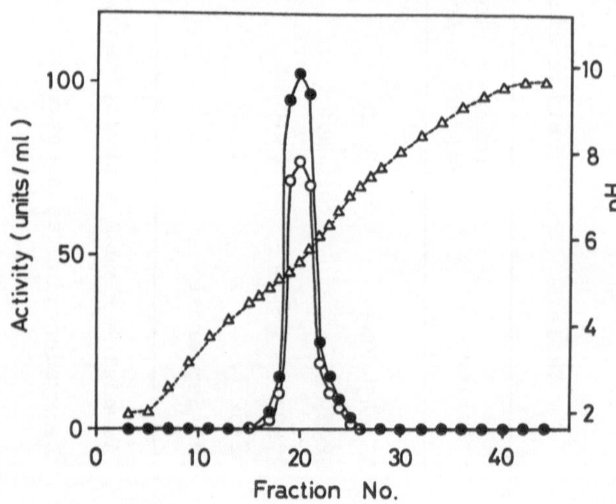

Fig. 2. Sucrose density gradient electrofocusing was carried out according to the method of Weller et al.[6] Fractions of 0.3 ml each were collected and assayed for adenosine (O) and deoxyadenosine (●) kinase activity. (Δ), pH.

Table 2. Substrate Specificy

Substrate	Concentration	Activity
	(mM)	(%)
d-Adenosine	1.0	100
Adenosine	0.025	72.5
	0.25	30.5
	1.0	15.0
d-Guanosine	0.25	0
Guanosine	0.25	0
d-Cytidine	1.0	0
Cytidine	1.0	0
d-Uridine	1.0	0
Uridine	1.0	0
d-Thymidine	1.0	0

Enzyme Properties

The final preparation was tested for purity and appeared to be homogeneous, since only a single protein band was observed by polyacrylamide gel electrophoresis both in the presence and absence of sodium dodecyl sulfate, and also by acrylamide gel electrofocusing.

When the native enzyme preparation was applied to a Sephadex G-100 column, enzyme activity towerd adenosine and deoxyadenosine was eluted from the column as a single peak and a molecular weight of 40,000 was estimated. The molecular weight of the kinase was also determined by electrophoresis in 7.5 % polyacrylamide gel carried out in the presence of 0.1 % sodium dodecyl sulfate. The kinase migrated as a single electrophoresis species and molecular weight of 40,000 was calculated. These results indicate that the kinase has a monomeric structure.

When the final preparation was assayed with the common nucleosides and deoxynucleosides as substrate, only adenosine and deoxyadenosine showed activity (Table 2). During purification, the ratio of the ability to phosphorylate adenosine to deoxyadenosine remain constant at a value of around 0.7 after gel filtration step (Table 1).

Sucrose density gradient isoelectric focusing with Ampholine, pH 3.5 to 10, gave only one peak of phosphorylating activity toward adenosine and deoxyadenosine (Fig. 2). From these experiments, pI value of the kinase was estimated to be pH 5.5.

Under the standard assay conditions, the optimum pH for deoxyadenosine was pH 7.5-8.0. Adenosine phosphorylation, however, was optimum at pH 6.0.

When adenosine phosphorylation was studied under the standard conditions, Km values for adenosine, ATP and $MgCl_2$ were <1 µM, 297 µM and 250 µM, respectively; deoxyadenosine as substrate, 357 µM, 5.4 µM and 250 µM for deoxyadenosine, ATP and $MgCl_2$, respectively.

The requirements for optimum activity are quite different between adenosine kinase and deoxyadenosine kinase activities. However, the present data strongly suggest that a single protein species is involved in the phosphorylation of adenosine and deoxyadenosine.

REFERENCES

1. A. Cohen, R. Hirschhorn, S. D. Horowitz, A. Rubinstein, S. H. Polmar, R. Hong and D. W. Martin, Jr., Deoxyadenosine triphosphate as a potentially toxic metabolite in adenosine deaminase deficiency, Proc. Natl. Acad. Sci. USA, 75:472 (1978).
2. J. Donofrio, M. S. Coleman, J. J. Hutton, A. Daoud, B. Lampkin and J. Dyminski, Overproduction of adenine deoxynucleosides and deoxynucleotides in adenosine deaminase deficiency with severe combined immunodeficiency disease, J. Clin. Invest., 62:884 (1978).
3. B. Ullman, L. J. Gudas, A. Cohen and D. W. Martin, Jr., Deoxyadenosine metabolism and cytotoxicity in cultured mouse T lymphoma cells: a model for immunodeficiency, Cell, 14:365 (1978).
4. B. S. Mitchell, E. Mejias, P. E. Daddona and W. N. Kelley, Purinogenic immunodeficiency diseases: selective toxicity of deoxyribonucleosides to T cells, Proc. Natl. Acad. Sci. USA, 75:5011 (1978).
5. D. G. Streeter, L. N. Simon, R. K. Robins and J. P. Miller, The phosphorylation of ribavirin by deoxyadenosine kinase from rat liver. Differentiation between adenosine and deoxyadenosine kinase, Biochemistry, 13:4543 (1974).
6. D. L. Weller, A. Heaney and R. E. Sjogren, A simple apparatus and procedure for electrofocusing experiments: pI of lactate dehydrogenase and isocitrate lyase, Biochim. Biophys. Acta, 168:576 (1968).

RADIOIMMUNOCHEMICAL ANALYSIS OF HUMAN ERYTHROCYTE ADENOSINE DEAMINASE

Peter E. Daddona, Michael A. Frohman and William N. Kelley

Departments of Internal Medicine and Biological Chemistry, Human Purine Research Center, University of Michigan Medical School, Ann Arbor, Michigan U.S.A.

Adenosine deaminase (adenosine aminohydrolase, EC 3.5.4.4.) catalyzes the irreversible hydrolytic deamination of adenosine to produce inosine and ammonia. Markedly reduced or absent adenosine deaminase activity in man is associated with an autosomal recessive form of severe combined immunodeficiency disease and the relationship appears to be causal.[1,2,3,4] Affected patients present a defect of both cellular and humoral immunity characterized clinically by severe recurrent infections with a uniformly fatal outcome if untreated.

Patients with adenosine deaminase deficiency and severe combined immunodeficiency disease have been reported to exhibit virtually no enzyme activity in their erythrocytes, while other tissues of these patients demonstrate variable amounts of residual adenosine deaminating activity.[5,6,7] However, the adenosine deaminase enzymatic assay used in all of these studies could only detect biologically active protein and so may have underestimated genetic expression of the adenosine deaminase alleles.

In a previous study of fibroblast cell strains from three adenosine deaminase deficient subjects, an immunoassay based on enzyme activity revealed that one cell strain with trace adenosine deaminase activity had immunoreactive protein (CRM) representing 35% of normal, while two other adenosine deaminase deficient cell strains demonstrated less than 5% of normal CRM.[8] In this report, we have described a newly developed specific radioimmunoassay for erythrocyte adenosine deaminase which is at least 100 times more sensitive than the previously published immunoassay.[8] In addition, we have used this immunoassay for the analysis of hemolysate from

9 heterozygotes and 4 homozygotes with adenosine deaminase deficiency, representing a combined total of 8 different families.

In developing a radioimmunoassay for adenosine deaminase, our goal was to make the assay as sensitive as possible while retaining a high degree of specificity. As shown in Figure 1, the adenosine deaminase radioimmunoassay was linear over a standard enzyme concentration range of 160 - 0.63 ng/ml with a slope of -1.00 ± 0.02 (15 determinations) and a calculated least detectable dose of 0.3 ng/ml. This assay would be sensitive to a calculated enzyme activity of 0.17 nmol/min/ml based on an absolute specific activity of 563 μmol/min/mg CRM (Table I). The specificity of the radioimmunoassay for the analysis of adenosine deaminase in hemolysate was verified by finding no detectable CRM in normal hemolysate pretreated with excess adenosine deaminase antiserum and by finding no detectable CRM in one of the homozygous adenosine deaminase deficient hemolysates (R.M. Jr.).

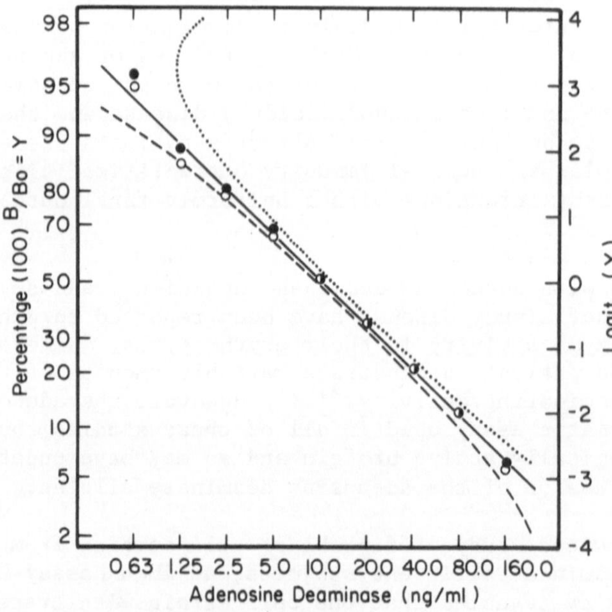

Figure 1. Adenosine deaminase radioimmunoassay. The standard curve was generated using various dilutions of either adenosine deaminase standard (●——●) or crude erythrocyte adenosine deaminase (o——o). Data are shown as the linearized logit-log transform of the response variate (100 x B/Bo) versus the log of the adenosine deaminase concentration (ng/ml) according to Rodbard.[11]

Table I

Characterization of Erythrocyte Adenosine Deaminase

Hemolysate*	Sp. Act. nmol/min/mg	ng CRM/mg	Abs. Sp. Act. μmol/min/mg CRM
Normal Adults (18)	0.88±0.24** (0.53 - 1.20)	1.67±0.60 (0.90 - 2.23)	563±46 (531 - 620)
Normal Children (5)	0.86±0.26 (0.62 - 1.20)	1.55±0.50 (1.07 - 2.22)	553±14 (540 - 572)
Heterozygote			
R.M.	1.35	2.40	562
D.M.	0.29	0.54	537
T.T.	0.74	1.47	500
E.B.	0.56	1.25	448
La.L.	0.68	1.16	586
L.L.	0.41	0.72	569
M.H.	0.56	0.94	595
F.H.	0.60	1.12	535
R.W.	0.33	1.02	320
Homozygote			
R.M., Jr.	$<0.8x10^{-3}$	<.001	-
B.W.	$<0.8x10^{-3}$	0.007	-
E.M.	$<0.8x10^{-3}$	0.052	-
M.H.	$<0.8x10^{-3}$	0.087	-

*Protein concentrations for hemolysates ranged from 287-156 mg/ml.
**Mean ± 1 S.D. range given in parentheses.

This radioimmunoassay was used in conjunction with a sensitive enzymatic assay to characterize adenosine deaminase in the hemolysate of normal subjects and of adenosine deaminase deficient subjects. The radioimmunoassay and the enzymatic assay for adenosine deaminase were equally sensitive capable of detecting 0.06% of normal mean CRM and 0.09% of normal mean enzyme activity, respectively.

Table I shows the characteristics of erythrocyte adenosine deaminase in hemolysates from normal, heterozygous and homozygous deficient subjects. Adenosine deaminase activity for 18 normal adults and 5 normal children was 0.88 ± 0.24 and 0.86 ± 0.26 nmol/min/mg hemolysate protein, respectively. Using the radioimmunoassay, the quantity of adenosine deaminase immunoreactive protein (CRM) was determined to be 1.67 ± 0.60 and 1.55 ± 0.50 ng CRM/mg hemolysate protein and the mean absolute specific activity was calculated to be 563 ± 46 and 553 ± 14 μmol/min/mg CRM, respectively, for normal adults and children. The calculated absolute specific activities compared well with the value of 515 and 538 μmol/min/mg previously reported for homogeneous preparations of erythrocyte adenosine deaminase.[9,10] The close agreement in the absolute specific activities provides further evidence for the accuracy of the adenosine deaminase radioimmunoassay used in conjunction with the enzymatic assay and confirms the absence of systematic error.

As noted in Table I, enzyme activity varied substantially from 0.53 - 1.20 nmol/min/mg hemolysate protein as did the level of CRM (0.9 - 2.23 ng/mg hemolysate protein) for the 18 normal adult hemolysates analyzed. However, using linear regression analysis, the level of enzyme activity and CRM for each sample correlated well ($r = 0.98$, 18 pairs) suggesting that the catalytic activity per molecule was similar while the concentration of enzyme varied in each hemolysate sample.

The specific activities of adenosine deaminase in the hemolysates of the heterozygous parents ranged from 0.29 to 1.35 nmol/min/mg with several samples falling within the normal adult range (Table I). The quantity of adenosine deaminase CRM among these samples also showed a wide range. However, the calculated absolute specific activity for adenosine deaminase (μmol/min/mg CRM) in the majority of hemolysates from heterozygotes appeared to fall within the normal range. Samples E.B., T.T. and R.W. were the exceptions with calculated absolute specific activities at least 1 S.D. below the normal adult mean. CRM in these three heterozygote hemolysates showed the same affinity toward adenosine deaminase antibody as did the normal erythrocyte enzyme. Further analysis of the adenosine deaminase from these subjects revealed a normal Km for adenosine and complete inhibition of enzyme activity in the presence of the competitive enzyme inhibitor erythrohydroxynonyl adenine (EHNA). Unfortunately, the hemolysates from the affected children of these heterozygote parents were either not available for analysis or were not suitable for analysis because of recent blood transfusions (with the exception of R.M., Jr.).

Variable genetic expression of the "silent" or apparently defective adenosine deaminase allele was even more apparent in the hemolysate of the homozygous deficient subjects. The hemolysates from homozygous adenosine deaminase deficient subjects all showed undetectable enzyme activity of <0.09% of the normal mean activity (representing the lower limit of our enzymatic assay) (Table I). Enzyme activity resulting from mixing normal and adenosine deaminase hemolysate was additive suggesting the absence of an adenosine deaminase inhibitor in these samples.

The hemolysate of R.M. Jr., the child of D.M. and R.M., failed to exhibit any detectable CRM, i.e., less than 0.06% of normal child mean CRM, consistent with the calculation of a normal absolute specific activity for adenosine deaminase in the hemolysate of both parents (R.M. and D.M.). In this family, the data suggested that only the "normal" adenosine deaminase allele was expressed in the hemolysate of the parents while the inherited "silent" or defective alleles did not produce a gene product immunologically detectable by this antiserum or alternatively produced a gene product which may have been rapidly degraded *in vivo*. The hemolysates of B.W., E.M. and M.H., however, all had easily detectable quantities of adenosine deaminase CRM ranging from 0.4% to 5.6% of the normal child mean CRM. From this type of analysis it was impossible to determine whether our CRM measurement reflected the expression of both adenosine deaminase alleles or only one producing an immunoreactive gene product.

In conclusion, heterozygous and homozygous hemolysate samples from 5 of the 8 total families analyzed revealed apparent variable genetic expression of the "silent" or defective adenosine deaminase allele(s). Further, hemolysate from some heterozygous subjects had quantities of the normal gene product which brought total enzyme quantities to a normal mean level suggesting possible autologous regulation of adenosine deaminase in the reticulocytes of these subjects.

REFERENCES

1. E. R. Giblett, J. E. Anderson, F. Cohen, B. Pollara and H. J. Meuwissen, Adenosine deaminase deficiency in two patients with severely impaired cellular immunity *Lancet* 2:1067-1069 (1972).

2. J. Dissing and B. Knudsen, Adenosine deaminase deficiency and combined immunodeficiency syndrome *Lancet* 2: 1316 (1972).

3. R. Parkman, E. W. Gelfand, F. S. Rosen, A. Sanderson and R. Hirschhorn, Severe combined immunodeficiency and adenosine deaminase deficiency *New Eng. J. Med.* 292:714-719 (1975).

4. H. J. Meuwissen, B. Pollara and R. J. Pickering, Combined immunodeficiency disease associated with adenosine deaminase deficiency *J. Pediatrics* 86:169-181 (1975).

5. S. H. Chen, C. R. Scott and K. R. Swedberg, Adenosine deaminase deficiency: expression of the enzyme in cultured skin fibroblasts and amniotic fluid cells, *Am. J. Hum. Genet.* 27:46-53 (1975).

6. R. Hirschhorn, N. Beratis and F. S. Rosen, Characterization of residual enzyme activity in fibroblasts from patients with adenosine deaminase deficiency and combined immunodeficiency: evidence for a mutant enzyme Proc. Natl. Acad. Sci. 73:213-217 (1976).

7. R. Hirschhorn, F. Martiniuk and F. S. Rosen, Adenosine deaminase activity in normal tissues and tissues from a child with severe combined immunodeficiency and adenosine deaminase deficiency Clin. Immunol. and Immunopath. 9:287-292 (1978).

8. D. A. Carson, R. Goldblum and J. E. Seegmiller, Quantitative immunoassay of adenosine deaminase in combined immunodeficiency disease J. Immunol. 118:270-273 (1977).

9. P. E. Daddona and W. N. Kelley, Human adenosine deaminase: purification and subunit structure J. Biol. Chem. 252:110-115 (1977).

10. W. P. Schrader, A. R. Stacy and B. Pollara, Purification of human erythrocyte adenosine deaminase by affinity column chromatography J. Biol. Chem. 251:4026-4032 (1976).

11. D. Rodbard, W. Bridson and P. L. Payford, Rapid calculation of radioimmunoassay results J. Lab. and Clin. Med. 74:770-781 (1969).

ADENOSINE DEAMINASE CONVERSION PROTEINS: A POTENTIAL ROLE

Paul P. Trotta and M. Earl Balis

Sloan-Kettering Institute for Cancer Research
1275 York Avenue, New York, New York 10021

Adenosine deaminase (ADase) catalyzes the deamination of adenosine. Its importance is underscored by clinical and *in vitro* evidence that supports a causal relationship between absence of ADase and one form of severe combined immunodeficiency disease (1). In addition, the metabolism of several anti-tumor and anti-viral nucleosides is regulated by this enzyme (2).

ADase is polymorphic in charge and molecular weight. The normal human enzyme is characterized predominantly by two cytosolic forms that differ in molecular weight: type A (mol wt ≥ *ca* 200,000) and type C (mol wt *ca* 35,000) (3). Studies in several laboratories have established the existence of a class of high molecular weight proteins, which we refer to as conversion proteins, that bind to the type C ADase with a high degree of specificity to produce the various type A enzymes. We describe here the tissue distribution of these proteins and ADase in the rabbit, and compare their physico-chemical properties from animal and human sources. Preliminary data suggest a role for the conversion proteins in anchoring adenosine deaminase to membranes.

As reported for other species, a wide range in tissue specific activity exists in the rabbit (Fig 1). Thus cells of the small intestine and lymphoid system have the highest activities while kidney, heart and lung are an order of magnitude lower. Employing gel filtration, we have characterized the percentage of ADase present as type A (high molecular weight) in each of these tissues (Fig 2). The highest percentages of type A enzyme were characteristic of tissues with low specific ADase activity. However, not all tissues with low ADase had a high percentage of type A enzyme;

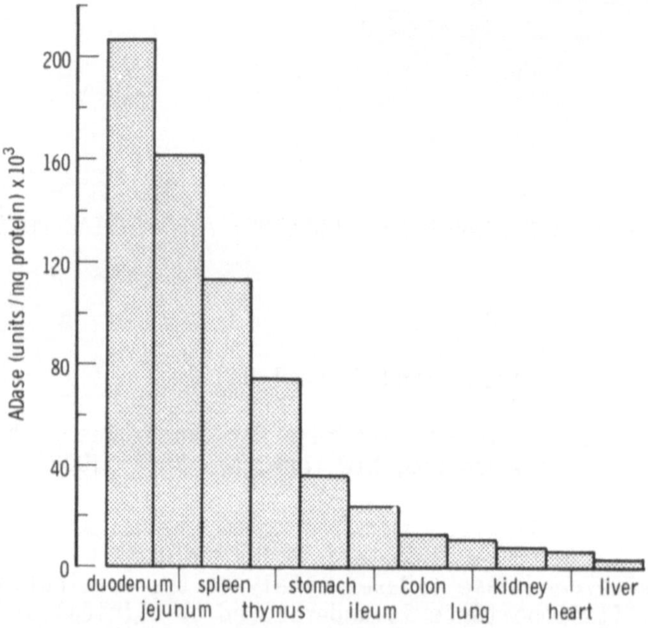

Figure I: Tissue distribution of rabbit adenosine deaminase activity. Activities were measured on 10,000 xg supernatants of 50% (w/v) tissue homogenates either spectrophotometrically or, for low activities, with radioactive substrate (4).

for example, stomach and colon. On the other hand, tissues with high specific ADase uniformly had little type A enzyme. These data may reflect a higher rate of synthesis of type C enzyme compared to conversion protein in these tissues.

The distribution of type A enzyme suggested that kidney, ileum and lung might represent good sources of conversion protein. The tissue distribution of conversion activity supports this conclusion (Table I). Kidney and lung have also been reported to represent good sources of human conversion protein (5). These data lend further support to the hypothesis that type A ADase in vivo is produced from the interaction of conversion protein with endogenous type C enzyme.

Sephadex G-200 chromatography of kidney conversion protein generally yielded one main peak of activity with ADase eluting somewhat earlier. The apparent molecular weights were ca 215,000 and 295,000 respectively. However, other protein with conversion activity were occasionally noted on gel filtration chromatography. These forms eluted at positions consistent with aggregation and/or dissociation of the major species.

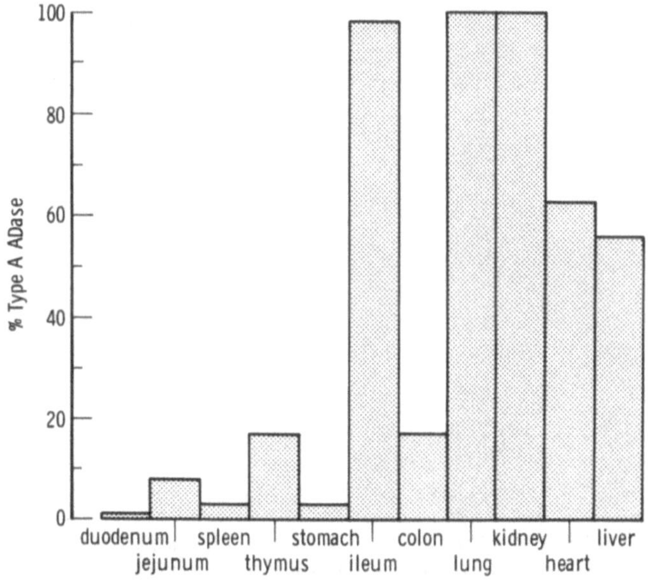

Figure 2: Distribution of rabiit high molecular weight ADase. ADase was quantitated by Sephadex G-100 chromatography tissue homogenates. Assay was either spectrophotometric or with radioactive substrate (4).

Gel filtration of human placental conversion proteins yielded several variants (Fig 3). Conversion protein "A" eluted in the excluded volume, followed by peaks "B" and "C", although the relative amounts of each varied from preparation to preparation. The molecular weight of form C was comparable to that of the rabbit preparation (<u>ca</u> 210,000) while that of form B was approxi-

Table 1

TISSUE	CONVERSION ACTIVITY (UNITS/MG PROTEIN)x10^2
Kidney	1.72
Ileum	0.52
Lung	0.32
Liver	0
Heart	0

<u>Tissue distribution of rabbit kidney ADase conversion activity</u>. conversion of calf intestinal ADase to type A enzyme was measured by gel filtration chromatography, followed by enzyme activity measurement of the amount of type A enzyme formed. These are the numbers which conform to our new definition of a conversion unit: the amount of conversion protein that produces one unit of type A ADase under the specified conditions of chromatography, etc.

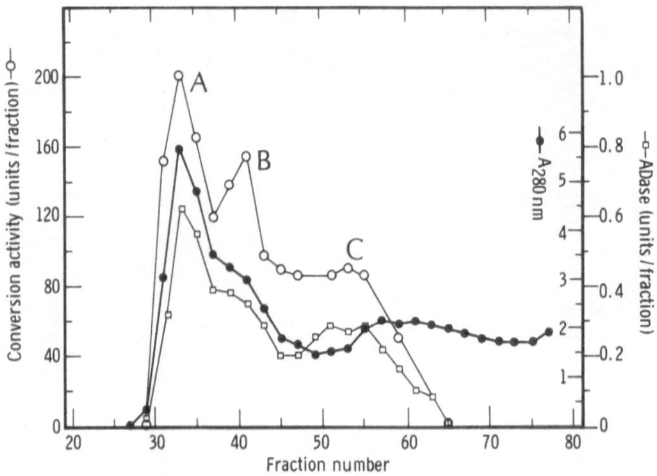

Figure 3: Sephadex G-200 gel filtration chromatography of human placental conversion proteins. Precepitate from 60% ammonium sulfate of human placental homogenate in 50 mM imidazole.HCl, 100 mM sodium chloride, pH 7 and applied to a Sephadex G-200 column. Fractions were collected and assayed for conversion and ADase activity.

mately twice this value. A constant fraction of ADase activity was consistently observed associated with each conversion protein peak. These data suggest that these various forms represent different states of aggregation of a common gene product although genetic polymorphism has not been rigorously eliminated.

We have further characterized rabbit kidney conversion protein eluted from Sephadex by isoelectric focusing. Four peaks of activity were detected with pI values of ca 5.56, 5.05, 4.50 and 4.15, respectively, while two main peaks of ADase had pI values of ca 4.85 and 4.35. The presence of a further heterogeneity was suggested by the unsymmetrical character of both conversion and ADase peaks. No ADase was detectable in the pI region of conversion protein I. All ADase was associated with some degree of conversion activity.

Isoelectric focusing of human placental conversion proteins also demonstrated a high degree of heterogeneity. Four peaks of activity were observed (pI ca 5.8, 5.4, 4.6 and 4.3), but only two of ADase. The broadness of the peaks and our inability to resolve them into sharp zones is again suggestive of a high degree of microheterogeneity. The reproducibility of this broad distribution was tested by repeating the focusing on fractions pooled from the major peaks. Under these conditions a similar profile of conversion activity was observed, thus minimizing convective disturbances as the cause of the apparent heterogeneity.

The difference in the number of conversion protein and ADase peaks may reflect a preferential binding of low molecular weight type C ADase to certain conversion protein variants. Alternatively, the isoelectric points of the ADase-conversion protein complexes may be so close as to render them unresolvable. In any case, the presence of large quantities of conversion protein with properties distinct from any ADase indicates a further inherent complexity to the system and does suggest a role for these proteins independent of the deamination of adenosine.

Although the basis for the multiple pI values of conversion activity has not been established, it is reasonable that the differences are produced by differences in carbohydrates, as noted in other systems. To test this we have determined the binding of these electrophoretic forms to concavalin-A (Con-A) and wheat germ lectin Sepharose resins. Wheat germ lectin binds sialic acid and N-acetylglucosamine resides while Con-A is specific for α-D-mannopyranosyl and α-D-glucopyranosyl residues. The high degree of binding to each resin (Table 2) indicates that the conversion proteins are glycoproteins. The variation of the binding of the pI of the variants shows a different carbohydrate content in each. Trends observed with the human placental conversion proteins were very similar in that each pI variant was also found to display a different affinity for each of these lectin resins.

Affinity chromatography with purified calf intestinal ADase is the most efficient step in conversion protein purification (6). By coupling this with gel filtration we have purified rabbit kidney conversion proteins _ca_ 1700 fold to apparent homogeneity. Gel electrophoresis in sodium dodecylsulfate indicates one component with a molecular weight of _ca_ 110,000. Rabbit conversion proteins are therefore most likely dimers.

Table 2

Conversion Protein Variant	Percentage of Conversion Activity Bound		
	Control Sepharose	Con-A	Wheat-Germ Lectin
I (pI 5.65)	14	94	81
II (pI 5.05)	16	92	69
III (pI 4.50)	44	88	88
IV (pI 4.15)	42	94	91
Total Eluant	17	54	58

Binding of rabbit kidney conversion proteins to lectin-Sepharose. Sepharose resins were incubated with sample containing 0.1 units of conversion activity in 50 mM imidazole.HCl, 100 mM sodium chloride, pH 7.0 with gentle shaking. After removal of the resin conversion activity was determined. Non-specific binding was determined with lectin-free Sepharose, "Control Sepharose".

Although no biological activity has been established for the conversion proteins, their heterogeneous carbohydrate content suggests a potential for interaction with membranes and subcellular particles. The conversion proteins that have been studied as soluble proteins may actually have derived from membrane structures. We have therefore extracted with detergent human placental plasma and microsomal membranes as well as subcellular membranes. The number of conversion units derived from membrane actually exceeded the units in the soluble (105,000 x$\underline{g}$) supernatnat. The specific activity in the plasma membrane-microsomal fraction was actually the highest of all (7). Other studies (8) have suggested an association between the cytosolic ADase and nucleoside transport in human erythrocyte plasma membrane. It is therefore a reasonable hypothesis that conversion proteins may function to anchor ADase to a membrane structure in order to preferentially channel any adenosine produced toward deamination. Potential sources of adenosine include the action of 5'-nucleotidase on AMP and entry via the nucleoside transport system.

REFERENCES:

1. H.J. Meuwissen, R.J. Pickering, B. Pollara and J.H. Porter (eds). Combined Immunodeficiency Disease and Adenosine Deaminase Deficiency. New York, Academic Press Inc, 1975.
2. R. Koshiura and G.A. Le Page. Some Inhibitors of Deamination of 9-β-D-arabinofuranosyladenine and 9-β-D-Xylofuranosyladenine. Cancer Research 28:1014, 1968.
3. P.F. Ma and T.A. Magers. Comparative Study of Human Adenosine Deaminases, Intern J Biochem 6:281, 1975.
4. P.P. Trotta and M.E. Balis. Structure and Kinetic Alterations in Adenosine Deaminase Associated with the Differentiation of Rat Intestinal Cells, Cancer Research 37:2297 1977.
5. H. Nishihara, S. Ishikawa, K. Shinkai and H. Akedo. Isolation and Properties of a Conversion Factor from Human Lung. Biochim Biophys Acta 302:429, 1973.
6. W.P. Schrader and A.R. Stacy. Purification and Subunit Structure of Adenosine Deaminase. J Biol Chem, 252:6409, 1977.
7. P.P. Trotta, S. Sur and M.E. Balis. Unpublished observations.
8. R.P. Agarwal and R.E. Parks, Jr. A Possible Association Between the Nucleoside Transport System of Human Erythrocytes and Adenosine Deaminase. Biochem Pharmacol 24:547, 1975.

REGULATORY PROPERTIES OF AMP DEAMINASE ISOZYMES

Nobuaki Ogasawara, Haruko Goto and Yasukazu Yamada

Department of Biochemistry, Institute for Developmental Research, Aichi Prefectural Colony, Kasugai, Aichi-480-03, Japan

INTRODUCTION

AMP deaminase catalyzes the deamination of AMP to form IMP and ammonia. The function of AMP deaminase in cellular metabolism remains obscure, but a number of physiological roles have been proposed. The enzyme may be responsible for the conversion of adenine nucleotide to inosine or guanine nucleotide[1], for stabilizing the adenylate energy charge[2], and the reaction of the purine nucleotide cycle[3].

Chromatographic, kinetic, and immunological data strongly suggested the existence of three parental isozymes in rat; isozyme A (muscle type), isozyme B(liver type) and isozyme C(heart type). All of them are allosteric enzyme regulated by alkali metal ions, nucleoside di- or triphosphates and inorganic phosphate.

In the present investigation, the effects of various monovalent cations, nucleotides and RNAs on the enzyme activity of three AMP deaminase isozymes were analyzed and the regulatory properties of isozymes were compared.

MATERIALS AND METHODS

Nucleotides and adenylate kinase were obtained from Boehringer, Mannheim. Other reagents were commercial preparations of the highest purity available.

AMP deaminase isozyme A and B were prepared from rat leg muscle and liver, respectively, to apparent homogeneity by the methods described previously[4]. Isozyme C was also purified from the rat heart extracts[5]. Ribosomal RNAs were prepared from rat liver ribosomal fraction by phenol extraction followed by centrifugation through sucrose density gradient. tRNA was also prepared from the pH 5 fraction of rat liver.

The deamination of AMP to IMP was measured by following the decrease in absorbance at 265 nm minus 300 nm on a 356 Hitachi two-wavelength double beam spectrophotometer at 25°C. Reaction mixtures typically contained 50 mM imidazol-HCl, pH 6.5, 100 mM KCl, AMP at the indicated concentrations, enzyme and water to a final volume of 0.3 ml. The reaction was carried out in a 3 ml quartz cuvette containing a silica insert to reduce the light path to 0.1 cm. One unit of deaminase activity is defined as the amount of enzyme which consumes 1 µmole of AMP in 1 min.

To test the effect of adenylate energy charge on the deaminase isozymes, the desired adenylate charge was obtained by the appropriate mixture of AMP and ATP. Adenylate kinase(3.6 units) was added to 0.2 ml of mixture, and the mixture was allowed to equilibrate for 15 min at 37°C. Following equilibration, AMP deaminase (0.03 unit) was added and the activity was assayed by determining ammonia formation in 5 min. Final reaction mixture contained 4 mM total adenine nucleotides, 8 mM $MgCl_2$, 50 mM cacodylate buffer, pH 6.6, 3.6 units of adenylate kinase and 0.03 unit AMP deaminase in a volume of 0.25 ml.

RESULTS AND DISCUSSION

In the presence of 100 mM KCl and 50 mM imidazol-HCl, pH 6.5, the Km values for AMP were 0.38, 1.09 and 0.13 mM for isozymes A, B and C, respectively. Thus, isozyme C has apparently a higher affinity for AMP than other type of deaminase, while isozyme B has lowest affinity of all.

As is the case for most preparations of AMP deaminase, three isozymes were activated by monovalent cations. Molovalent cations act as typical allosteric activators to reduce apparent Km values for AMP. The activating effect of each of the monovalent cations determined at a fixed concentration of AMP(0.1 mM) is shown in Table 1. When isozyme A was tested, highest activity was observed in the presence of K^+ and Na^+, and Li^+, NH_4^+ and Rb^+ were about half as effective as K^+. To activate isozyme B, Na^+ and Li^+ were more effective than K^+, and the activating effect was in the order of; $Na^+ > Li^+ > K^+ > NH_4^+ > Rb^+$. Isozyme C was activated by the monovalent cations, in the order of; $K^+ > Li^+ > Na^+ = NH_4^+ = Rb^+$. Cs^+ exerts either the lowest activation or no effect. Thus, three isozymes are activated by the monovalent cations, but each isozyme has the specific monovalent cations for activation.

Table 1. Effect of Monovalent Cations[a]

Ions	Isozyme from Muscle	Liver	Heart
Li	62	189	92
Na	105	267	74
K	100	100	100
NH_4	45	75	76
Rb	63	56	68
Cs	5	22	5

[a]Reaction mixture contained 100 mM monovalent cations, 0.1 mM AMP and 50 mM imidazole buffer, pH 6.5.

Effect of various ATP concentrations on AMP deaminase isozymes was studied in the presence of 50 mM imidazol-HCl buffer, pH 6.5, 0.1 mM AMP and 100 mM KCl. Fig. 1 shows that under these conditions 12.5 µM ATP causes an inhibition of 65 % in isozyme A. When the ATP concentration is raised over the range from 25 to 250 µM the inhibitory effect in isozyme A is partially removed. ATP inhibited isozyme A at low concentration, but in contrast, isozyme B and C were predominantly activated by ATP.

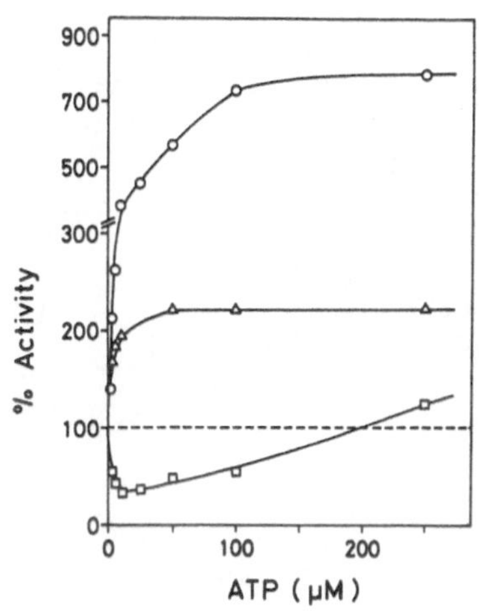

Fig. 1. Activity of AMP deaminase isozymes as a function of ATP concentration. Reaction mixtures contained 50 mM imidazol-HCl, pH 6.5, 100 mM KCl, 0.1 mM AMP and various concentrations of ATP. (□),isozyme A; (O), B; (Δ), C.

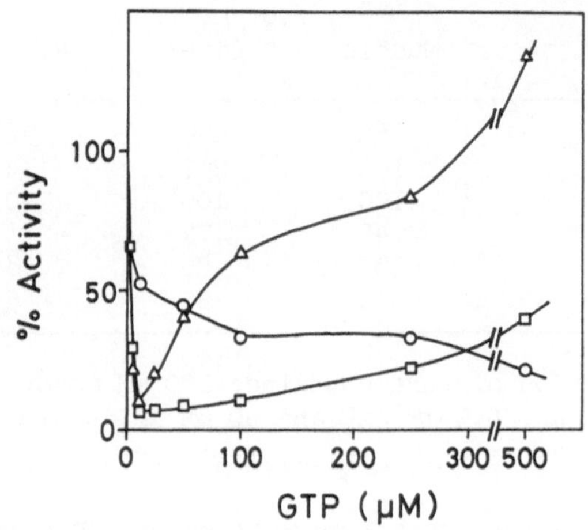

Fig. 2. Activity of AMP deaminase isozymes as a function of GTP concentration. Reaction mixtures contained 50 mM imidazol-HCl, pH 6.5, 100 mM KCl, 0.1 mM AMP and various concentrations of GTP. (□), isozyme A; (○), B; (△), C.

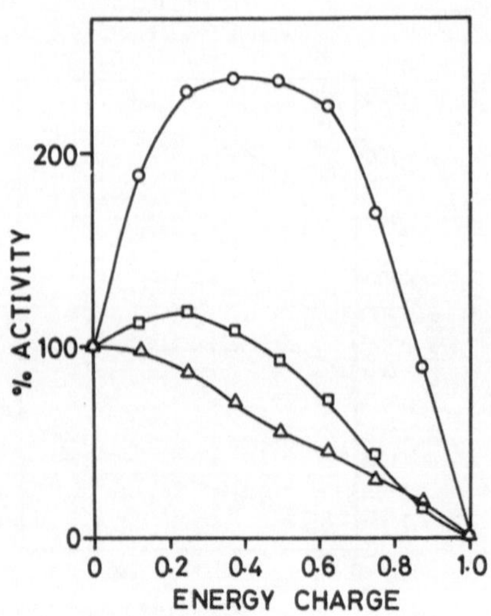

Fig. 3. Response of AMP deaminase isozymes to variations in the adenylate energy charge at a total adenine nucleotide concentration of 4 mM. (□), isozyme A; (○), B; (△), C.

The effect of GTP on enzyme activity was also examined (Fig. 2). Low concentration of GTP inhibited isozyme A and C, with maximum inhibition at approximately 12.5 μM. By a further increase of GTP concentration the enzyme activity of isozyme A and C increased. Thus, the effects of GTP on isozyme A and C were biphasic. Isozyme B was also inhibited by GTP, but its effect was not biphasic.

Effect of CTP and UTP was also tested. They were effective inhibitors for three isozymes at low concentration(25 μM). At higher concentration(1 mM) the enzyme activities were higher than those observed at 25 μM concentration. ADP at any concentrations activated all three isozymes.

One of the physiological functions suggested for this enzyme is to regulate the adenylate energy charge in the cell. In Fig. 3 is shown the response of three isozymes as a function of alterations in the adenylate energy charge. The activity of isozyme A increased linealy as the energy charge decreased from 1.0 to 0.3, and then decreased slowly showing maximum activity at energy charge value of 0.3. Isozyme B showed rapid increase of the enzyme activity as the decrease of energy charge from 1.0 to 0.6, maximum activity at 0.3-0.5 and then rapid decrease of activity at lower energy charge values. The activity of isozyme C was highest at an energy charge value of 0.

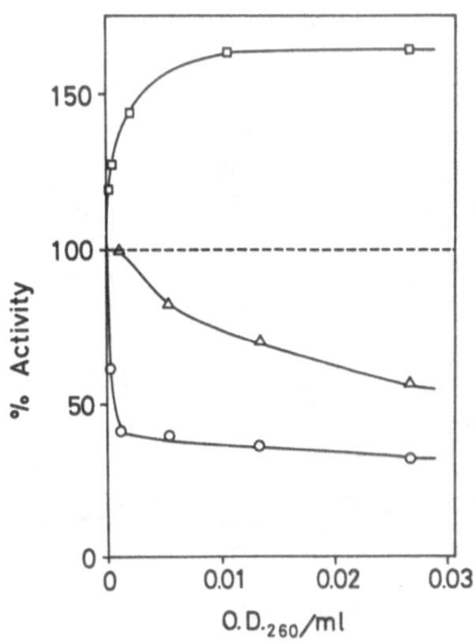

Fig. 4. Activity of AMP deaminase isozymes as a function of 28 S RNA concentrations. Reaction mixtures contained 50 mM imidazol-HCl, pH 6.5, 100 mM KCl, various concentrations of 28 S RNA, and 0.3, 1.0 and 0.1 mM AMP for isozyme A(□), B(○) and C(△), respectively.

Since a number of nucleotides are effective in inhibiting or activating AMP deaminase, the effect of RNAs on the enzyme activity of three AMP deaminase isozymes was tested. In Fig. 4 is shown the effect of 28 S ribsomal RNA on isozymes in the presence of AMP concentrations required to achieve half maximal velocity. Isozyme A was activated by 28 S RNA, with half maximum activation at 0.001 OD_{260} unit/ml. In contrast to isozyme A, isozyme B and C were inhibited by RNA. Isozyme B was inhibited at lower concentrations of RNA than those required to inhibit isozyme C. The concentration required to give half maximal inhibition was less than 0.0005 OD_{260} unit/ml for isozyme B and 0.01 OD_{260} unit/ml for isozyme C. Other RNAs such as 18 S, 5 S and tRNA from rat liver were also effective in activating isozyme A and inhibiting isozyme B and C. The macromolecular structure of RNAs are essential for activation or inhibition of the enzyme, since the effect of RNAs disappeared after RNAase T_1 or T_2 treatment.

To test further whether the enzyme binds RNA, sedimentation velocity experiments were performed using isozyme A purified from rat muscle, and rat liver 28 S ribosomal RNA. A complex was formed when a RNA and isozyme A were mixed in 20 mM imidazol-HCl, pH 6.5, and 100 mM KCl at excess molar ratio of RNA. Free AMP deaminase showed 11.5 S, while the faster peak of enzyme activity, which

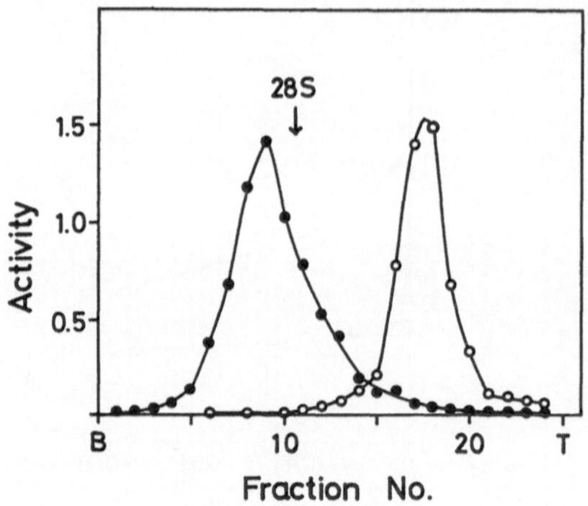

Fig. 5. Sucrose density gradient centrifugation of isozyme A and a mixture of isozyme A and 28 S RNA. Isozyme A(○) or a mixture (●) of isozyme A and 28 S RNA was placed on a linear 5-25% sucrose gradient in 20 mM imidazol-HCl, pH 6.5, and 100 mM KCl. Centrifugation was carried out for 3 hrs at 8°C and 45,000 rpm in the Hitach RPS-50 rotor.

appears to represent a complex between the enzyme and 28 S RNA, has a S value of greater than 28 S (Fig. 5). When ATP, GTP, CTP, UTP, ADP and PPi were added at the final concentration of 1 mM, the peak corresponding to RNA-enzyme complex was no longer present. Clearly, the addition of the ligands to a solution containing the RNA-enzyme complex results in the dissociation of the complex.

REFERENCES

1. A. Askari and S. N. Rao, Regulation of AMP deaminase by 2,3-diphosphoglyceric acid: A possible mechanism for the control of adenine nucleotide metabolism in human erythrocytes, Biochim. Biophys. Acta, 151:198 (1968).
2. A. G. Chapman and D. E. Atkinson, Stabilization of adenylate energy charge by adenylate deaminase reaction, J. Biol. Chem., 248:8309 (1973).
3. J. M. Lowenstein, Ammonia production in muscle and other tissues: The purine nucleotide cycle, Physiol. Rev., 52:382 (1972).
4. N. Ogasawara, H. Goto, Y. Yamada and M. Yoshino, Subunit structures of AMP deaminase isozymes in rat, Biochem. Biophys. Res. Commun., 79:671 (1977).
5. N. Ogasawara, H. Goto and T. Watanabe, Isozymes of rat AMP deaminase, Biochim. Biophys. Acta, 403:530 (1975).

HUMAN ADENOSINE DEAMINASE: STOICHIOMETRY OF THE LARGE FORM COMPLEX

Peter E. Daddona and William N. Kelley

Department of Internal Medicine and Biological Chemistry, Human Purine Research Center, University of Michigan Medical School, Ann Arbor, Michigan U.S.A.

Adenosine deaminase (adenosine aminohydrolase, ADA, EC 3.5.4.4.) is widely distributed in human tissues. In some tissues ADA exists exclusively as the small molecular form (36,000-38,000) while in other tissues the large molecular form of the enzyme predominates (298,000).[1] The small form of ADA has been purified to homogenity from human erythrocytes and was shown to be a single polypeptide of molecular weight 38,000.[2,3] Physical and kinetic characteristics of small form ADA from various tissues appeared to be identical to the purified erythrocyte enzyme.[1,2] The small form of the enzyme could be converted to the large form ADA by incubation with a specific ADA binding protein (BP) (also termed complexing protein or conversion factor).[1,4,5] This purified binding protein from human kidney was shown to be a dimer of identical subunits with a native molecular weight of 190,000-200,000.[4,5] The large form ADA produced _in vitro_ was physically and kinetically indistinguishable from the native large form ADA from various tissues.[1,4]

In this report we have analyzed the stoichiometry of the large form ADA (ADA-BP complex) by binding measurements, chemical cross-linking and sedimentation equilibrium analyses.

The stoichiometry of the large form ADA was first analyzed by direct binding experiments. Initially a quantitative assay was developed to detect the binding of purified ^{125}I-small form ADA to its binding protein. This binding reaction was found to be complete within 10 min at 37°C, to proceed over a broad pH range of 5 to 8 and to be unaffected by changes in ionic strength (0-200 mM KCl), azide (0.02%), thiols (5 mM), divalent metal ions (5 mM) and bovine serum albumin (1-25 mg/ml) After incubating purified ^{125}I-labeled small form ADA and highly pure binding protein, the reaction mixture containing the newly produced ^{125}I-labeled large form ADA (ADA-BP complex) and unreacted ^{125}I-small form ADA was separated by polyacrylamide gel electrophoresis and quantitated by gamma counting. As shown in Fig. 1, small form ^{125} I-small form ADA migrated with an Rf of 0.61 while the ^{125}I-labeled ADA-BP complex (large form) migrated with an Rf of 0.14.

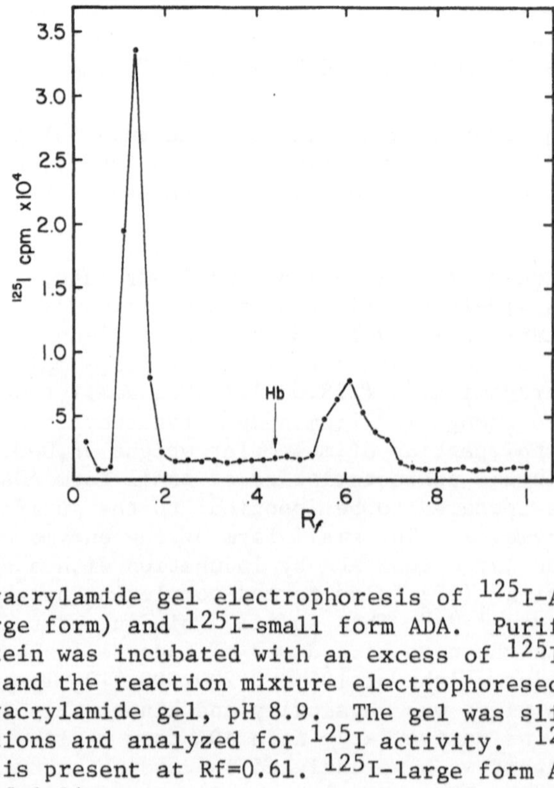

Fig. 1. Polyacrylamide gel electrophoresis of ^{125}I-ADA-BP complex (large form) and ^{125}I-small form ADA. Purified binding protein was incubated with an excess of ^{125}I-small form ADA and the reaction mixture electrophoresed on a 7.5% polyacrylamide gel, pH 8.9. The gel was sliced into 2 mm sections and analyzed for ^{125}I activity. ^{125}I-small form ADA is present at Rf=0.61. ^{125}I-large form ADA is present at Rf=0.14.

In the direct binding experiment, a fixed concentration of binding protein was incubated with varying concentrations of ^{125}I-small form ADA. As shown in Fig. 2, the binding was saturable. A Scatchard-type plot of these data (Fig. 2, inset) indicates that 2.15 moles of small form ADA are bound to 1 mole of binding protein under the conditions of our assay.

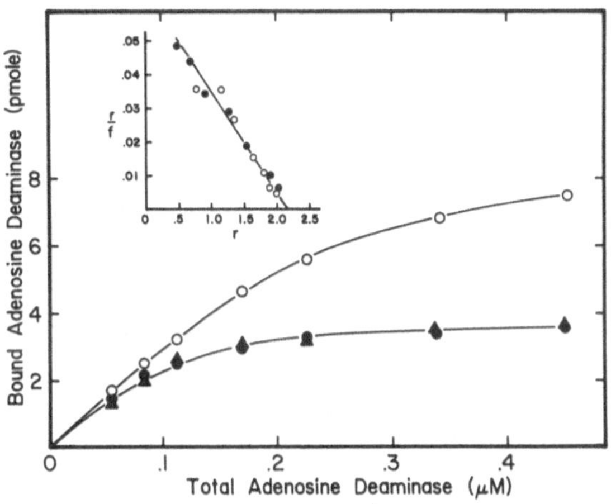

Fig. 2. Direct binding of small form ADA to binding protein. Highly purified binding protein (0.0514 μM, ●—●; 0.1028 μM, o—o) was added to ^{125}I-small form ADA (3.6×10^9 dpm/pmol) diluted serially from 0.45 μM to 0.05 μM. ^{125}I small form ADA (1.8×10^9 dpm/pmol) prepared by dilution with nonradioactive ADA, was also incubated with 0.0514 μM of highly purified binding protein (▲—▲) in a similar manner. The molecular weight of small form ADA and binding protein used in these calculations was assumed to be 38,000 and 213,000, respectively, based on sedimentation equilibrium analyses data. The inset shows a Scatchard-type plot of these data.

Another approach used to determine the stoichiometry of large form ADA was chemical cross-linking. Successful chemical cross-linking of a native protein is dependent upon both the accessibility of the necessary reactive group on the subunits of the protein and the ability of the bifunctional cross-linking reagent to bridge between the reactive groups on the adjacent subunits. If both conditions are satisfied, then the number of cross-linked species produced for ADA should theoretically equal the maximum number of different combinations of its subunits.

If large form ADA were composed of 2 moles of small form ADA (A,A) and 1 mole of binding protein (BB) then 8 possible cross-linked combinations would be theoretically possible: BBAA, BBA, BAA, BB, BA, B, AA and A. When the ^{125}I-labeled ADA-BP complex and the ^{125}I-labeled BP-ADA complex were cross-linked with glutaraldehyde, 6 different radioactive peaks were consistently observed (see Fig. 3). Radioactive peaks at Rf=0.49 and 0.31 presumably corresponded to binding protein monomer (B) and dimer (BB), respectively, while the peak at Rf=0.88 was most probably monomeric small form ADA (A). The remaining radioactive peaks at Rf=0.40,

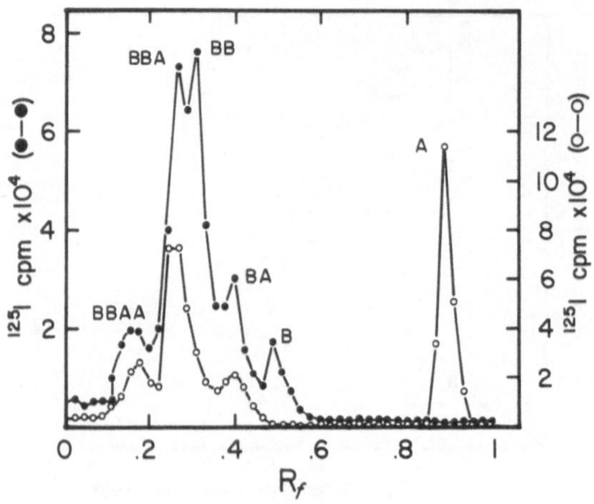

Fig. 3. SDS polyacrylamide gel profiles of the cross-linked ^{125}I-ADA-BP complex. Samples were cross-linked individually with glutaraldehyde for 1 1/2 h and electrophoresed on 3-10% linear gradient cylindrical SDS polyacrylamide gels. Gels were sliced into 1.5 mm sections and analyzed for ^{125}I radioactivity: Cross-linked ^{125}I BP-ADA complex (●—●) and cross-linked ^{125}I ADA-BP complex (o—o). The putative cross-linked species are indicated by the appropriate letter abreviations.

0.27 and 0.18 appeared to represent combinations of binding protein and ADA subunits.

Since the mobility of each radioactive cross-linked species on SDS gels should roughly correlate to its molecular weight, this relationship was used to limit the possible combinations of A and B subunits contained in the 3 coincident radioactive peaks present in Fig. 3. The radioactive peak at Rf=0.40 which migrated between binding protein monomer (B́, Mr=106,000, Rf=0.49) and dimer (BB, Mr=213,000, Rf=0.31) would presumably be a BA combination of expected molecular weight 144,000. The peak at Rf=0.27 present just after the binding protein dimer peak (BB, Mr=213,000, Rf=0.31), could possibly be a BBA combination of expected molecular weight 251,000 and finally the peak at Rf=0.18 could reflect the cross-linked native complex, AABB of expected molecular weight 289,000. A plot of the expected molecular weight for the proposed cross-linked species versus the log of their Rf values was found to be linear. Noticebly absent in the radioactive gel profile were peaks corresponding to an AA species and an AAB species. This suggested that the distance between the two ADA subunits in the native large form ADA molecule was too great for successful cross-linking under these conditions. If the stoichiometry of large form ADA were 1 mole of small form ADA and 1 mole of binding protein then five cross-linked combinations of the subunits would theoretically be possible, two of which would contain A and B subunit combinations. Our data clearly showed more cross-linking than the 1:1 model would predict and thus strongly favored a minimum 2:1 stoichiometry model.

Sedimentation equilibrium analyses revealed native molecular weights of 37,500 for small form ADA, 213,000 for binding protein, and 300,000 for large form ADA (ADA-BP complex). The molecular weight data would also be most consistent with the proposed stoichiometry of 2 moles of small form ADA and 1 mole of binding protein for the large form complex.

The molecular weight of the large form ADA prepared *in vitro* compared well with our previous estimated molecular weight of large form ADA from several different tissue sources.[1] This suggested that the stoichiometry of the large form complex may be constant regardless of the tissue source of the enzyme.

REFERENCES

1. M.B. Van der Weyden and W.N. Kelley, Human adenosine deaminase: distribution and properties. *J. Biol. Chem.* 251: 5448-5456 (1976).

2. P.E. Daddona and W.N. Kelley, Human adenosine deaminase: purification and subunit structure. *J. Biol. Chem.* 252: 110-115 (1977).

3. W.P. Schrader, A.R. Stacy and B. Pollara, Purification of human erythrocyte adenosine deaminase by affinity column chromatography. *J. Biol. Chem.* 251: 4026-4032 (1976).

4. P.E. Daddona and W.N. Kelley, Human adenosine deaminase binding protein: assay, purification and properties. *J. Biol. Chem.* 253: 4617-4623 (1978).

5. W.P. Schrader and A.R. Stacy, Purification and subunit structure of adenosine deaminase from human kidney. *J. Biol. Chem.* 252: 6409-6415 (1977).

GUANASE FROM HUMAN LIVER — PURIFICATION AND CHARACTERISATION

Rudolf Kuzmits, Heinrich Stemberger, Mathias M. Müller
Dept. of Medical Chemistry, Institute for Specific Prophylaxis and Tropical Medicine, 2nd Dept. of Medicine, University of Vienna

INTRODUCTION:

Guanine aminohydrolase (E.C.3.5.4.3.) plays an important role in the catabolism of purines, catalyzing the formation of xanthine from guanine.

The aim of the study was to examine the organ distribution of guanase in man and to purify and characterize the enzyme from human liver.

MATERIALS AND METHODS:

The determination of guanase activity was carried out with a new radiochemical method according to van Bennekom (1). Livers and organ samples were taken from patients without signs of liver dysfunction shortly after death. 1g of human tissue was homogenized in 2 ml 49mM ammonium phosphate buffer, pH 5.0 for 60 seconds at 4° C. Homogenates were centrifuged at 39000 x g for 20 minutes. Guanase activities were determined in the supernatant.

The fractionation of human liver tissue was carried out according to the technique described by Berthelet (2). Enzyme activities and protein content were determined in the large particle fraction, in the nuclear, the microsomal and the soluble fraction.

Enzyme purification from human liver: 200 g human liver was cut in slices and homogenized in 400ml ammonium phosphate buffer, pH 5.0 for 60 seconds. All operations were performed at 4°C. To remove membrane fragments and cell stroma the homogenate was centrifuged at 39000 x g for 20 minutes. The supernatant, contain-

ing guanase activity, was used for further purification.
a) Ammonium sulphate fractionation: The supernatant was brought to 0.4 saturation, stirred for 1 hour and then centrifuged for 30 minutes at 1400xg. To this supernatant ammonium sulphate was added to gain a 0.75 saturation. After 2 hours this solution was centrifuged for 30 minutes at 1400xg. The supernatant was discarded, the precipitate was dissolved in a minimal volume of 50 mM ammonium acetate buffer, pH 5.8 and dialysed against this buffer for 2 hours. Saturated ammonium sulphate solution was added to bring the solution to 0.75 saturation. After centrifuging for 45 minutes the sediment was resuspended in a minimal volume of ammonium phosphate buffer, pH7.3 and dialyzes against the buffer (19 mmol/l) overnight.
b) DEAE-cellulose column chromatography: The dialyzed liver extract was run through a DEAE-cellulose column (5cmx50cm). Elution was carried out with 500ml ammonium phosphate buffer, 19 mmol/l, pH 7.3, followed by 1300ml ammonium phosphate buffer, 120 mmol/l, pH 7.3. Guanase was eluted as a single peak, fractions with guanase activity were pooled and concentrated to a volume of 7ml.
c) Isoelectric focusing according to Svensson (3): Isoelectric focusing was carried out over a pH-range of 2.5-6.0 (LKB Ampholine Ampholite 2.5-4.0 and 4.0-6.0 1:1). Concentration of ampholyte was 2.36%. At the beginning of the separation procedure, which extended over 72 hours, voltage was 400V at a current of 6mA. Voltage was steadily increased, until a constant current of 2.8 mA was reached at 1000V. After focusing 5ml fractions were collected and dialysed against ammonium phosphate buffer, 19 mmol/l, pH7.3. Enzyme activities were eluted in 2 distinct peaks which were concentrated to a volume of 0.5 ml.
d) Sephacryl chromatography: The Sephacryl column (2.5 cm x 80 cm) was calibrated with a o.5 ml sample containing 10 mg bovine serum albumine, 10mg highly purified rabbit g-globuline, 5mg cytochrome c and 5mg blue dextran. The two guanase fractions obtained from the isoelectric focusing were run separately through the Sephacryl column for further purification and molecular weight determination. Elution was performed with ammonium phosphate buffer, 19 mmol/l, pH 7.3. Fractions with guanase activity were pooled and concentrated to a volume of 1 ml.

RESULTS

Organ and intracellular distribution: Guanase is

mainly located in brain, liver, gut and kidney but nearly absent in all other human tissues.

Table 1. Distribution of guanase activity in human tissue

Tissue		Guanase activity nmol/mg prot./hr	Tissue	Guanase activity nmol/mg prot./hr
Cerebral	cortex	726	Muscle	38
	medulla	830	Testes	29
Cerebellum		40	Pancreas	12
Liver		453	Prostate	5
Gut		169	Thyroid gland	2
Kidney	cortex	151	Adrenal tissue	2
	medulla	108	Lung	1
Epididymis		85	Spleen	0.5
Heart		50		

In human liver the greatest part of enzyme activity is found in the soluble fraction (87% of total activity). However, guanase activity was also present in the microsomal fraction (11%).
Enzyme purification: The elution profile from the DEAE-cellulose column offered a single guanase peak after an elution volume of 1150 ml.

Table 2. Purification of guanase from human liver

Fraction	Spec.activity ymol/mg prot/hr	Tot.prot. mg	Vol. ml	Recovery %
Homogenate 39.000xg (supernatant)	0.453	14.750	500	100
0.75 Ammonium sulphate saturation I	1.428	2.900	200	62
0.75 Ammonium sulphate saturation II	1.632	2.480	200	60
DEAE-cellulose chromatography	34.500	46.2	15	24

Fraction	Spec.activity ymol/mg prot/hr	Tot.prot. mg	Vol. ml	Recovery %
Isoelectric focusing frac-tion "A"	198.463	2.27	21	7
Isoelectric focusing frac-tion "B"	125.277	1.99	21	4
Sephacryl column fraction "A"	278.102	0.210	1	0.9
Sephacryl column fraction "B"	217.066	0.082	1	0.3

Isoelectric focusing reproducibly separated the guanase activity into two distinct peaks.

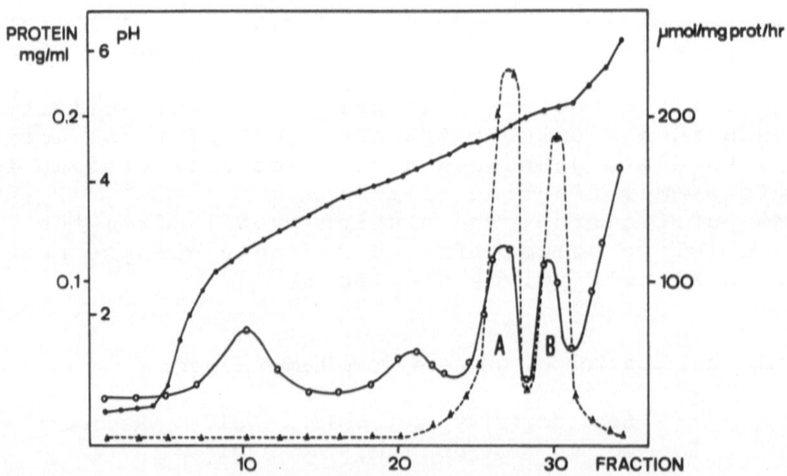

Figure 1. Preparative isoelectric focusing of guanase. o–o–o Protein concentration, ▲--▲ Enzyme activity, •–•–• pH-gradient.

The two isoenzymes obtained were called "A" and "B" according to their migration to the anode; their isoelectric points were 4.7 and 5.15 respectively. The final step of the purification performed separately for each fraction with a Sephacryl column resulted in the elution of further two minor proteins besides the isoenzyme "A" and one protein beside the isoenzyme "B". For the isoenzyme "A" as well as for the isoenzyme "B" a molecular weight

of 100 000 was determined.
Properties of guanase isoenzymes: The characteristics of the two guanase isoenzymes are summarized in tab. 3.
Both isoenzymes retained full activity after heat treatment at 45° C for 30 minutes. In contrast at 56° C a decrease of activity of about 20% for isoenzyme "A" and a loss of activity of approximately 10% for isoenzyme "B" due to denaturation was observed.

Table 3. Characteristics of guanase isoenzymes.

	Isoenzyme "A"	Isoenzyme "B"
Molecular weight	100.000	100.000
pH-optimum	7.4	9.0
pK_i	4.70	5.15
K_m (umol/l)	0.56	0.26
V_{max} (ymol/mg prot./hr)	236	200

DISCUSSION

Human liver guanase has not been purified and characterized previously. With the purification procedure described guanase of human liver was purified and separated into two isoenzymes, which showed electrophoretically homogeneity. Two isoenzymes of guanase were distinguished in rat liver, rat brain and mouse brain, whereas in mouse liver the existence of guanase isoenzymes could not be demonstrated (4). Farkas (5) reported on the separation of two guanase isoenzymes from rat and mouse erythrocytes. The preparation of guanase was carried out by these authors mainly by means of ammonium sulphate fractionation, ion exchange chromatography on DEAE-cellulose and gel-filtration on Sephadex- G-200. The two isoenzymes separated from human liver showed pH-optima at 7.4 and 9.0 respectively. Guanase purified from animal liver tissue was reported to have pH-optima at 9.0, 6.8 or 7.7. We presume, that the purification of guanase in these cases resulted in the separation of mainly one actual isoenzyme. This might explain the different pH-optima for preparations from the same source.
Both isoenzymes offered a high catalytic activity with guanine as substrate, the K_m value for the isoenzyme "B" was about two times as high as the K_m value for isoenzyme "A". It is noteworthy that guanase purified from human liver shows a ten times higher affinity for guanine than guanase from animal tissues.
On basis of the results presented in this study the existence of two guanase isoenzymes in human liver was demonstrated. At pre-

sent it is not possible to elucidate the origin of the two isoenzymes. The same molecular weight favours the assumption that posttranscriptional modulation of the synthesized guanase results in two isoenzymes with different catalytic and physical properties. Because of the main localization in the cytoplasmatic fraction the two isoenzymes do not represent guanase from different cellular compartments. Furthermore guanase does not catalyze the transformation of xanthine to guanine. Therefore the polyisocymic hypothesis for unilocular enzymes is not valid for the explanation of guanase isoenzymes described. (6).

REFERENCES

1. van Bennekom, C.A., van Laarhoven, J.P., de Bruyn, C.H.M.M. and Oei, T.L.: A simple and sensitive radiochemical assay for plasma guanase. J.Clin.Chem.Clin.Biochem. 16: 245-248 (1978).

2. Berthelet, J. and de Duve, C.: Tissue fractionation studies. The existence of a mitochondria-linked, enzymatically inactive form of acid phosphatase in rat-liver tissue. Biochem. J.50: 174-181 (1952).

3. Svensson,H.: Isoelectric fractionation, analysis and characterization of ampholytes in natural pH-gradients. III.Description of apparatus for electrolysis in columns stabilized by density gradients and direct determination of isoelectric points. Arch. Biochem. Biophys. Suppl. 1: 132-138 (1962)

4. Sree Kumar, K., Sitaramayya, A. and Krishnan, P.S.: Guanine deaminase in rat liver and mouse liver and brain. Biochem. J. 128: 1079-1088 (1972).

5. Farkas, W.R. and Singh, R.D.: Guanine aminohydrolase in rat and mouse red cells: A potent inhibitor of guanylation of tRNA. Biochim. Biophys. Acta 377: 166-173 (1975).

6. Ureta, T.: In Current Topics in Cellular Regulation, (Horecker, B.L. and Stadtmann, E.R.), Vol 3., pp 233-258, Academic Press, New York (1978).

NICOTINAMIDE AND LIVER XANTHINE OXDIDASE

Di Stefano A.[1], Pizzichini M.[1] and Marinello E.[2]

1 Dept. Biological Chemistry, University Siena, Italy

2 Dept. Biological Chemistry, E.U.L.O. Brescia, Italy

INTRODUCTION

When rats receive injection of high doses of nicotinamide, there is a temporary 10 fold increase in the diphosphyridine nucleotide content of the liver (1).

Maximal synthesis of NAD is accompanied by no detectable changes in the amount of adenine nucleotides (other than NAD), ribonucleic acid or deoxiribonucleic acid of the liver (2). The increase in liver NAD would therefore require either mobilization of adenine from sources outside the liver or new synthesis.

No increase of the incorporation of labeled glycine into urinary allantoin has been demonstrated in such conditions (3). However, Shuster and Goldin (4), have shown a net increased synthesis of purine bases-specifically of adenine- in the liver of mices treated with nicotinamide.

These results indicate that the liver of rats treated with nicotinamide may be used as a model for studying increased purine biosynthesis; they also suggest that nicotinamide may have a role in regulating such process,on enzymatic level.

With the present experiments, we have taken in consideration this last problem, and specifically if nicotinamide has a role in regulation "in vivo" or "in vitro", of xanthine oxidase (E.C.1.2.3.2.) the enzyme which, in mammalian, controls the last step of free purine formation.

Xanthine oxidase activity can be induced in protein depleted mice (5), by repeated injections of xanthine, and in chicks (6) by administration of inosine.

Xanthine oxidase activity is increased in the liver of gouty patients, a situation in which there is a net increase of purine biosynthesis. RNA load or hypoxanthine, 2-ethylamino-1,3,4, thiadiazole (EA-TDA), fructose load, in normal subjects causes liver xanthine oxidase levels to rise to values that are two to fourfold higher than those for the control group (7).

Two major forms of xanthine oxidase have been demonstrated in the rat liver by Stirpe and Della Corte (8): one can react with O_2 as electron acceptor and has been interpreted as an oxidase " O form", whereas the other form shows an activity detectable on addition of NAD, and has been interpreted as a dehydrogenase " D form ".

Free sulfhydryl groups are essential for activity of "D form". "D form" can be readly converted into "O form" and this conversion has been called, by Stirpe and Della Corte "activation", and is afforded by various treatments of rat liver supernatant, storage at -20°C, (9), treatment with sulfhydryl reagents (10), incubation in air in the presence of particulate subcellular fractions, preincubation under anaerobic condition heating at 37°C (8). Most of these treatments oxidize the -SH groups of the "D form" to the corresponding disulfides with loss of dehydrogenase activity and net increase of oxidase activity. Heating at 37°C alters the conformation of the enzyme, abolishing the dehydrogenase activity but does not oxidize the -SH groups; trypsinization removes a peptide which is rich in -SH groups and essential for dehydrogenase activity.

We have evaluted the activity of rat liver enzyme, taking 2-amino-4-hydroxypteridine as a substrate: in such condition, the "O form" of the enzyme is evaluted.

The effect of nicotinamide - which is not a substrate for the enzyme - has been evaluted "in vitro", in the presence of various concentrations of nicotinamide and "in vivo" after nicotinamide injection.

MATERIALS AND METHODS

Our experiments were carried out on adult Wister strain male albino rats maintained on standard laboratory diet.

The nicotinamide solution was given intraperitoneally (100 mg /100 g b.w.) and the rats were killed by decapitation at various times after treatments. The liver was removed within 1 min after the death of the animals, cut into small pieces and homogenized with 0.2 M phosphate buffer (pH 7.6) in a Potter-Elvehiem glass homogenizer. 3% homogenates were centrifuged at 3,000 x g for 10 min and xanthine oxidase activity was assayed in the homogenate by the fluorimetric method of Burch et al. (11).

Incubation mixtures contained 1×10^{-5}M 2-amino-4-hydroxypteridine and liver tissue samples in a concentration of 2 mg/ml of substrate. The incubation was carried out at 37°C for 30 min; readings were taken at zero time and every 5 min in a Turner Fluorimeter model 110 at 395 nm (filter 7-60) and an emission wavelenght of 460 nm (filter 2A). The sensitivity was reduced 100 time with a filter Kodak n.96.

A blank consisting of 4 ml of 0.2M phosphate buffer (pH 7.6), was run in parallel, with a quantity of tissue equivalent to that of the samples. Enzymatic activity was expressed as mμmoles of substate transformed in 60 min for mg of tissue.

Aliquots of rat liver homogenates (12%) in 0.2 M phosphate buffer (pH 7.6) were boiled for 10 min at 100 C and centrifuged at 3,000 x g for 10 min. The supernatant was taken and the precipitated resuspended was again centrifuged. In the joined supernatants the assay of oxypurines was performed according to Fried and Fried (12).

RESULTS AND DISCUSSION

When nicotinamide is added "in vitro" to incubation mixtures containing rat liver homogenate, inibition of xanthine oxidase activity is observed, at concentrations 7.5×10^{-3} M -- 2.5×10^{-3} M (table 1).

Also nicotinic acid, which is produced in the liver by nicotinamide, has an inhibitory effect on the enzyme at lower concentrations than nicotinamide (table 2).

The inhibitory concentrations of nicotinamide and of nicotinic acid observed "in vitro" are not reached "in vivo". Ricci (13) has studied the hepatic levels of nicotinic acid and of nicotinamide in the rat following nicotinamide administation.

After nicotinamide administration, some properties of the enzyme seem to modified. The well known "activation" of xanthine oxidase, the conversion of "D form" to "O form" in phosphate buffer (pH 7.6) at 0 C, is decreased in rat liver homogenate taken at the 1st hour after intraperitoneal administration of nicotinamide (14).

Table 1. Inhibition of rat liver xanthine oxidase by nicotinamide. The reported values represent the percent of inhibition, at the indicated final concentrations of 2-amino-4-hydroxypteridine and nicotinamide.

2-amino-4-hydro-xypteridine	nicotinamide		
	7.5×10^{-3}M	5×10^{-3}M	2.5×10^{-3}M
1×10^{-5}M	38	34	15
5×10^{-6}M	34	30	17
2.5×10^{-6}M	50	45	18
1.3×10^{-6}M	75	50	25
0.6×10^{-6}M	100	67	34

Table 2. Inhibition of rat liver xanthine oxidase by nicotinic acid. The reported values represent the percent of inhibition, at the indicate final concentrations of 2-amino-4-hydroxypteridine and nicotinic acid.

2-amino-4-hydro-xypteridine	nicotinic acid		
	2.5×10^{-3}M	1.25×10^{-3}M	0.6×10^{-3}M
1×10^{-5}M	37	27	25
5×10^{-6}M	63	50	28
2.5×10^{-6}M	60	55	38
1.3×10^{-6}M	67	68	66
0.6×10^{-6}M	68	69	67

Results are shown in Fig.1,where a low effect of nicotinamide added "in vitro" on the "activation" is evident also.

In diluted homogenates, the "activation" of the enzyme has the same degree of controls (table 3).

These results suggest that some cofactor is responsible in nicotinamide treated rat homogenates for the low activation observed at 0°C. In order to elucidate the nature of this cofactor, boiled or dyalyzed extracts from treated rat livers were added to incubation mixtures containing homogenates from normal rat liver for assay of xanthine oxidase.

It appears by table 4 that no difference in thermostable cofactors is evident in treated rat in comparison to the controls. The dyalyzed extracts do not exert any effect on enzyme activity.

Since oxypurines are known to inhibit xanthine oxidase (15) and even to interfer in the "activation" of the enzyme (8), their liver content has been investigated.

However, since their hepatic level is very low and their esti-

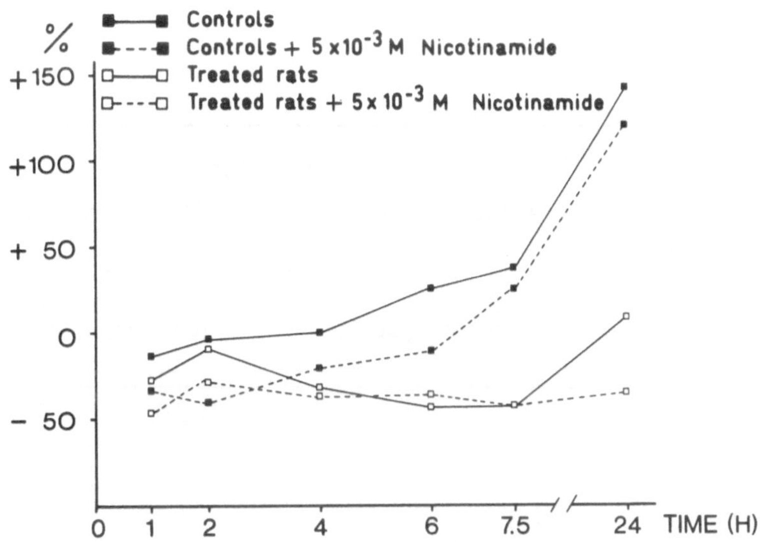

Fig. 1. Activity of xanthine oxidase (from normal and treated rat) after storage at different times in the presence or absence of added nicotinamide.-

Table 3. Xanthine oxidase activity of 3% and 1.5% rat liver homogenates (H) stored at 0 C. The reported values represent the percent of variation at the different times of storage.

	normal rat		treated rat	
times	3% H	1.5% H	3% H	1.5% H
0	0	0	0	0
2 h	- 10	- 24	- 25	- 25
6 h	+ 34	+ 10	- 11	0
24 h	+ 173	+ 115	+ 17	+ 106

Table 4. Effect of boiled extract from normal (N) and nicotinamide treated (T) rat livers added to normal rat liver homogenates. Enzyme activity is expressed as mμmoli/60 min/mg tissue. Additions are referred to 1 ml incubation mixtures.

	additions			
	N boiled extract		T boiled extract	
mg hepatic tissue	enzyme activity	% inhibition	enzyme activity	% inhibition
0	3.70	-	3.70	-
2 mg	3.70	0	3.50	5
10 mg	2.96	20	3.14	15
20 mg	1.85	50	1.85	50
30 mg	1.48	60	1.85	50
40 mg	1.37	63	1.48	60

mation is considerably difficult, results obtained are not definitive.

There is no evidence at the moment for a different oxypurines content in treated rats in comparison to the controls.

The possibility is considered also that the low activation at 0°C of xanthine oxidase induced by nicotinamide load is due to modification of enzyme conformation.

This conformational change would have effect on the conversion of "D form" in "O form" and on other properties, i.e., fractional turnover rate of the enzyme.

REFERENCES

1. A. Bonsignore and C. Ricci, Sulla natura del composto dosabile come N_1 - metilnicotinamide nel fegato di ratti trattati con nicotinamide. Identificazione con il DPN. Boll. Soc. It. Biol. Sper. 25:710 (1949).
2. L. Shuster, T.A. Langan, Jr N.O. Kaplan, A. Goldin, Significance of induced in vivo synthesis of diphosphate pyridine. Nature 182:512 (1958).
3. A. Di Stefano, M. Pizzichini and E. Marinello, Incorporazione di glicina - U - ^{14}C nell'allantoina urinaria del ratto trattato con nicotinamide. Boll. Soc. It. Biol. Chem. LIV:425 (1978).
4. L. Shuster and A. Goldin, The effect of nicotinamide on incorporation in vivo of formate - ^{14}C. J. Biol. Chem. 230:878 (1958).
5. A. Mangoni, V. Pennetti and M.A. Spadoni, Aumento adattativo di xantina ossidasi in topini alimentati con dieta con diverso contenuto proteico. Boll. Soc. Biol. Sper. 31:1397 (1955).
6. E. Della Corte and F. Stirpe, Regulation of xanthine dehydrogenase in chick liver. Biochem. J. 102:520 (1967).
7. R. Marcolongo, E. Marinello, G. Pompucci and R. Pagani, The role of xanthine oxidase in hyperuricemic states. Arthritis and Rheumatism 17:430 (1974).
8. F. Stirpe and E. Della Corte, The regulation of rat liver xanthine oxidase. J. Biol. Chem. 244:3855 (1969).
9. E. Della Corte and F. Stirpe, Regulation of rat liver xanthine oxidase. Biochem. J. 108:349 (1968).
10. E. Della Corte and F. Stirpe, The regulation of rat liver xanthine oxidase. Biochem. J. 126:739 (1972).

11.H.B. Burch, O.H. Lowry, A.M. Padilla and A.M. Combs, Effects of riboflavin deficiency and realimentation on flavin enzyme of tissues. J. Biol. Chem. 233:29 (1956).

12.R. Fried and L. W. Fried, in H.U. Bergmeyer Methodem der enzymatischen Analyse Verlag Chemie, Weinheim (1970).

13.C. Ricci, Regolazione ormonale della biosintesi epatica del nicotinamide-adenina-nucleotide, relazione tenuta al XIV convegno della salute, Ferrara (27-28 maggio 1967).

14.A. Di Stefano, M. Pizzichini and E. Marinello, Regolazione della xantina ossidasi epatica. Atti del congresso S.I.B., comunicazione VII:16" (1978).

15.P.B. Rowe and J.B. Wyngaarden, The mechanism of djetary alteration in rat hepatic xanthine oxidase level. J. Biol. Chem. 241:5571 (1976).

XANTHINE OXIDASE ACTIVITY IN HUMAN INTESTINES. HISTOCHEMICAL AND RADIOCHEMICAL STUDY

C. Auscher, N. Amory, P. van der Kemp and F. Delbarre

Institut de Rhumatologie, Unité n° 5 INSERM - ERA 337 CNRS, Hôpital Cochin, Paris, France

Xanthine oxidase (xanthine oxygen-oxidoreductase EC 1.2.3.2.) catalyses oxidation of hypoxanthine and xanthine to uric acid. In man, small intestines and liver are the only tissues that normally show abundant xanthine oxidase (XO) activity. But localization, level, form of catalytic activity and physiological function in the different tissues are not clearly known. In xanthinuria, characterized by a gross deficiency of XO in the tissues it is not known whether the variations in the level of residual activity of XO represent methodologic limitations or heterogeneity of the genetic and molecular defect.

The present report describes the cellular localization and enzyme characteristics of XO in human intestines. It was performed on tissue by biopsies obtained from duodenum (perorally), rectal part of the large intestine and jejunum (surgically).

HISTOCHEMICAL LOCALIZATION

XO activity was detected on unfixed sections of frozen tissue by the reduction of nitro-blue-tetrazolium (NBT) in insoluble formazan with hypoxanthine as substrate as previously reported (1, 2).

As shown on fig. 1 (a, b) XO was normally located in the absorptive cells of the epithelium covering the free surface of the villi of the duodenum and jejunum segments. None activity was demonstrated in the mucosa of the large intestine and its rectal part (fig. 1 c).

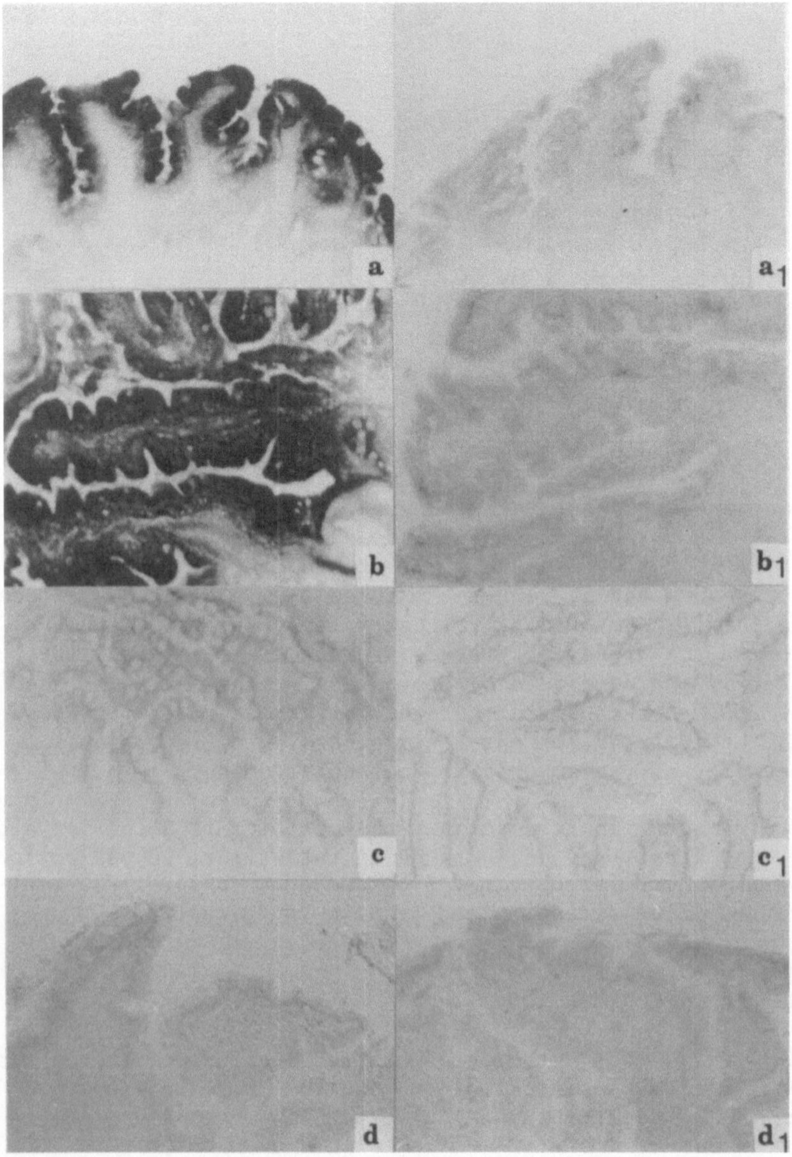

Fig. 1 : Reaction G x 125 Control (Allopurinol)

Small intestine : a Duodenum ; b Jejunum
Large intestine : c Rectal part
Xanthinuria : d Duodenum.

Therefore the catalytical activity of XO may be studied on biopsies obtained from the duodenum mucosa, which is much easier to obtain than jejunum one.

ENZYME CHARACTERISTICS

A sensitive radiochemical method was developped in order to study the catalytic activity of XO in the surnageant of crude homogenate of tissues obtained by biopsies from human duodenum mucosa with low level (0.010 mM) of 8-C14 hypoxanthine (3).

The study of the two steps of oxidation of hypoxanthine and xanthine in uric acid was carried out as follow : aliquotes of the supernatant (after centrifugation at 13.000 g) of crude homogenate of tissue (54 to 130 μg protein per 100 μl of reaction mixture) was incubated at 37°C in phosphate buffer pH 7.8 with 1.0 nmole of 8-C14 hypoxanthine (47 mCi/mM). The reaction was stopped by perchloric acid at 15 %. The radioactive hypoxanthine, xanthine and uric acid were detected and quantified directly on the plates by a chromato-scanner actigraph III Nuclear Chicago and a Victor computer after their separation by thin layer chromatography on cellulose plate. The time course of hypoxanthine oxidized, xanthine and uric acid formed in the reaction mixture were plotted as indicated on fig. 2.

Activity of XO expressed in hypoxanthine oxidized and estimated from the initial rate of oxidation of hypoxanthine within the first 20 minutes of reaction was 9.6 $\pm$ 3.5 nmol/h/mg protein (n = 12) while the rate of uric acid linearly formed was slower : 3.5 $\pm$ 1.5 nmol/h/mg protein (n = 12).

An amount up to 2.6 $\pm$ 0.3 μM (n = 12) of xanthine was necessary for uric acid to be formed. Then xanthine increased up to a steady state of 4.1 $\pm$ 0.1 μM (n = 8).

The Km value with hypoxanthine as substrate was 2.65 $\pm$ 0.86 μM (n = 6). It was estimated from a single progress curve of hypoxanthine (4) after geing assured at several enzyme concentration that the enzyme was not inactivated or inhibited during assay (5). It indicated a great affinity of the enzyme for hypoxanthine in the same magnitude as that found for milk XO (6).

Accumulation of xanthine in the medium was also found with milk XO and spectrophotometric methods (7, 8), but up to now no quantitative data were given concerning the amount of xanthine required in the reaction mixture for uric acid formation.

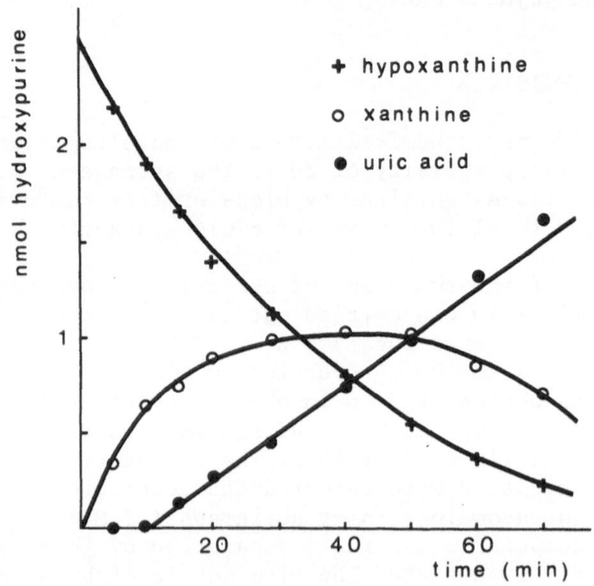

Fig. 2 : time course of hydroxypurine in the reaction mixture : 2.5 nmol hypoxanthine ; 130 µg of protein of homogenate in 250 µl (final vol.) phosphate buffer pH 7.8.

None activity of XO was detected with both histochemical method and radiochemical method in duodenum biopsies of a xanthinuric patient (normal histological villi) (fig. 1 d).

ACKNOWLEDGMENTS

We would like to acknowledge Miss J. Chevallier for the preparation of this manuscript.

REFERENCES

(1). C. Auscher and N. Amory, The histochemical localization of xanthine oxidase in the rat liver, Biomed. 25 : 37 (1976).

(2). N. Amory, F. Delbarre and C. Auscher, Localisation par voie histoenzymatique de la xanthine oxydase dans les tissus chez le rat et chez l'homme, C.R. Acad. Sc. Paris t. 287 série D.: 1007 (30 oct. 1978).

(3). C. Auscher, N. Amory and P. van der Kemp, Radiochemical assay of xanthine oxidase in biopsies of human duodenum with hypoxanthine as substrate (to be published).

(4). Shyun-Long Yun and C.H. Sueller, A simple method for calculating Km and V from a single enzyme reaction progress curve, Biochim. Biophys. Acta 480 : 1 (1977).

(5). M.J. Selwyn, A simple test for inactivation of an enzyme during assay, Biochim. Biophys. Acta 105 : 193 (1965).

(6). F. Bergman and L. Leven, Oxidation of N methyl substitued hypoxanthines xanthines purines 6-8 diones and the corresponding 6 thioxo derivatives by bovine milk xanthine oxidase, Biophys. Biochim. Acta 429 : 672-688 (1976).

(7). D.G. Priest and J.R. Fisher, Substrate activation with a xanthine oxidase reaction. An alternative to dismutation as an explanation in the chicken liver system, Europ. J. Biochem. 10 : 439 (1969).

(8). M.J. Jezewska, Xanthine accumulation during hypoxanthine oxidation by milk xanthine oxidase, Europ. J. Biochem. 36 : 385 (1973).

PURINE AND PYRIMIDINE METABOLISM IN HEREDITARY OROTICACIDURIA DURING A 15 YEAR FOLLOW-UP STUDY

D. R. Webster, H. A. Simmonds, C. F. Potter and
D. M. O. Becroft
Purine Laboratory, Guy's Hospital, London;
Princess Mary Hospital, Auckland, New Zealand.

Pyrimidines and purines are vital to the body; each has many important independent functions, as well as being essential components of DNA and RNA.

Hereditary oroticaciduria is the only recorded disorder of pyrimidine metabolism. The accumulation, or excretion of orotic acid results from the inhibition of two tandem enzymes of pyrimidine biosynthesis, orotate phosphoribosyltransferase (OPRT: EC 2.4.2.10) and orotidine-5-phosphate decarboxylase (ODC: EC 4.1.1.23). The defect may be primary as in hereditary oroticaciduria where orotic acid is excreted almost exclusively[1]. Alternatively, it may be secondary to the use of pharmacological agents, such as the pyrimidine analogues, 6-aza-uridine or 5-aza-orotic acid (oxonic acid - a uricase inhibitor), or the purine analogue allopurinol used in gout[2]. In this case, inhibition primarily of the ODC enzyme by nucleotide analogues results in the excretion of orotidine as well as orotic acid.

Increased PP-ribose-P levels are noted in subjects deficient in the purine salvage enzyme, hypoxanthine guanine phosphoribosyltransferase, where gross stimulation of <u>de novo</u> purine biosynthesis is considered a consequence of the raised PP-ribose-P levels. Despite the obvious dependence of <u>de novo</u> pyrimidine as well as purine biosynthesis on PP-ribose-P availability, PP-ribose-P levels are surprisingly reputedly normal in hereditary oroticaciduria and purine metabolism has not been studied in depth.

A 15 year clinical follow-up study of the longest surviving case of oroticaciduria on chronic uridine supplementation provided the opportunity to study both purine and pyrimidine metabolism in this defect.

In view of previous studies which have documented the reversal of both drug-induced (allopurinol) as well as hereditary oroticaciduria by dietary nucleic acid[3], the study included a diet both poor and rich in nucleoprotein with and without allopurinol.

METHODS

Clinical

The patient was investigated over four consecutive periods on a caffeine-free diet including three days on low or high purine diet, with and without allopurinol. Throughout the study the patient (D.G.) 54 kg in weight, continued the daily uridine supplement (150 mg/kg: 0.5 mmol/kg; 32.6 mmoles uridine per day); drugs were given as thrice daily supplements with meals.

Biochemical

Purine and pyrimidine metabolites were estimated in 24 h urine samples by methods reported in detail previously[4,5].

RESULTS

Pyrimidine metabolism

1. Basal levels and the effect of diet: The results obtained in the low and high purine diet are summarised in Fig. 1. Uracil and uridine were both abnormally elevated in the urine; they were excreted in roughly comparable amounts and represented together 2-7% of the daily uridine. Their excretion was apparently reduced on the high purine diet. Orotic acid excretion totalled approximately 7 mmoles daily and was essentially unaltered by the change to a high purine diet.

Purine metabolism

2. Basal levels and effect of diet: Urinary purine metabolites during the different periods of investigation are summarised in Fig. 2. Uric acid levels in plasma were low for a male of this age on a low purine diet, while urine uric acid levels were high (4.3-6.5 mmoles) (up to three-fold normal on a low purine diet) and were surprisingly essentially unaltered after the change to the high purine diet. Xanthine and hypoxanthine (which together with uric acid make up the urinary 'total oxypurines') were also within the normal limits for either dietary regime. Uric acid clearances were up to four times the normal mean (approximately 30 ml/min).

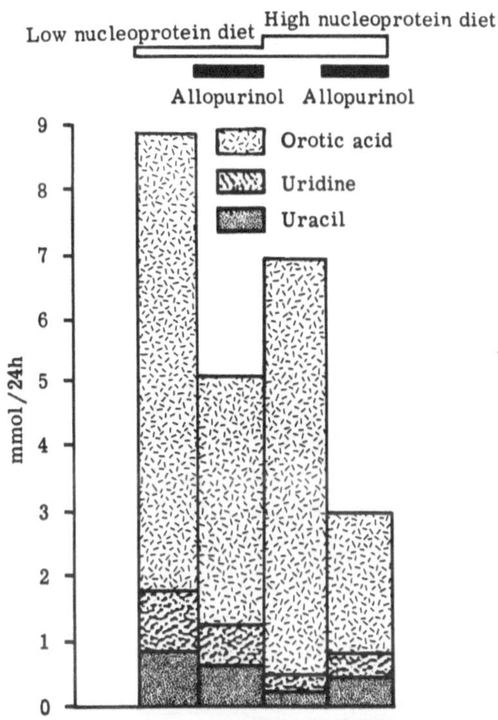

Fig. 1. Urinary pyrimidine excretion in hereditary oroticaciduria on low and high nucleoprotein diets with and without allopurinol.

3. Effect of allopurinol therapy: The effect of allopurinol on pyrimidine metabolism and purine metabolism is also given in Fig. 1 and 2. The dramatic effect of allopurinol in reducing both orotic acid and uric acid excretion concomitantly is clearly evident (Fig. 3). The dramatic reduction in uric acid excretion is mirrored by the reduction in orotic acid excretion. The metabolism of allopurinol was similar to that in the normal situation but the total metabolites were 10-30% less suggesting impaired absorption from the gastrointestinal tract.

DISCUSSION

Considering first basal pyrimidine metabolism, it is interesting that despite the thrice daily uridine supplements amounting to

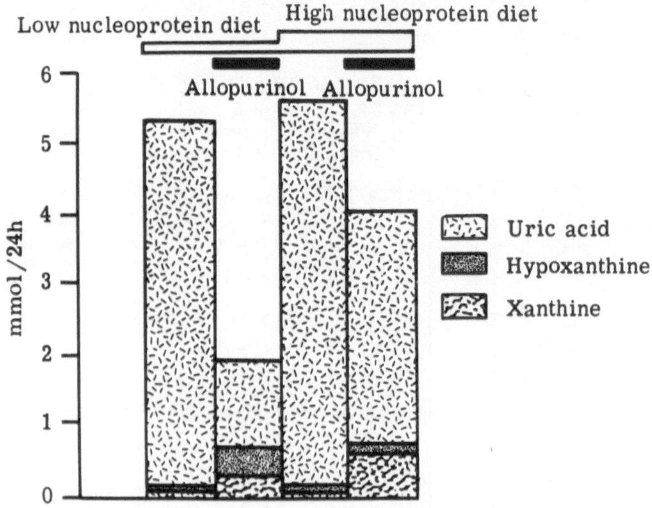

Fig. 2. Urinary oxypurine levels (xanthine + hypoxanthine + uric acid) in the same subject on low and high nucleoprotein diets with and without allopurinol.

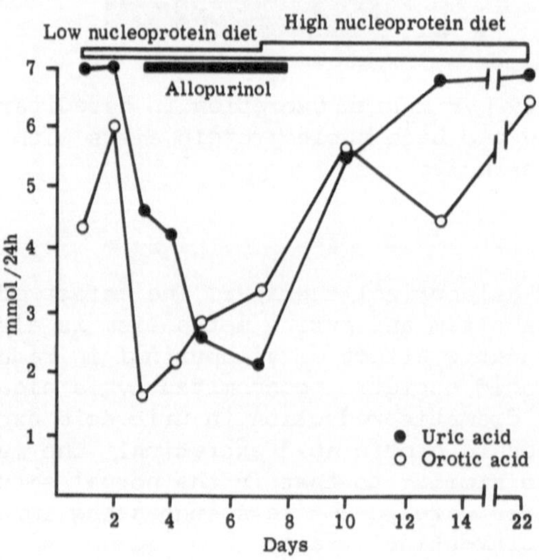

Fig. 3. Relationship between uric acid and orotic acid excretion during the different dietary regimes.

a total intake of approximately 33 mmoles only traces of the oral uridine (≃3%) were excreted unchanged in the urine, or as the pyrimidine base uracil. The lack of any effect of the high nucleoprotein diet on orotic acid levels in this instance is in accord with the patients clinically normal status on uridine supplementation. In this study, no orotidine was detected in the urine at any time in contrast to an earlier report where allopurinol produced a considerable orotidinuria without alteration in orotic acid excretion[6]. The difference between these two studies is even more striking when the dramatic orotic acid-lowering effect produced by allopurinol in this case is considered; yet the dosage was comparable (≃5 mg/kg) and uridine therapy was continued throughout in both instances. This effect was instantaneous and no apparent increase in enzyme levels in erythrocyte lysates was noted. Some orotidinuria might also have been anticipated if enzyme induction were the explanation in our case.

Allopurinol is known to reduce the absorption of purine bases and nucleosides but no effect on pyrimidine absorption has been noted[7]. However, competitive transport across the interstitial mucosa has been demonstrated between purine and pyrimidine bases; this may also apply to nucleosides and purine and pyrimidine analogues. Purine metabolism, and the effect thereon of allopurinol was equally remarkable. On a low purine diet urinary uric acid levels were 2-3fold normal. Low plasma uric acid levels were observed throughout and were thus associated with a urate clearance 3-4 times normal; a result presumably related to the well-documented uricosuric effect of orotic acid. The absence of any significant increment in plasma or urinary uric acid levels, as seen in normal subjects, on a high purine diet was also unusual. It is possible that the daily uridine supplement may prevent the absorption of dietary purine (as discussed above). Alternatively the normal daily extra-renal uric acid loss via the gastrointestinal tract may be abnormal in some way in oroticaciduria.

The reduction in total oxypurine excretion (xanthine and hypoxanthine and uric acid) by more than 50% even on the low purine diet during allopurinol therapy was dramatic and in direct contrast to the normal situation, where relatively little change in total oxypurine excretion is observed on a low purine diet (although a significant decrease is frequently found when a high purine diet is combined with allopurinol) (Van Acker et al - this symposium). The fact that allopurinol in this case of hereditary oroticaciduria produced a considerable reduction in total oxypurine excretion levels to a near normal total, irrespective of diet, could be considered evidence for purine over-production in hereditary oroticaciduria. However, equally convincing arguments against this hypothesis may be produced. The effect may represent altered absorption/secretion at the intestinal level, or effects thereon by allopurinol as already discussed.

The metabolic studies in this patient are interesting when considered in relation to the patients essentially normal clinical status on oral uridine. The total urinary load of potentially insoluble purines (uric acid, 6 mmoles) and pyrimidines (orotic acid, 7 mmoles) totalling more than 12 mmoles, was grossly in excess of that sometimes tolerated for uric acid alone, yet even maximal urinary concentrating capacity, the most sensitive indication of early renal dysfunction in gouty subjects, was normal.

This 15 year follow-up study in a case of hereditary oroticaciduria on chronic uridine therapy has demonstrated excessive urinary levels of orotic acid and uric acid, together with a close quantitative relationship between their excretion levels. The unexpected effect of allopurinol in reducing orotic acid in parallel with uric acid excretion was in direct contrast with its induction of mild oroticaciduria in the normal situation[2] and is unexplained.

The dramatic uric acid lowering effect of allopurinol was accompanied by only a small increment in the precursors xanthine and hypoxanthine, while diet alone was without effect on uric acid levels. These findings suggest the operation of two alternative mechanisms in concert: competition between exogenous purines and pyrimidines for intestinal absorption on the one hand, interaction between endogenous purine and pyrimidine biosynthetic pathways on the other.

REFERENCES

1. D. M. O. Becroft, L. I. Phillips and H. A. Simmonds, Hereditary oroticaciduria: Long-term therapy with uridine and a trial of uracil, J. Pediat., 75:885 (1969).
2. R. M. Fox, M. H. Wood and W. J. O'Sullivan, Studies on the coordinate activity and lability of orotidylate phosphoribosyltransferase and decarboxylase in human erythrocytes and the effects of allopurinol, J. Clin. Invest., 50:1050 (1971).
3. N. Zöllner and W. Gröbner, Influence of oral ribonucleic acid on oroticaciduria due to allopurinol, Z. Ges. Exper. Med., 156: 317 (1971).
4. H. A. Simmonds, Two-dimensional thin-layer high-voltage electrophoresis and chromatography for the separation of urinary purines, pyrimidines and pyrazolopyrimidines, Clin. Chim. Acta, 23:319 (1969).
5. L. E. Rogers and F. S. Porter, Hereditary oroticaciduria. II. A urinary screening test, Paediatrics, 42:423 (1968).
6. T. D. Beardmore and W. N. Kelley, Mechanism of allopurinol-mediated inhibition of pyrimidine biosynthesis, J. Lab. Clin. Med., 78:696 (1971).
7. H. A. Simmonds, P. J. Hatfield, J. S. Cameron, A. S. Jones and A. Cadenhead, Metabolic studies of purine metabolism in the pig during the oral administration of guanine and allopurinol, Biochem. Pharmacol., 22:2537 (1973).

EFFECT OF ALLOPURINOL ON PYRIMIDINE METABOLISM IN HUMAN WHITE BLOOD CELLS. ROLE OF THE SALVAGE PATHWAY.

P. Banholzer, W. Gröbner, N. Zöllner

Medical Policlinic, University of Munich

Pettenkoferstraße 8a, Munich

The administration of allopurinol leads to an increase of urinary excretion of orotic acid and orotidine (Fox et al. 197o) due to an inhibition of orotidine-5-phosphate decarboxylase (ODC).

Defects at this site can result in "pyrimidine starvation" from depletion of the intracellular pool of pyrimidine nucleotides as seen in hereditary orotic aciduria. Side effects during allopurinol treatment which can be attributed to pyrimidine starvation are not known until now. Nelson et al. (1973) showed that administration of even 2o mg/kg body weight allopurinol produces only a transitory decrease of uridinephosphates in rat liver and kidney.

We studied the effect of allopurinol and oxipurinol on the de novo and the salvage pathway of the pyrimidine metabolism by measuring the activity of ODC and uridinekinase in human leucocytes in vitro and in vivo. We also investigated the incorporation of labelled orotic acid and uridine into nucleic acids of cultured lymphoblasts.

METHODS

Preparation of leucocytes:

2o ml heparinized venous blood was mixed with 4 ml 5% Dextran T 5oo in physiological NaCl and allowed to sediment by gravity for 45 min. The leucocytes in the supernatant were washed twice with o,87% NH_4Cl-solution, then with physiological NaCl. The pellet was suspended in 2 ml o,o1 M Tris HCl-buffer pH 7,4 and sonified with a Branson Desintegrator. After centrifugation with 5oo x g at room temperature the supernatant was used for assays.

Lymphocytes cultures:

Heparinized venous blood was centrifugated over a Ficoll gradient. The interphase containing lymphocytes was washed and then suspended in RPMI-Medium (Gibco No. 164o) containing fetal calf serum 5o U/ml penicillin 5o µg/ml of streptomycin and 2,4 µg/ml phytohemagglutinine (Fa. Wellcome).

Incorporation of orotic acid-6-^{14}C and uridine-2-^{14}C into nucleic acids.
(Isotopes had a specific activity 5o mCi/mmol):

Cultured lymphoblasts were incubated at 37°C. After 48-72 hours uridine-2-^{14}C (o,o4 µCi) or orotic acid-6-^{14}C (2 µCi) and oxipurinol o,5 mM was added to 5oo ul of lymphoblasts suspension containing $1o^5$-$1o^6$ cells. Controls contained no oxipurinol. After 9o minutes the reaction was stopped by adding ice cold PBS (NaCl 8 g/l, KCl 2 g/l, Na_2HPO_4 x $2H_2O$ 1,45 g/l, KH_2PO_4 o,2 g/l). Nucleic acids were precipitated with trichloroacetic acid at a final concentration of 1o%. At last the acid insoluble material was dissolved in 1 ml o,1 n NaOH and counted in a Packard Liquid Scintillation counter with an efficiency of 9o%.

Enzyme assays:

Uridinekinase was assayed by using a modification of the method of Westwick et al. (1974). Sonified leucocytes were incubated for 2o minutes at 37°C in a medium containing 5 mM $MgCl_2$, 4 mM adenosine-triphosphate, o,1 M Tris-HCl-buffer pH 8, 7o mM uridine-2-^{14}C (o,4 uCi) in a final volume of 1oo µl. The reaction was stopped with 2o µl 2 mM EDTA. 2o µl of the reaction mixture was spotted on a Whatman No. 3 MM chromatography paper in the presence of UMP, UDP and UTP as carriers and separated by high voltage electrophoresis. The uridinephosphate spots were visualized by UV-light, cut out and counted in a Packard Liquid Scintillation counter at 9o% efficiency.

ODC-activity was measured according to Kelley and Beardmore (197o). Protein was determined by the method of Lowry et al. (1971). All enzyme assays were linear with regard to protein and time. Uric acid was assayed enzymatically.

RESULTS

Table 1 shows the influence of oxipurinol, Mg-PRPP or oxipurinol and Mg-PRPP on the activity of ODC and uridinekinase in lysed human leucocytes. Oxipurinol produces no change of ODC activity. In the presence of oxipurinol and Mg-PRPP ODC is inhibited. Oxipurinol does not change uridinekinase activity with or without Mg-PRPP.

Table 1: Lysed leucocytes were incubated for 2o minutes at 37°C in the presence of o,5 mM oxipurinol and Mg-PRPP. Then the enzymes were assayed as described above.

	ODC activity (per cent of control)	Uridinekinase activity (per cent of control)
None	100	100
Oxipurinol	101	104
MgPRPP	131	134
MgPRPP + Oxipurinol	67	141

When 9oo mg of allopurinol was orally administered in the morning as a single dose to healthy volunteers no change of uridinekinase activity in leucocytes was found within 1o hours as compared to controls.

A decrease of uridinekinase activity during the afternoon was observed in both groups, controls and allopurinol treated volunteers. The decrease of serum uric acid after allopurinol administration indicated allopurinol absorption from the gut.

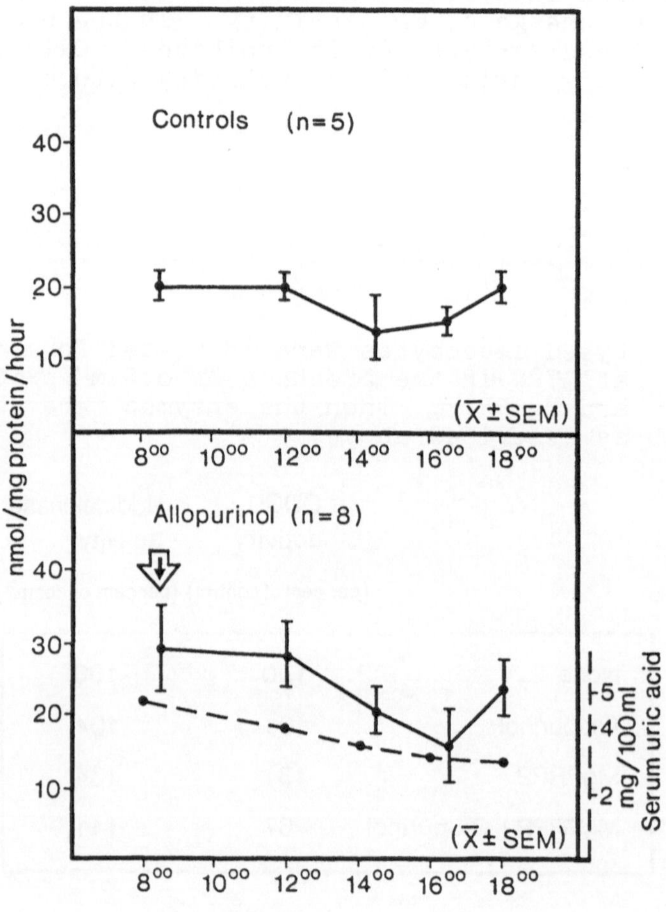

Fig. 1: 8 healthy volunteers received 9oo mg allopurinol as a single oral dose at 8^{00} a. m. Uridinekinase activity of leucocytes was measured up to 1o hours. Controls did not receive allopurinol.

Table 2: Lymphocytes were cultured for at least 48 hours. 5oo µl of the cell suspension containing $1o^5$-$1o^6$ cells were incubated for 9o minutes in the presence of o,5 mM oxipurinol and orotic acid-6-^{14}C (o,o4 uCi) or uridine-2-^{14}C (2 uCi). Isotope incorporation was measured as described in methods.

	Orotic acid incorporation (per cent of control)	Uridine incorporation (per cent of control)
None	100	100
Oxipurinol 0,5 mM	52	102
	46	110
	60	97
	64	95

Orotic acid-6-^{14}C incorporation into nucleic acids of human lymphocytes in culture was diminished after 9o minutes incubation with o,5 mM oxipurinol at 37°C, whereas uridine incorporation remained unchanged under the same conditions (table 2).

DISCUSSION

The results show that incubation of lysed human peripheral leucocytes with oxipurinol and Mg-PRPP results in a decrease of ODC activity. This is in agreement with investigations of Beardmore et al. (1971) with lysed human erythrocytes and Becker et al. (1974) with human lymphocytes in culture. The inhibition of ODC is attributed to the ribonucleotides of allopurinol and oxipurinol (Gröbner and Kelley, 1975; Becker et al. 1974). The activity of uridinekinase in peripheral white blood cells is not changed by oxipurinol in vitro. In addition a single dose of allopurinol as well as daily administration of 300 - 450 mg allopurinol did not influence uridinekinase activity in leucocytes. The decrease of uridinekinase activity during the afternoon is not clearly understood. Possible reasons are nutritional and/or hormonal influence or a circadian rhythm of the enzyme activity as known for some digestive enzymes. The increased ODC activity in erythrocytes during a long term allopurinol therapy is attributed to enzyme stabilisation. Incorporation of uridine into nucleic acids of lymphocytes cultures was not measurably affected by oxipurinol while orotic acid incorporation was decreased under these conditions. These data agree with those of Becker et al. (1974). In fibroblasts Kelley (1971) found a slightly but not significant enhanced incorporation of uridine into nucleic acids in the presence of oxipurinol, while incorporation of orotic acid was diminished.

The amount of orotic acid excreted during allopurinol therapy equals approximately 1-5% of that found in children with complete deficiency of ODC. We assume therefore that the allopurinol induced inhibition of ODC in vivo amounts to only 1-5%. This small amount must not necessarily lead to a decrease in uridinephosphates.

ACKNOWLEDGEMENT

This work was supported by the Deutsche Forschungsgemeinschaft.

REFERENCES

Beardmore, T. D., Kelley, W. N., 1971, Mechanism of allopurinol-mediated inhibition of pyrimidine biosynthesis, J. Lab. clin. Med. 78:696

Becker, M. A., Kent, F. A., Fox, R. M., and Seegmiller, E., 1974, Oxipurinol Assoziated Inhibition of Pyrimidine Synthesis in Human Lymphoblasts, Molec. Pharm., 10:657.

Fox, R. M., Royse-Smith, D., and O'Sullivan, W. J., 197o, Orotidinuria induced by allopurinol., Science, N. Y., 168:861.

Gröbner, W., and Kelley, W. N., 1975, Effect of Allopurinol and its Metabolic Derivatives on the Configuration of Human Orotate Phosphoribosyltransferase and Orotidine 5'-Phosphate Decarboxylase. Biochem. Pharm., 24:379.

Kelley, W. N., Beardmore, T. D., Fox, I. H., and Meade, J. C., 1971, Effect of Allopurinol and Oxipurinol on Pyrimidine Synthesis in Cultured Human Fibroblasts., Biochem. Pharm. 2o:1471.

Kelley, W. N., and Beardmore, T. D., 197o, Allopurinol: Alteration in Pyrimidine Metabolism in Man. Science, N. Y., 168:861

Lowry, O. H., Rosebrough, N. J., Farr, A. L., and Randall, R. J., 1951, Protein Measurement with the Folin Phenol Reagent, J. Biol. Chem., 193:265.

Nelson, D. J., Buggé, C. J. L., Krasny, H. C., and Elion, G. B., 1973, Formation of Nucleotides of ($6-^{14}C$) Allopurinol and ($6-^{14}C$) Oxipurinol in Rat Tissues and Effects on Uridine Nucleotide Pools. Biochem. Pharm. 22:2oo3.

Westwick, W. J., Allsop, J., and Watts, W. E., 1974, The Effect of Gold Salts on the Biosynthesis of Uridinnucleotides in Human Granulocytes., Biochem. Pharm. 23:153.

KINETICS AND COMPARTMENTATION OF ERYTHROCYTE PYRIMIDINE METABOLISM.

E. H. Harley, P. Zetler and S. Neal.

Department of Chemical Pathology,
University of Cape Town,
Cape Town, South Africa.

Purine metabolism and transport have been extensively studied in the mammalian erythrocyte ; much less, however, is known about corresponding aspects of the metabolism of pyrimidines and their derivatives, or even whether pyrimidine nucleotides have any physiological role in the erythrocyte. Until quite recently it was considered, from studies on stored blood, that negligible metabolism of pyrimidine nucleotides took place in the erythrocyte [1]. However, if fresh red cells are used it can be shown that both orotate and uridine can be actively taken up and converted to uridine nucleotides ; the rate of nucleotide formation from uridine circulating in the plasma has been estimated to be about 0.8 nmole/min 10^{10} RBC [2]. The rate of uridine nucleotide degradation back to uridine by the pyrimidine-specific 5' nucleotidase of erythrocytes is of a comparable order, if the properties of this enzyme in cell-free preparations [3,4] apply in the intact cell. Our previous studies on red cell uridine nucleotide degradation after pulse labelling normal and pyrimidine 5' nucleotidase deficient red cells with (^{3}H) uridine, confirm that rapid conversion to uridine occurs [2]. The presence, therefore, of an ATP-dependent cycling of pyrimidines between the nucleosides in plasma and the non-diffusable nucleotides in the red cell, mediated by uridine (cytidine) kinase and by pyrimidine 5' nucleotidase, and with a flux sufficient to turn over the circulating plasma uridine at least once per hour, strongly suggests a physiological role for this energy dependent cycle.

Uridine nucleotides synthesised via the de novo pathway do not necessarily have the same metabolic fate as those derived from the

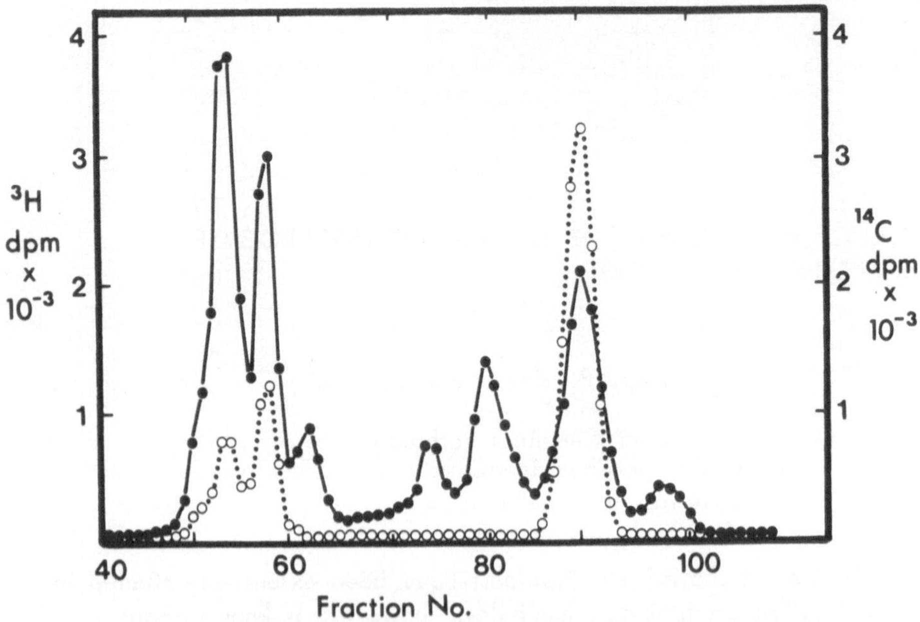

Fig.1. Biogel P-2 column chromatographs of labelled acid-soluble components from erythrocytes labelled for 1 h. at 37° with 1 µc/ml (6-^{14}C) orotate and 5 µc/ml (5-^{3}H) uridine. The ^{3}H labelled peaks correspond, from left to right, to the elution positions, on this or similar chromatographs, of markers of UDP, UMP, β alanine, unidentified, ^{3}HHO, uridine and uracil respectively. The ^{14}C-labelled peaks co-elute with markers of UDP, UMP and orotate [6]. O-----O, ^{14}C dpm; ●——●, ^{3}H dpm.

salvage pathway : utilising a line of cultured rat hepatoma cells, free of mycoplasma, which retains the ability of its tissue of origin to take up both uridine and orotate, we were able to study the comparative metabolism of both de novo and salvage pathway derived pyrimidine ribonucleotides by following the fate of (^{14}C) orotate and (^{3}H) uridine taken up from the culture medium. Differences in the ratios of ^{14}C : ^{3}H between the CTP and UTP of the acid soluble nucleotide pool, and between the CMP and UMP of hydrolysed RNA from cells labelled in this way provided evidence for compartmentation of de novo from salvage pathway derived uridine nucleotides. Orotate was predominantly directed to a UTP pool drawn on directly for RNA synthesis and uridine was directed to a UTP pool destined primarily for CTP synthesis [5].

This double-label approach was therefore used to compare orotate and uridine metabolism in normal human erythrocytes. The methods and

Table 1. ^{14}C : ^{3}H ratios in UDP and UMP from erythrocytes labelled with (6-^{14}C) orotate and (5-^{3}H) uridine as for Fig.1 and then washed and chased in unlabelled medium for varying periods of time.

Duration of Chase (min)	^{3}H UDP/UMP	^{14}C UDP/UMP	UDP $^{14}C/^{3}H$	UMP $^{14}C/^{3}H$
0	2.75	2.23	0.93	1.21
30	1.45	1.15	0.90	1.12
60	1.23	0.82	0.88	1.19
120	0.85	0.67	0.89	1.11

conditions for separating erythrocytes from leukocytes, washing and labelling them and the preparation and chromatographic analysis of the labelled pyrimidine derivatives, have been described previously[2].

Washed erythrocytes were labelled with (^{3}H) uridine and (^{14}C) orotate, and the chromatograph of the neutralised acid-soluble supernatant from these cells is shown in Fig.1. Both labels have been incorporated into two species corresponding to the elution positions of unlabelled markers of UDP and UTP respectively. Since labelled UDP is derived directly from UMP it would be expected, given a constant rate of labelling of UMP from both sources, that the ratio of ^{14}C : ^{3}H in UDP would reflect that in UMP. In fact, the ratio in UDP was found to be markedly different from that in UMP, the values being 0.22 ± 0.01 and 0.40 ± 0.02 respectively. This difference was found to be a reproducible feature in many subsequent experiments. Measurement of these ratios after different periods of labelling showed no significant delay in the appearance of label in UMP from orotate as compared with uridine, and the ratios of ^{14}C : ^{3}H in UMP and UDP remained almost constant over a four hour period (Table 1) despite a fall in the proportion of labelled UDP relative to UMP in the cells with time of incubation.

Labelling erythrocytes with the labels reversed, i.e. using (5-^{3}H) orotate and (2-^{14}C) uridine, resulted in a similar order of difference in the ratios of ^{14}C : ^{3}H between UDP and UMP. Values were 0.040 ± 0.001 and 0.020 ± 0.001 for UDP and UMP respectively. The proportions were reversed as compared to the results from the experiment illustrated in Fig.1, confirming that there was no selective isotope effect on UMP kinase.

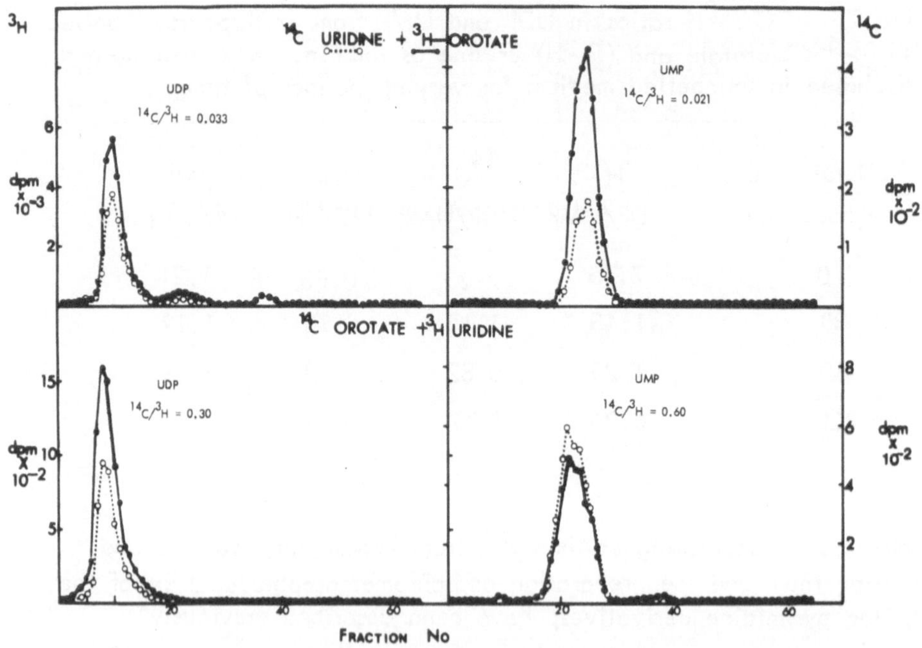

Fig.2. Paper chromatographs of UDP and UMP fractions, separated on preparative Biogel P-2 columns, from erythrocytes co-labelled as in Fig.1 with either (5-^{3}H) orotate and (2-^{14}C) uridine (above) or (6-^{14}C) orotate and (5-^{3}H) uridine (below). O ---- O ^{14}C dpm ; ● —— ●, ^{3}H dpm.

It was possible that this difference in ratios between UDP and UMP could be caused by a single-labelled species co-migrating with either UDP or UMP. Orotidine monophosphate was not a candidate for such a species since it elutes slightly earlier than UDP on Biogel P-2 [6] and also since any contribution of its label to counts measured in the UDP peak would cause the ratio to be the reverse of that found. The UDP and UMP peak fractions from typical Biogel P-2 separations of doubly-labelled erythrocytes were therefore lyophylised to concentrate them and chromatographed on a second system with different separation characteristics [2,6]. The labels migrated with R_fs corresponding to UDP and UMP respectively [6] and the ^{14}C and ^{3}H remained together in single peaks (Fig.2) which maintained the disparity in their respective ^{14}C : ^{3}H ratios, rendering the possibility of co-migrating single labelled species very unlikely.

It was therefore difficult to escape the conclusion that there is compartmentation between orotate and uridine derived pyrimidine nucleotides. Two different models can be proposed : firstly, the compartmentation might be between two red cell populations (e.g. young and old cells). This could account for the findings if one population had both lost the ability to phosphorylate uridine by uridine kinase and also had a markedly lowered intracellular ratio of UDP to UMP. This would result in enhanced orotate-derived label in UMP relative to UDP in the whole red cell extract. Secondly, there might be compartmentation of orotate from uridine derived uridine nucleotides in each red cell similar in principle to that described in nucleated mammalian cells [5]. This is feasible, despite the absence of obvious morphological compartments, if one (or both) compartments consists of a membrane or enzyme bound nucleotide complex. This model is perhaps not so unlikely as might at first appear ; we have observed that labelled orotate is not washed out of erythrocytes as rapidly as uridine, and it is well known that most of the pyrimidine de novo pathway enzymes in nucleated cells exist as multienzyme complexes [7]. The second model may have a parallel with the demonstration of compartmentation of uridine and UMP between cell membrane and cytosol in rat liver cells [8]. Experiments are therefore in progress to determine, by physical separation of these compartments, which of these two models is correct.

This investigation was supported by grants from the S.A. Medical Research Council, the Atomic Energy Board, the Harry Crossley Foundation and the Cancer Research Trust.

REFERENCES.

1. T.S. Lieu, R.A. Hudson, R. K. Brown and B.C. White : Transport of pyrimidine nucleotides across human erythrocyte membranes, Biochim. Biophys. Acta 241 : 884 (1971).
2. E.H. Harley, A. Heaton and W. Wicomb : Pyrimidine metabolism in hereditary erythrocyte pyrimidine 5' nucleotidase deficiency, Metabolism 27 : 1743 (1978).
3. D.E. Paglia and W.N. Valentine : Characteristics of a pyrimidine specific 5' nucleotidase in human erythrocytes, J. Biol. Chem. 250 : 7973 (1975).
4. J.D. Torrance, D. Whittaker, and E. Beutler : Purification and properties of human erythrocyte pyrimidine 5' nucleotidase, Proc. Natl. Acad. Sci. USA 74 : 3701 (1977).
5. M.J. Losman and E.H. Harley : Evidence for compartmentation of uridine nucleotide pools in rat hepatoma cells, Biochim. Biophys. Acta 521 : 762 (1978).

6. M. J. Losman and E. H. Harley : Chromatographic separation of pyrimidine derivatives, S. Afr. J. Sci. 72 : 343 (1976)
7. W.T. Shoaf and M.E. Jones : Uridylic acid synthesis in Ehrlich ascites carcinoma, Biochemistry 12 : 4039 (1973).
8. J.K. Tseng and E. Gurpide : Compartmentalization of uridine and UMP in rat liver slices, J. Biol Chem. 248 : 5634 (1973).

SIMULTANEOUS DETERMINATION OF RATES OF PURINE AND PYRIMIDINE SYNTHESIS IN CULTURED HUMAN LYMPHOBLASTS AND FIBROBLASTS

William H. Huisman, Kari O. Raivio and Michael A. Becker

Veterans Administration Medical Center and
University of California, San Diego, La Jolla,
California, 92161, USA

INTRODUCTION

Measurement of rates of incorporation of radiolabeled precursors into pathway intermediates and end products has commonly been employed in the study of rates of specific biosynthetic processes in intact cells. We report here the development and validation of an isotopic method which exploits the shared requirement of the pathways of both purine and pyrimidine nucleotide synthesis de novo for one molecule of CO_2 per base moiety to provide simultaneous estimates of rates of these pathways in cultured human lymphoblasts and fibroblasts.

Purine nucleotide synthesis de novo proceeds by a sequence of reactions in which the purine base moiety is synthesized on a ribose-5-phosphate backbone donated by 5-phosphoribosyl 1-pyrophosphate (PRPP). In contrast, the synthesis of pyrimidine nucleotides de novo involves formation of the base orotate prior to conversion of this compound to the first nucleoside monophosphate, OMP, in a reaction catalyzed by orotate phosphoribosyltransferase (E.C. 2.4.2.10). By means of the ion-exchange chromatographic procedure described here, all intermediates and products of pyrimidine biosynthesis, except carbamyl phosphate, can be recovered and separated from purine compounds, which include all stable pathway intermediates distal to CO_2 incorporation and all completed purine products.

METHODS

Cultured fibroblasts and lymphoblasts were harvested during log-phase growth. Fibroblasts were detached from culture vessels by treatment with trypsin as previously described (1). Cells were washed once in modified-autoclavable Eagle's minimum essential medium supplemented with 0.4% bovine serum albumin, 2 mM glutamine, 1 mM sodium pyruvate, non-essential amino acids (0.1 mM each), and 2.6 mM trisodium phosphate (pH 7.3) and were resuspended in this medium at the appropriate cell densities (0.5 to 10 x 10^6 per ml). Cell suspensions in a volume of 1 ml were incubated in covered test tubes for 30 min at 37° after addition of sodium bicarbonate (final concentration 15 mM). At this point, sodium [^{14}C] bicarbonate (20 µl; 50 µCi/µmol) was added, and incubation was continued for 60 min. Reactions were terminated by addition of 0.1 ml 6 N HCl and unfixed CO_2 was trapped in center wells containing hyamine hydroxide. Extracts were heated for 60 min at 100° to convert purine nucleotides to purine bases and pyrimidine nucleotides to pyrimidine nucleoside monophosphates. Heated extracts were then diluted with 0.1 N HCl and chromatographed on Dowex 50 x 8 (200 - 400 mesh) ion-exchange columns equilibrated with 0.1 N HCl. This procedure separated pyrimidine compounds (eluted with 0.1 N HCl) from purine bases (eluted with 6 N HCl) at an efficiency of greater than 98%. Pyrimidine and purine-containing fractions were mixed with 10 ml Aquasol scintillation mixture and counted on a Beckman LS-230 scintillation counter at 75% efficiency.

RESULTS

Incorporation of [^{14}C] bicarbonate into purine and pyrimidine compounds by intact lymphoblasts and fibroblasts was linear for up to 120 min (Figure 1) and was proportional to cell density over ranges from 0.15 to 6.0 x 10^6 per ml for lymphoblasts and from 0.2 to 5.0 x 10^6 per ml for fibroblasts. Over a period of months, considerable variation was found in rates of [^{14}C] bicarbonate incorporation into the purine and pyrimidine compounds of individual lymphoblast lines. This variation appeared to be attributable to periodic changes in growth rate (which occurred despite apparently unaltered culture conditions), since for each cell line nearly linear relationships were observed between rates of [^{14}C] bicarbonate incorporation into purine and pyrimidine compounds and the log phase growth rate of the culture at the time of study (Figure 2). Growth rates of fibroblast strains were similar for all 9 strains studies (from 0.053 to 0.038 doublings per hour) between the 7th and 16th passage in culture. Correlations of rates of label incorporation and growth rate therefore, appeared unnecessary for fibroblast strains. Table 1 shows rates of [^{14}C] bicarbonate incorporation determined in 4 lymphoblast lines and 9 fibroblast strains, certain of which bear genetic abnormalities in enzymes of purine metabolism. As will be discussed below, under specific conditions, values for labeled bicarbonate incorporation into purine and pyrimidine compounds appear to provide valid estimates of relative rates of the respective biosynthetic pathways in individual cell strains. Extension of this incorporation data to the comparison of rates of nucleotide synthesis between cell strains is, however, currently unwarrented even though rates of incorporation of label into purines appear to reflect the in vivo increased uric acid production of individuals with HGPRT-deficiency or excessive PRPP synthetase activity.

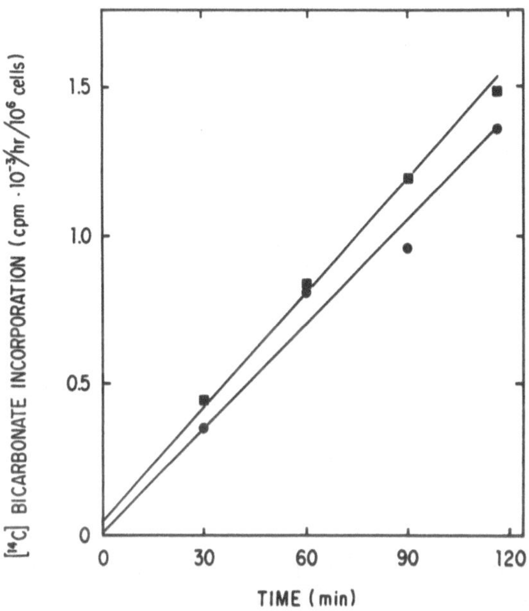

Figure 1

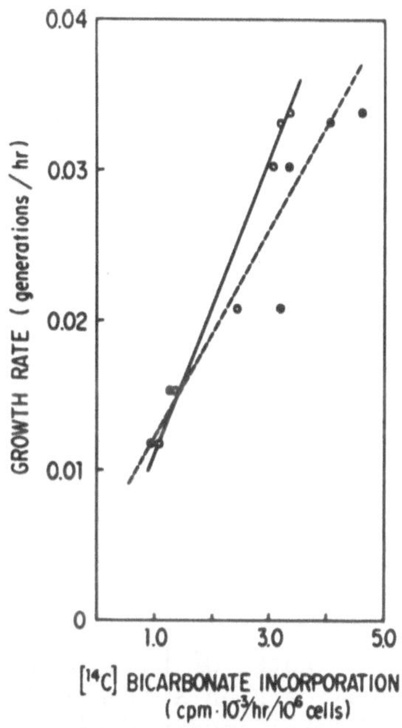

Figure 2

Figure 1. Incorporation of [^{14}C] Bicarbonate into pyrimidine (■) and purine (●) compounds of lymphoblast lines RS-760 as a function of time at a cell density of 1.0 x 10^6.

Figure 2. Relationships between rates of [^{14}C] bicarbonate incorporation into pyrimidine (●) and purine (○) compounds of lymphoblast line RS-760 and the growth rate of the culture. Cells were harvested during log phase growth at the rates indicated on the ordinate. [^{14}C] Bicarbonate incorporation into pyrimidine and purine compounds was determined as described in "Methods". Linear regression coefficients of 0.95 and 0.98, respectively were calculated from the results of the 6 experiments shown here.

Table 1. Relative Rates of [^{14}C] Bicarbonate Incorporation in Cultured Human Lymphoblast Lines and Fibroblast Strains

Lymphoblast Line	Phenotype	Relative Rate of Incorporation into Pyrimidines	Relative Rate of Incorporation into Purines
Wl-L2	Normal	1.00	1.00
RS-760*	Normal	1.11	1.16
Ag^{r}9 CL35SCl	HGPRT-deficient	0.98	2.68
MTI^r 107a	Adenosine Kinase-deficient (9)	1.26	0.76
Fibroblast strain			
MG*	Normal	1.00	1.00
JL*		0.97	1.10
ON*		1.07	1.09
WO*		0.98	0.83
S.Ma.*	HGPRT-deficient	1.18	1.61
RA	Superactive PRPP synthetase	2.04	1.78
HB		1.55	1.51
TC		1.82	1.60
SM		1.56	1.78

Rates of [^{14}C] bicarbonate incorporation were determined in duplicate on at least 3 occasions. For lymphoblast lines, rates of label incorporation were extrapolated to an arbitrary 30-hour doubling time from linear regression values calculated for the relationship between incorporation of label and growth rate. Rates are expressed relative to values of 1.00 for incorporation of [^{14}C] bicarbonate into pyrimidine and purine compounds in normal cells, corresponding in lymphoblasts to 4015 and 2880 cpm/h/10^6 cells, respectively for Wl-L2; and in fibroblasts to 5600 and 4800 cpm/h/10^6 cells, respectively for M.G.

*These lymphoblast lines and fibroblast strains were obtained from Dr. J. E. Seegmiller.

Studies were undertaken to evaluate the bicarbonate incorporation method in light of results obtained with prior methods for separate estimation of the rates of nucleotide synthesis _de novo_. Acceleration of purine and pyrimidine synthesis in mammalian lymphocytes during exposure to phytohemagglutinin (PHA) provided one such model for comparison. The increments in rates of [^{14}C] bicarbonate incorporation into purine compounds in human peripheral blood lymphocytes incubated 72 hours with PHA (1 μg/ml) were very similar to the increments in rates of purine synthesis under these conditions estimated either by [^{14}C] formate incorporation into FGAR or total purines,or by [^{14}C] glycine incorporation into purines (2). In addition, stimulation of [^{14}C] bicarbonate incorporation into pyrimidine compounds during PHA treatment closely approximated a previous estimate of the increment of pyrimidine synthesis under these circumstances (3).

The effects on [^{14}C] bicarbonate incorporation of several compounds shown by previous methods to alter rates of purine and pyrimidine synthesis (4,5) were tested in normal (WI-L2) (6) and HGPRT-deficient (Ag^{r}9C135SC1) (7) lymphoblast lines. Incubation of lymphoblasts for 30 min with 10^{-4} M adenine, adenosine, hypoxanthine or inosine prior to addition of label significantly decreased incorporation of labeled bicarbonate into the purine compounds of WI-L2. Adenine and adenosine, but not hypoxanthine or inosine, had similar inhibitory effects on incorporation of [^{14}C] bicarbonate into the purine compounds of the HGPRT-deficient cell line. Addition of 10^{-4} M uridine decreased incorporation of label into pyrimidine compounds in both cell lines, while both showed significantly increased rates of incorporation of label into pyrimidine compounds and precursors after addition of 10^{-4} M 6-azauridine.

Although the results of the above studies agree well with previous observations utilizing other isotopic methods for the study of rates of purine or pyrimidine synthesis, rates of incorporation of a radiolabeled precursor into pathway intermediates and products do not necessarily constitute valid estimates of the rates of synthesis of these compounds. Dilution of label by branching pathways, delayed equilibration of label with intracellular pools, and variations in assay conditions which may alter pool sizes or specific activities of pathway intermediates must all be considered. With the current method, one potential source of discrepancy between rates of label incorporation and pyrimidine synthesis is the undetermined pool of carbamyl phosphate. Carbamyl phosphate contents in cultured cells were, therefore, estimated after enzymatic conversion of carbamyl phosphate into [^{14}C] citrulline by incubation of neutralized acid-soluble cell extracts with excesses of ornithine carbamyltransferase (E.C. 2.1.3.3) and [1-^{14}C] ornithine (7). Labeled citrulline was separated from [1-^{14}C] ornithine by chromatography on Dowex 50 x 8 (200 - 400 mesh) ion-exchange columns equilibrated with 50 mM sodium citrate (pH 5.3). Under baseline incubation conditions, carbamyl phosphate contents in lymphoblasts and fibroblasts ranged from 0.4 to 2.1 pmol per 10^6 cells and were not significantly altered by incubation for an additional 60 min or by the addition of the purine base or purine and pyrimidine nucleoside effectors of bicarbonate incorporation described above. The baseline values determined for carbamyl phosphate content represent less than 0.1% of the hourly flux of [^{14}C] bicarbonate into pyrimidine compounds, suggesting a very rapid turnover of the carbamyl phosphate pool. Thus equilibration of the specific activity of the carbamyl phosphate pool with that of the intracellular bicarbonate pool should be quite rapid and the potential contribution of retarded equilibration to a discrepancy between rate of incorporation of label and rate of pyrimidine synthesis should be negligible.

The specific activity of the intracellular bicarbonate pool thus appears to be the critical factor in attempting to relate rates of [^{14}C] bicarbonate incorporation to rates of synthesis of both purines and pyrimidines. Although direct determination of the bicarbonate pool was not possible, our findings suggest that the specific activity of this pool remained essentially constant for each cell strain under the experimental conditions described. First, incorporation of [^{14}C] bicarbonate was linear for 120 min in the presence or absence of effectors. Second, effector compounds which increased or decreased rates of incorporation of label appeared to do so by altering rates of synthesis rather than the specific activity or size of the bicarbonate pool, since each compound tested had constant effects on rates of incorporation into each class of compounds regardless of the order of addition of [^{14}C] bicarbonate and effector.

These findings support the view that under the conditions specified, rates of [^{14}C] bicarbonate incorporation into purine and pyrimidine compounds provide valid relative values for rates of synthesis by the respective pathway in individual cell strains. Since, however, potential differences in the bicarbonate pool sizes and specific activities of different strains have not been evaluated, differences in rates of incorporation between strains (Table 1) cannot be directly attributed to differences in rates of nucleotide synthesis <u>de novo</u>. Despite this limitation, the current method should prove useful in the study of the coordinate control of purine and pyrimidine biosynthetic pathways in individual cell strains.

ACKNOWLEDGMENTS

This work was supported in part by the Medical Research Service of the Veterans Administration and Grant AM-18197 from the National Institutes of Health.

REFERENCES

1. W.H. Huisman, K.O. Raivio, and M.A. Becker, J.Biol.Chem. In Press.
2. T. Hovi, A.C. Allison, K.O. Raivio, and A. Vaheri, in: "Purine and Pyrimidine Metabolism," K.Elliott and D.W. Fitzsimons, eds., Ciba Foundation Symposium 48 (new series): 225, Elsvier, Amsterdam (1977).
3. K. Ito and H. Uchino, J.Biol.Chem. 246:4069 (1971).
4. F.F. Snyder, M.S. Hershfield, and J.E. Seegmiller, Adv.Exp. Med.Biol. 76A:30 (1976).
5. F.F. Snyder and J.E. Seegmiller, FEBS Letters 66:102 (1976).
6. J. Levy, M. Virolainen, and V. Defendi, Cancer 22:517 (1968).
7. G. Nuki, J. Lever, and J.E. Seegmiller, Adv.Exp.Med.Biol. 41A:255 (1974).
8. W.H. Huisman and M.A. Becker, submitted for publication.
9. M.S. Hershfield, E. Spector, and J.E. Seegmiller, Adv.Exp.Med.Biol. 76A:303 (1976).

ACTIVITIES OF ENZYMES OF PURINE AND PYRIMIDINE METABOLISM IN NINE MYCOPLASMA SPECIES

M. Hamet, C. Bonissol, and P. Cartier

Laboratoire de Biochimie, CHU Necker-Enfants Malades,

and Laboratoire des Mycoplasmes, Institut Pasteur, Paris.

INTRODUCTION

Cultured fibroblasts are commonly used for studies of purine and pyrimidine metabolism and may be helpful for the prenatal diagnosis of enzyme deficiencies. Mycoplasmas, a common contaminant of cell cultures, are unable to realize "de novo" synthesis of purines and pyrimidines. It is necessary to add to the medium precursors such as free bases (guanine, uracil, thymine) or corresponding nucleosides or nucleotides in order to obtain a good growth in vitro. The study of the best medium composition shows that mycoplasmas possess active pathways of nucleotides degradation and bases salvage (1).

It is well known that the different species of mycoplasmas have their own and diversified metabolic pathways. We thought important to establish the enzymatic profile of the most commonly isolated mycoplasmas in cell cultures, in order to determine to what extent they might interfere with enzymatic determination in fibroblasts.

Nine mycoplasma species were investigated : M. hominis, M. buccale, M. salivarium, M. arginini, M. pneumoniae, M. hyorhinis, M. fermentans and Acholeplasma laidlawii.

METHODS

The reference strains were grown in the classical broth medium of Hayflick containing horse serum (20 %), yeast extract (10 %), glucose (0.1 %) and penicillin. After 3 successive transfers at 37°C, the cells were washed three times with isotonic NaCl, resuspended in distilled water and disintegrated with ultrasound. The final protein concentration was 2 to 10 mg/ml determined by the method of Lowry et al. (1951). All extraction and centrifugation procedures were performed at 4°C.

Enzyme assays

The following enzymatic activities were measured : adenosine phosphorylase, adenosine deaminase, adenosine kinase, adenase, APRT, AMP deaminase, 5'-nucleotidase, PNP, HGPRT, Inosine kinase, 5'-pyrimidine nucleotidase, uridine phosphorylase, cytidine deaminase and uracil phosphoribosyltransferase. All the activities were determined at 37°C and assayed with radioactive substrates by measuring the amount of radioactive products formed per minute and mg protein.

RESULTS

1. We found that some enzymes had very little or undetectable activity in all species tested : adenosine deaminase, adenase, adenosine kinase and uridine kinase. Enzymatic activities are gathered in table I, for mycoplasmas hydrolysing arginine, and in table II, for fermenting mycoplasmas.

Table I : Mycoplasmas hydrolysing arginine

	Hominis	Buccale	Orale	Salivarium	Arginini
n =	4	3	4	3	4
5'-Nucleotidase	0.22	92,9	0.10	0.38	0.25
Ado phosphorylase	387	158	285	258	229
Ino phosphorylase	21.9	23.5	34.8	23.5	20.0
APRTase	0.48	1.06	0.36	2.70	0.24
HxPRTase	0.20	0.80	0.10	0.62	0.09
GPRTase	0.98	2.75	0.30	2.17	0.46
5'Pyr-Nucleotid	0.46	155.3	0.16	0.65	0.92
Ur phosphorylase	41.7	<0.05	30.3	11.6	14.3
Cytidine deaminase	7.3	0.26	6.1	8.5	15.1
Uracil PRTase	2.9	5.7	0.58	1.38	6.2

Table II : Fermenting mycoplasmas (glucose)

	Pneumoniae	Hyorhinis	Fermentans	Acholeplasma laidlawii
n =	5	4	4	5
5'-Nucleotidase	0.20	8.3	0.50	1.7
Ado phosphorylase	129	868	364	362
Ino phosphorylase	17.7	114.9	22.6	71.0
APRTase	0.45	5.3	2.7	14.5
HxPRTase	0.13	2.8	22.5	17.4
GPRTase	0.41	13.9	22.3	17.0
5'Pyr Nucleotidase	1.2	13.0	0.30	62.2
Ur phosphorylase	26.7	24.6	27.1	68.8
Cytidine deaminase	<0.05	7.1	27.3	<0.05
Uracil PRTase	0.36	2.1	14.7	6.6

2. Some enzymes were found in all the mycoplasmas tested, some at a high level such as adenosine phosphorylase, others at a lower level such as 5'-nucleotidase.

3. Some mycoplasma species were found to have a peculiar enzymatic pattern : for exemple, M. buccale shows high levels of 5'-purine and pyrimidine nucleotidase activities, but undetectable uridine phosphorylase. Acholeplasma laidlawii has high levels of purine phosphoribosyltransferases but undetectable cytidine deaminase activity. So, according to the strain of contaminating mycoplasma, an alteration of the cell metabolism can be expected.

4. Adenosine phosphorylase activity is very high in all species of mycoplasmas studied, contrary to mammalian cells which have no or very weak activity. The determination of this enzyme therefore is an excellent test for evaluation of contamination in cells culture. (3). In Fig. 1 a skin fibroblast culture of an APRT deficient child has been contaminated by M. Hyorhinis. The values of adenosine phosphorylase and APRT which were found at different stages of contamination are represented. A non negligible APRT activity appeared in the same time as adenosine phosphorylase (which expresses the degree of contamination).

Table III : Some enzyme activities of mycoplasmas and normal human fibroblasts

Fibroblasts		Orale	Arginini	Hyorhinis	Acholeplasma laidlawii
	Contamination (2) frequency in :				
	cultures	35 %	22 %	19 %	7 %
	Bovine serum	-	37 %	-	43 %
14.8	Ino phosphorylase	34.8	20.0	114.9	71.0
3.2	APRTase	0.36	0.24	5.3	14.5
1.8	HxPRTase	0.10	0.09	2.8	17.4
2.3	GPRTase	0.30	0.46	13.9	17.0

Units = nmoles.mg protein^{-1}.min^{-1}

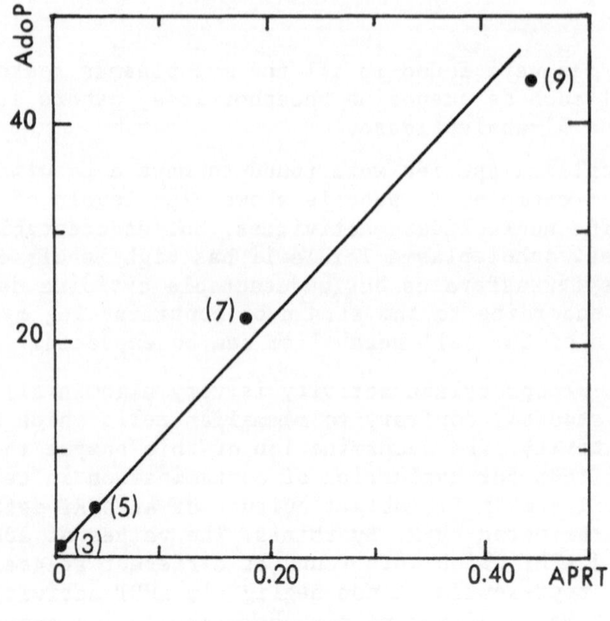

Fig. 1

REFERENCES

1. S. Razin : the mycoplasmas. Microbiol. Rev. 42:414 (1978).
2. M. Barile : Mycoplasmal contamination of cell cultures. In Contamination in tissue culture. J. Fogh, ed. Academic Press, New York (1973).
3. M. Hatanaka, R. Del Giudice, C. Long : Adenine formation from adenosine by mycoplasmas : adenosine phosphorylase activity. Proc. Nat. Acad. Sci. 72:1401 (1975).

INCREASES IN PURINE EXCRETION AND RATE OF SYNTHESIS BY DRUGS INHIBITING IMP DEHYDROGENASE OR ADENYLOSUCCINATE SYNTHETASE ACTIVITIES

R. C. Willis and J. E. Seegmiller

University of California San Diego

La Jolla, California USA 92093

Hershfield reported an assay system devoid of hypoxanthine for evaluating the rate of purine synthesis _de novo_ based on rates of [^{14}C]-formate incorporation into intracellular purine components of lymphoblasts and into purines excreted into the culture medium. Using this system normal and _in vitro_ selected, biochemically-characterized HPRT* deficient lines were found to have similar rates of total purine synthesis; however, these cell lines differed markedly in the excretion of newly-formed purine into the medium (1,2).

Normal lines cultured in media initially free of purines excrete as hypoxanthine 3-5% of the total purine synthesized. All HPRT-deficient lines excrete as hypoxanthine 15 to 30% of total purine synthesized. A Puo phosphorylase deficient line (3), like HPRT-deficient lines, excretes as inosine 20-30% of total purine synthesized. Lines characterized by feedback-resistant PRPP synthetase (4) excrete as hypoxanthine 15-20% of the total purine synthesized. In summary, cultures of lines established from patients overproducing purines as defined by three different aberrations of purine synthesis and reutilization all show a characteristic increase in [^{14}C]-formate incorporation into hypoxanthine and inosine excreted into the media rather than markedly increased synthesis _de novo_ of intracellular purines.

To date only two well documented enzyme defects of purine synthesis and reutilization have been identified among hyperuricemic patients; these are HPRT deficiency and increased activity of PRPP

*Abbreviations used. Hypoxanthine-guanine phosphoribosyl transferase (HPRT). Purine nucleoside phosphorylast (Puo phosphorylase). Ribose-5-phosphate pyrophosphokinase (PRPP synthetase). Glutamine amidophosphoribosyl transferase (PRPP amidotransferase). Adenylosuccinate synthetase (ASMP synthetase).

synthetase (4). These genetic aberrations account for less than 10% of all hyperuricemia patients who overproduce uric acid. Reports of pharmacologically-induced hyperuricemia which were traced to inhibition of IMP dehydrogenase (5) suggested to us that aberrations in,or deficiencies of, IMP dehydrogenase or ASMP synthetase activity could be identified in the undefined population of purine overproducers.In order to identify these individuals using lymphoblast culture system we are determining the patterns of purine synthesis and excretion for which to search.

For models of deficiencies of IMP dehydrogenase we have used the inhibitor mycophenolic acid and for deficiencies of ASMP synthetase, alanosine. The studies were performed with normal and HPRT-deficient lines derived from spleen and from peripheral blood. All studies were performed in RPMI 1640 medium supplemented with 0.1% fetal calf serum (2,6).Two methods were used to study the effects of the drugs on the various cell lines: I. An analysis of purine synthesis and excretion by [^{14}C]-formate incorporation (1,2), and II. An analysis of nucleotide content of cells and nucleoside and base content of culture medium by high-pressure liquid chromatography (1,2,6).

The most obvious effects of inhibition of ASMP synthetase (Table 1) are increased excretion of hypoxanthine; the second, elevations - which can be marked in HPRT deficient lines - in the total rate of purine synthesis. Adenylate synthesis is reduced to less than 5% of controls; guanylate synthesis is relatively unaffected; and IMP, recognized as cellular hypoxanthine, is elevated 5- to 10-fold. The rate of total purine synthesis of the normal lines induced by alanosine is increased up to 30% while the HPRT-deficient lines is consistently increased by 70 to 80% when adenylate synthesis is inhibited.

Table 1. Effect of 10μM alanosine on purine synthesis, base distribution, and excretion by line of lymphoblasts derived from spleen.

		[^{14}C]-FORMATE INCORPORATION INTO PURINES										
		CELL				MEDIUM						
			% Distribution				% Distribution					
Cell Line	Alanosine	Rate*	A	Hx	G	Rate*	A	Hx	G	Total Rate* (Rel.to control)	% Excreted
Control WI-L2	-	7.8	53	2	42	0.3	2	86	2	8.1 (1.0)	4
	+	5.1	4	39	53	5.8	<1	97	2	10.9 (1.35)	53
HPRT$^-$ TGR-729/6	-	9.3	55	3	38	3.0	3	92	3	12.3 (1.0)	24
	+	5.8	2	33	60	14.8	<1	98	1	20.6 (1.68)	72

* cpm x 10^{-3}/hr per 2 x 10^5 cells A=Adenine Hx=Hypoxanthine G=Guanine

Table 2. Effect of 10μM mycophenolic on purine synthesis, base distribution, and excretion by lymphoblasts derived from peripheral blood.

		[^{14}C]-FORMATE INCORPORATION INTO PURINES									
		CELL				MEDIUM					
			% Distribution				% Distribution				
Cell Line	Mycophenolic Acid	Rate*	A	Hx	G	Rate*	A	Hx	G	Total Rate* (rel.tc control)	% Excreted
Control 681	-	23.5	51	2	48	0.7	7	48	10	24.2 (1.00)	3
	+	11.2	87	8	3	25.7	<1	98	<1	36.9 (1.52)	70
HPRT$^-$ 681-TGR	-	20.3	49	2	49	4.6	2	77	14	24.9 (1.00)	19
	+	7.0	83	16	1	31.4	<1	98	<1	38.4 (1.54)	82

* cpm x 10^{-3}/hr per 2 x 10^5 cells A=Adenine Hx=Hypoxanthine G=Guanine

The most obvious effects of inhibition of guanylate synthesis (Table 2) are increased excretion of hypoxanthine and a marked elevation of the total rate of purine synthesis. Guanylate synthesis is reduced to less than 3% of controls; adenylate synthesis is marginally reduced, and IMP is elevated 3- to 5-fold. In contrast to inhibitions of ASMP synthetase, no consistent differences in total rates of purine synthesis were observed between normal and HPRT-deficient lymphoblasts when IMP dehydrogenase was inhibited.

The drug-induced inhibition of IMP dehydrogenase or of ASMP synthetase produces a similar effect on rates of purine synthesis, distribution, and excretion. The rate of IMP metabolism by the conjugate side of the branch point is initially maintained at a rate approximating the control. The excess IMP synthesized accumulates and is excreted into the media as inosine and hypoxanthine. The overall rate of purine synthesis of the culture is stimulated. The average stimulation observed with normal cell lines is 120% of control when adenylate synthesis is inhibited and 170% of control when guanylate synthesis is inhibited. The average stimulation of HPRT-deficient lines is 170% with either adenylate or guanylate synthesis inhibited. These results are consistent with a loss of feedback control by adenylates or guanylates on early steps of the _de novo_ pathway; for example, PRPP synthetase or PRPP amidotransferase (4,7). This presumably could result from diminished adenylate or guanylate pools (5,8,9).

Inhibition of AMP synthesis from IMP diminishes the ATP content (Fig. 1) as well as ADP and AMP. The GTP content increases transiently and then decreases with continued exposure of cells to the drug. The UTP and CTP contents decrease slowly with time. In these studies the IMP pool continued to increase over 4 hrs. from initial values <30 pmol/10^6 cells to values of 200 pmol/10^6 cells.

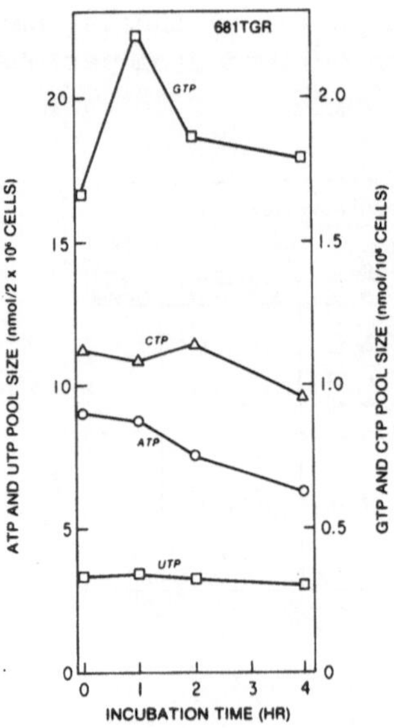

Figure 1. The effect of alanosine on nucleoside triphosphate content of lymphoblasts.

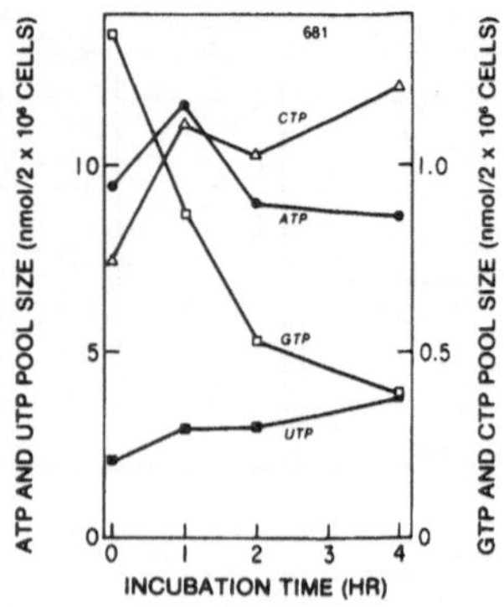

Figure 2. The effect of mycophenolic acid on nucleoside triphosphate content of lymphoblasts.

During the initial 120 minutes of drug exposure the increase in IMP content is concommitant with increased rates of purine synthesis. The inhibition of GMP synthesis from IMP rapidly diminishes the GTP content (Fig. 2). The ATP content increases transiently. The UTP and CTP contents also increase with time. The IMP content rapidly increased during the initial 120 mins of exposure to values of 180 pmol/10^6 cells and slowly increases for the remainder of the study.

In order to define further the expected characteristics of cultured lymphoblast with these deficiencies, long-term (14 days) culture studies of the effect of drug inhibition were performed. Using drug concentrations which provide <20% increase in culture doubling times, we observed by high-pressure liquid chromatography of medium samples a hypoxanthine output (per 10^6 cell increment) of 7-14 nmole with 0.1μM alanosine and 70-140 nmol with 0.3μM mycophenolic acid. Treatment with alanosine produced a 40-fold increase in IMP to approximately 1 mM, reduced ATP to 66% and increased GTP to 158% of control values. Treatment with mycophenolic acid increased the IMP content over 10-fold with a reduction of ATP to 80% and GTP to 17% of control values.

Therefore, we would propose that genetic aberrations leading to a deficiency of either IMP dehydrogenase or ASMP synthetase activity potentially could cause purine overproduction or hyperuricemia, which by analogy is recognized as elevated excretion of hypoxanthine by the cultured lymphoblast. Our support for the possible existence of these aberrances in the purine overproducing population of hyperuricemia patients is strengthened by many similarities between our studies and the studies of the ASMP synthetase mutant line selected by Ullman, et al. (10) in a mouse T-lymphoma line.

1. M.S. Hershfield and J.E. Seegmiller, Regulation of de novo purine synthesis in human lymphoblasts. J. Biol. Chem. 252: 6002 (1977).
2. R.C. Willis, D.A. Carson, and J.E. Seegmiller, Adenosine kinase initiates the major route of ribavirin activation in a cultured human cell line. Proc.Natl.Acad.Sci.USA 75: 3042 (1978).
3. L.H. Siegenbeek van Heukelom, J.W.N. Akkerman, G.E.J. Staal. C.H.M.M. DeBruyn, J.W. Stoop, B.J.M. Zegers, P.K. deBreu and S.K. Wadson, A patient with purine nucleoside phosphorylase deficiency: enzymological and metabolic aspects. Clin. Chim. Acta. 74: 271 (1977).
4. M.A. Becker, K.O. Raivio, and J.E. Seegmiller, Synthesis of phosphoribosylpyrophosphate in mammalian cells. in: "Advances in Enzymology", A. Mesiter, ed.,John Wiley & Sons, Inc. (1979).
5. A.J. Nelson, L.M. Rose, and L.L. Bennett, Jr., Mechanism of action of 2-amino-1,3,4-thiadiazole (NSC 4728). Cancer Res. 37: 182 (1977).
6. R.C. Willis and J.E. Seegmiller, Defining serum requirements of culture media for use in studies of physiological effects of various genetic diseases. in: "Monographs in Human Genetics", O.Sperling, A. deVries and P. Tiqva, ed., S. Karger,Basel (1978).
7. W.N. Kelley, E.W. Holmes, and M.B. Van Der Weyden, Current concepts on the regulation of purine biosynthesis de novo in man. Arth. & Rheum. 18: 673 (1975).
8. J.C. Graff and P.G.W. Plagemann, Alanosine toxicity in Novikoff rat hepatoma cells due to inhibition of the conversion of inosine monophosphate to adenosine monophosphate. Cancer Res. 36: 1428 (1976).
9. J.K. Lowe, L. Brox, and J.F. Henderson, Consequences of inhibition of guanine nucleotide synthesis by mycophenolic acid and virazole. Cancer Res. 37: 736 (1977).
10. B. Ullman, S.M. Clift, A. Cohen, L.J. Gudas, B.B. Levinson, M.A. Wormsted and D.W. Martin, Jr., Abnormal regulation of de novo purine synthesis and purine salvage in a cultured mouse T-cell lymphoma mutant partially deficient in adenylosuccinate synthetase. J. Cell. Physiol. 99: 139 (1979).

ACKNOWLEDGEMENTS

This work was supported in part by NIH grants GM17702, AM13622, AM07318 and The Kroc Foundation.

POSSIBLE ROLE FOR 5'-NUCLEOTIDASE IN DEOXYADENOSINE SELECTIVE TOXICITY TO CULTURED HUMAN LYMPHOBLASTS

Robert L. Wortmann, Beverly S. Mitchell, N. Lawrence Edwards, and Irving H. Fox

Human Purine Research Center
Departments of Internal Medicine and Biological Chemistry, University of Michigan, Ann Arbor, Michigan, 48109

Adenosine deaminase deficiency and purine nucleoside phosphorylase deficiency are relatively specific for causing a disorder of immune function. Several in vitro observations provide clues concerning the basis for this specificity. The addition of deoxyadenosine reduces the response of peripheral blood lymphocytes to mitogen stimulation when adenosine deaminase is inhibited[1,2]. The combination of deoxyadenosine and adenosine deaminase inhibition is also cytotoxic to T-lymphoblasts but not B-lymphoblasts[3-5]. Deoxyadenosine mediated cytoxicity in T-lymphoblasts is accompanied by increased concentrations of dATP.

Experiments were designed to study deoxyadenosine metabolism in cultured cells in order to elucidate the biochemical basis for the sensitivity of T-lymphoblasts and resistance of B-lymphoblasts to deoxyadenosine toxicity[6].

Deoxyadenosine metabolism was investigated in an in vitro model of adenosine deaminase deficiency using cultured human lymphoblasts from T- and B-cell lines. A block of adenosine deaminase was created by the addition of 5 μM EHNA, an adenosine deaminase inhibitor. Phosphorylation of deoxyadenosine to deoxyadenosine nucleotides was studied (Figure 1).

The most rapid synthesis of deoxyadenosine nucleotides was observed in the T-lymphoblasts. Using 58 μM deoxyadenosine, T-lymphoblasts synthesized 396 pmol of deoxyadenosine nucleotides per 10^6 cells in 5 minutes as compared to 9 pmol per 10^6 cells for

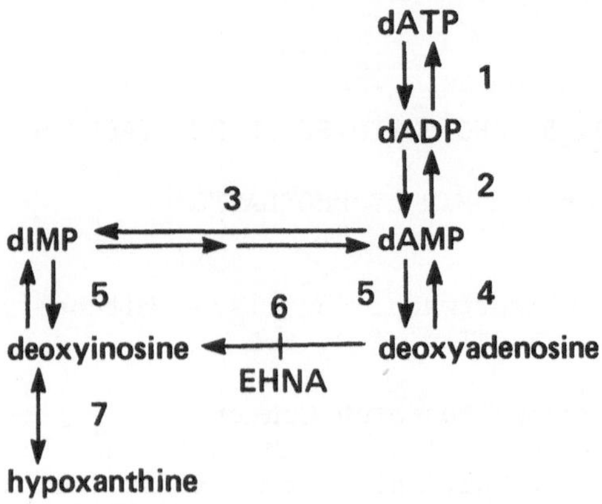

Fig. 1. Abbreviated diagram of deoxyadenosine metabolism. Enzymes displayed are: 1, deoxyadenosine diphosphate kinase; 2, deoxyadenylate kinase; 3, deoxyadenosine monophosphate deaminase; 4, deoxyadenosine kinase; 5, 5'-nucleotidase; 6, adenosine deaminase (this enzyme step is inhibited by EHNA); 7, purine nucleoside phosphorylase.

B-lymphoblasts (Figure 2). Other studies using 50-300 μM deoxyadenosine confirmed findings that T-lymphoblasts have a 20- to 45-fold greater capacity to synthesize deoxyadenosine nucleotides than B-lymphoblasts do.

The individual deoxynucleotides synthesized from deoxyadenosine were also measured; they demonstrate another difference between T- and B-lymphoblasts (Figure 3). While low levels of dAMP and dIMP accumulate in each cell line, there is a striking difference in the rate of dADP formation. In the first 10 minutes of the incubations with 58 and 300 μM deoxyadenosine, the concentration of dADP exceeded the dATP concentration in T-lymphoblasts. After ten minutes, the dADP levels reached a plateau, while dATP concentrations continued to increase linearly. In B-lymphoblasts, the

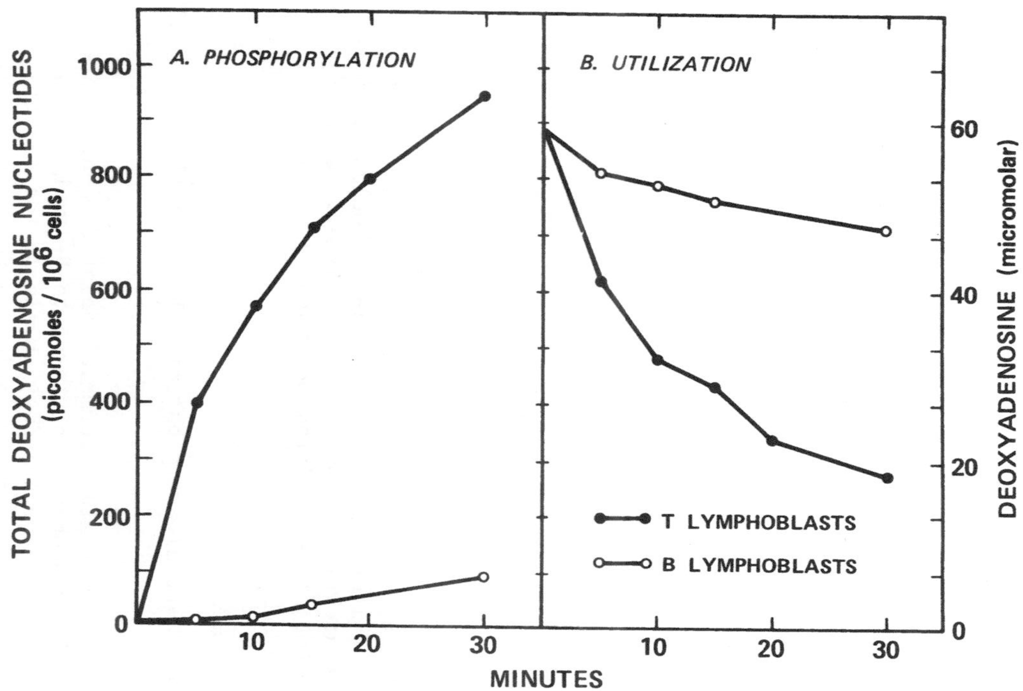

Fig. 2. (U-^{14}C)Deoxyadenosine metabolism during adenosine deaminase inhibition by EHNA (5 μM) in continuous Molt-4 T-lymphoblast (●) and MGL-8 B-lymphoblast (o) cell lines. (A) Quantities of total deoxyadenosine nucleotides formed by the phosphorylation of deoxyadenosine; (B) concomitant substrate utilization. The major pathway of deoxyadenosine metabolism is deamination to deoxyinosine despite inhibition of adenosine deaminase by EHNA (From Wortmann, et al[6]).

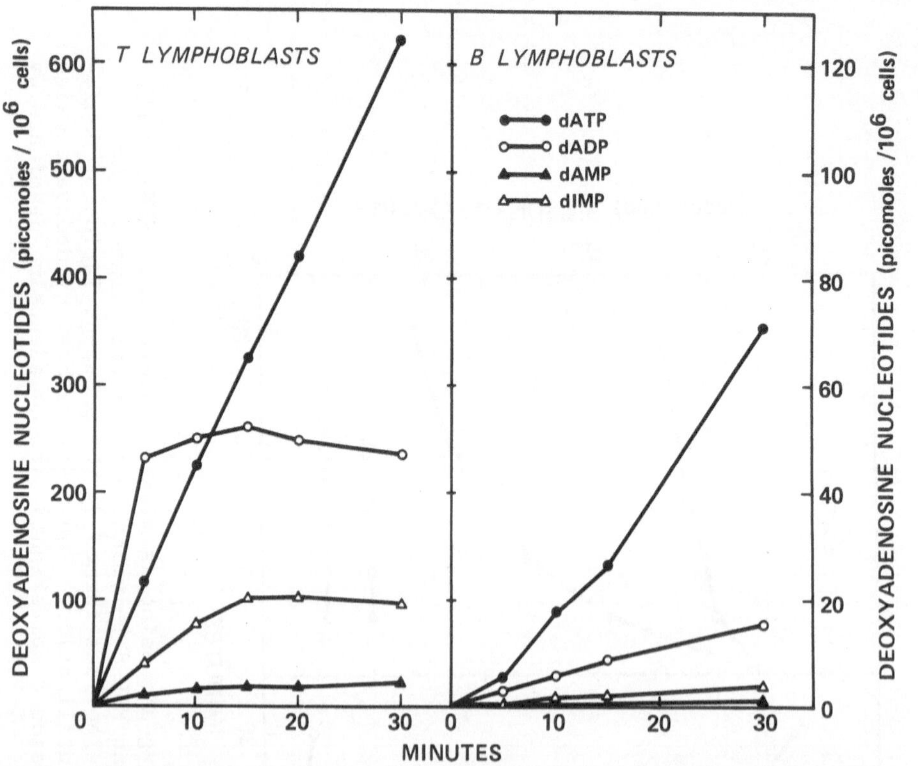

Fig. 3. Synthesis of deoxyadenosine nucleotides from (U-^{14}C)deoxyadenosine with adenosine deaminase inhibition by EHNA (5 μM) in (A) T-lymphoblasts and (B) B-lymphoblasts. Note the different vertical axis for each panel. ●, dATP; o, dADP; ▲, dAMP; Δ, dIMP (From Wortmann, et al[6]).

dADP concentrations were always substantially lower than the dATP concentrations.

The observations of greater accumulation of deoxyadenosine nucleotides in the T-lymphoblasts compared to B-lymphoblasts could be explained by either an increased synthesis in the T-cells or an increased degradation by dephosphorylation in the B-cells (Figure 1). Assessment of individual enzymes was undertaken to explain the increased synthesis of dADP observed with T-lymphoblasts as compared to B-lymphoblasts. The possibilities investigated were: (a) Increased deoxyadenosine kinase activity in T-cells compared to B-cells, (b) increased deoxyadenylate kinase activity in the T-cell line, and (c) increased 5'-nucleotidase activity in the B-cell line.

TABLE 1. PROPERTIES OF CULTURED HUMAN LYMPHOBLAST ENZYMES
(From Wortmann, et al[6])

Enzyme Activity	Specific Activity: T-Lymphoblasts (MOLT-4)	Specific Activity: B-Lymphoblasts (MGL-8)
Deoxyadenosine Kinase (nmole/hr/mg)	142	101
Deoxyadenylate Kinase (nmole/hr/mg)	702	665
5'-Nucleotidase ($nmole/hr/10^6$ cells)		
dAMP Substrate	1.3	9.15
AMP Substrate	1.1	48.5

Deoxyadenosine kinase and deoxyadenylate kinase activities were similar in B- and T-lymphoblasts. However, 5'-nucleotidase activity is distinctly different in the two cell types. Compared to the T-lymphoblasts, the B-lymphoblasts have an average of seven times as much ecto-5'-nucleotidase activity when deoxy-AMP is the substrate and 35 times as much when AMP is the substrate. The Michaelis constant for deoxyadenosine or for dAMP is similar for these three enzymes.

The interaction of 5'-nucleotidase activity and deoxyadenosine nucleotide synthesis may be based upon the rapid dephosphorylation and removal of dAMP when substantial 5'-nucleotidase activity is present. Figure 4 illustrates this proposed relationship. With only a small quantity of 5'-nucleotidase activity in the T-lymphoblasts, dAMP may be rapidly phosphorylated to dADP and dATP. The accumulation of these deoxynucleotides leads to cytotoxicity. The B-lymphoblasts, on the other hand, may only form small quantities of dADP and dATP because of the highly active and competing dephosphorylation of dAMP catalyzed by 5'-nucleotidase. The enhanced ability of B-lymphoblasts to degrade deoxynucleoside triphosphates as compared to T-lymphoblasts[7] supports this concept. Therefore, this model proposes that selective toxicity of deoxynucleosides to the immune system involves those cells that possess deoxynucleoside kinase to phosphorylate deoxynucleosides and a low activity of 5'-nucleotidase to allow deoxynucleotides to accumulate. This model is useful in that it suggests experiments to test its validity.

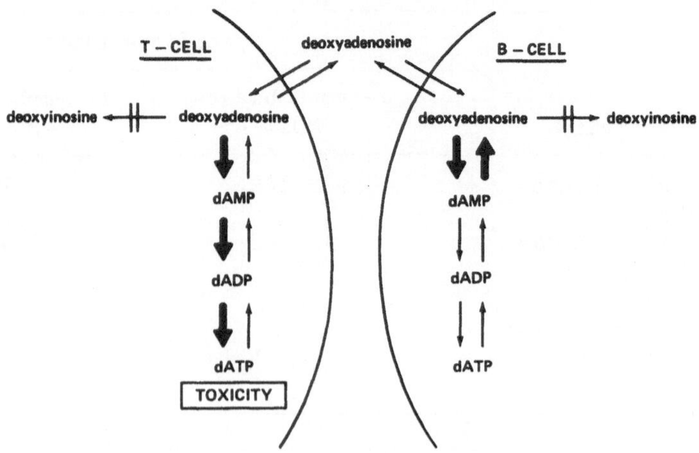

Fig. 4. Model for differential toxicity of deoxyadenosine to cultured lymphoblasts during adenosine deaminase inhibition. In T-lymphoblasts, the small amount of 5'-nucleotidase leaves dAMP available for further conversion to dADP and dATP. In B-lymphoblasts, the high activity of 5'-nucleotidase reconverts dAMP to deoxyadenosine, preventing its conversion to dADP and dATP. dATP is thought to be responsible for the cytotoxic effects (From Wortmann, et al[6]).

Whether cultured human lymphoblasts can serve as a relevant model for normal immunocompetent cells is not yet known. However, if the hypothesis relating deoxynucleoside-induced cytotoxicity to 5'-nucleotidase activity is applicable to the normal immune system, then similar changes in 5'-nucleotidase in man must be demonstrated. Normal lymphoid cells from peripheral blood, tonsil, and thymus were examined for 5'-nucleotidase activity to see if comparable variations of enzyme activity exist in man (Table 2). Intact peripheral mononuclear cells were examined prior to and after depletion of E-rosette forming lymphocytes. The E-RFC from thymus had the same low level of 5'-nucleotidase activity as the unfractionated thymocyte preparation and the T-lymphoblast in culture. Other immune cells studied had about an order of magnitude more 5'-nucleotidase than the thymocytes. If 5'-nucleotidase activity regulates sensitivity to deoxynucleoside toxicity, then these

TABLE 2. ACTIVITY OF 5'-NUCLEOTIDASE IN CELLS OF THE HUMAN IMMUNE SYSTEM (From Edwards, et al[8])

Cell Source	5'-Nucleotidase[a] (nmol/hr/10^6 cells) Unfractionated Cells	E-Rosette Forming Cells	Non-E-Rosette Forming Cells
Peripheral Blood Mononuclear Cells (9)[b]	21.9 (12.8-47.1)	21.2 (11.9-33.6)	24.0 (12.0-43.2)
Tonsil (5)	25.9 (13.0-34.4)	11.3 (5.9-23.5)	34.5 (15.8-55.8)
Thymus (4)	1.7 (1.4-2.1)	1.7 (1.4-2.1)	----
Cultured T-Lymphoblast (5)	1.8 (1.2-2.0)	----	----
Cultured B-Lymphoblast (5)	26.1 (6.7-47.2)	----	----

a. Enzyme levels are the mean values for multiple determinations with the range of values in parentheses.
b. Number of individual determinations performed in duplicate. In cultured lymphoblasts, five different cell lines were assayed.

findings suggest that human thymocytes may have an increased risk for toxicity when deoxynucleosides accumulate, as seen with cultured T-lymphoblasts.

To summarize, our studies show that T-lymphoblasts accumulate deoxyadenosine nucleotides at a faster rate than B-lymphoblasts and accumulate larger quantities of dADP. The differences observed between T- and B-lymphoblasts may be accounted for by the greater activity of 5'-nucleotidase in the B-cell line and may provide a biochemical basis for the increased sensitivity of the T-cell line to deoxyadenosine toxicity. Assessment of normal human lymphocyte subpopulations demonstrated low 5'-nucleotidase activity in thymocytes, similar to the value observed in T-lymphoblasts. If the differential sensitivity to deoxyadenosine toxicity observed in T- and B-lymphoblasts is applicable to human immunocompetent cells, then human thymocytes may be highly susceptible to the toxic properties of deoxynucleosides.

ACKNOWLEDGMENTS

The authors wish to thank Jumana Judeh for her excellent typing of the manuscript. This work is supported by USPHS grants AM19674 and 5M01RR42 and a grant from The American Heart Foundation (77-849) and The Michigan Heart Association. B.S.M. is a recipient of a National Institutes of Health Clinical Investigator Award. N.L.E. is the recipient of a Clinical Associate Physician Award for the General Clinical Research Center Branch of the National Institutes of Health.

REFERENCES

1. H. A. Simmonds, G. S. Panay, and F. Corrigali, A role for purine metabolism in the immune response: Adenosine deaminase activity and deoxyadenosine catabolism, Lancet 1:60 (1978).
2. D. A. Carson, J. Kaye, and J. E. Seegmiller, Lymphospecific toxicity in adenosine deaminase deficiency and purine nucleoside phosphorylase deficiency: Possible role of nucleoside kinase(s), Proc. Natl. Acad. Sci. USA 74:5677 (1977).
3. B. S. Mitchell, E. Mejias, P. E. Daddona, and W. N. Kelley, Purinogenic immunodeficiency diseases: Selective toxicity of deoxyribonucleosides for T-cells, Proc. Natl. Acad. Sci. USA 75:5011 (1978).
4. D. A. Carson, J. Kaye, and J. E. Seegmiller, Differential sensitivity of human leukemic T-cell lines and B-cell lines to growth inhibition by deoxyadenosine, J. Immunol. 121: 1726 (1978).
5. E. W. Gelfand, J. L. Lee, and H. M. Dosch, Selective toxicity of purine deoxynucleosides for human lymphocyte growth and function, Proc. Natl. Acad. Sci. USA 76:1998 (1979).
6. R. L. Wortmann, B. S. Mitchell, N. L. Edwards, and I. H. Fox, Biochemical basis for differential deoxyadenosine toxicity to T- and B-lymphoblasts: A role for 5'-nucleotidase. Proc. Natl. Acad. Sci. USA 76:2434 (1979).
7. D. A. Carson, J. Kaye, S. Matsumoto, J. E. Seegmiller, and L. Thompson, Biochemical basis for the enhanced toxicity of deoxyribonucleosides toward malignant human T-cell lines. Proc. Natl. Acad. Sci. USA 76:2430 (1979).
8. N. L. Edwards, E. W. Gelfand, L. Burk, H. M. Dosch, and I. H. Fox, Distribution of 5'-nucleotidase in human lymphoid tissue. Proc. Natl. Acad. Sci. USA 76:(In Press) (1979).

CYCLIC NUCLEOTIDE LEVELS AND MECHANISM OF INHIBITION OF LEUCOCYTE FUNCTION BY ADENOSINE DEAMINASE INHIBITION

Allen D. Meisel, Chandrasek Natarajan, Gary Sterba, and Herbert S. Diamond

State University of New York Downstate Medical Center, 450 Clarkson Avenue, Box #42, Brooklyn, New York 11203

Supported by a grant from the New York Chapter of the Arthritis Foundation and from the Irma T. Hirschl Career Scientist Award (HSD)

INTRODUCTION

The association of adenosine deaminase (ADA) deficiency and severe combined immunodeficiency syndrome has focused attention upon purine metabolism as a modulator of the immune response (1-3). In in vitro studies, adenosine deaminase appears to be necessary for normal lymphocyte blastogenesis in response to phytomitogens and for mononuclear cell maturation (4-6). Various mechanisms of potential cellular toxicity have been proposed including accumulation of adenosine (4, 5), pyrmidine starvation (7, 8), accumulation of deoxyadenosine (9), inhibition of S-adenosylmethionine (SAM) dependent methylation reactions (10), and accumulation of intracellular cyclic AMP (4, 11).

Although attention has been focused upon lymphocytes and lymphoblastoid cells because of the apparent defects in both B and T cell function, these patients with severe combined immunodeficiency syndrome secondary to ADA deficiency are subject to recurrent bacterial infections (1-3), and adenosine deaminase appears to be required for normal leucocyte function (12). This report documents that adenosine deaminase activity is required for normal human polymorphonuclear cell function, and that adenosine deaminase deficiency appears to modulate cellular function by leading to an accumulation of intracellular cyclic AMP.

METHODS

Blood from normal volunteers was collected with 25 units of sodium heparin/ml of blood and allowed to sediment at 37°C. The leucocyte rich plasma was diluted with TC 199 medium (pH 7.4) containing Hepes Buffer 0.62%, penicillin 100 units/ml and streptomycin 100 mg/ml, and layered over Ficoll-Paque™ by the method of Boyum (13). The separated PMNs were washed twice with TC 199, counted, and resuspended at a concentration of 2.5×10^5 PMN/10 ul in TC 199 alone or as appropriate.

Adenosine Deaminase Assay:

ADA activity was measured by conversion of adenosine to uric acid which was detected by the spectrophotometric change in absorbance at 293 nm.

Chemotaxis Under Agarose:

Agarose plates were prepared by the method of Nelson, Quie and Simmons (14) using TC 199, heat inactivated fetal calf serum with the pH adjusted to 7.4. The center well of each series of three received 10 ul cell suspensions containing 2.5×10^5 PMNs after 30 minutes preincubation. The inner well received 10 ul of the medium TC 199 and the outer well received 10 ul of endotoxin activated serum or zymosan activated serum. The plates were incubated for 3 hours, after which the cells were fixed and stained. The chemotactic differential is calculated by subtracting the random migration toward the well containing TC 199 from directed migration toward the well containing the chemotactic factor.

Isolated leucocytes (2.5×10^7 cells/ml) were incubated with varying concentrations of adenosine in TC 199, erythro-9-(2-hydroxy-3-nonyl) adenine hydrochloride (EHNA) (from Dr. Gertrude B. Elion of the Wellcome Research Laboratories, Research Triangle Park, N.C.), deoxyadenosine (1 mM), L-homocysteine thiolactone (1 mM) or theophyllin (1 mM).

In the case of a mixture of two agents, separate solutions were prepared in TC 199 having twice the concentration of each. When the total volume in which the cells were to be suspended was calculated, half the volume of each was used to make up the final suspension of 2.5×10^7 cells/ml. The cell suspension was then incubated at 37°C for 30 minutes in a CO_2 incubator before placing them in the wells for chemotaxis.

RESULTS

ADA activity in human polymorphonuclear cells was 324 ± 123

units/h/10^7 cells and was significantly less than ADA activity in mononuclear cells (4185 $\pm$ 1253 units/h/10^7). When leucocytes were incubated with radiolabelled adenosine, adenosine was incorporated into acid precipitable material at 37°, but only negligibly at 0°. Adenosine incorporation reached a plateau at 10 minutes of incubation, and did not change over 30-60 minutes.

Effect of ADA Inhibition on Chemotaxis

At concentrations of less than 15 ug/ml or 50 uM, the adenosine deaminase inhibitor, EHNA, had no effect on leucocyte chemotaxis while resulting in up to a 60% decrease in ADA activity. Increasing concentrations of EHNA resulted in inhibition of ADA activity in both intact cells and cell lysates and corresponded to increased inhibition of leucocyte chemotaxis. At EHNA concentrations of 25 ug/ml or 80 uM, there was a significant decrease in leucocyte chemotaxis of 21% (Figure 1). At 80 uM, EHNA decreased chemotaxis from 10.6 $\pm$ 0.7 mm to 8.6 $\pm$ 0.8 mm, but EHNA had no effect on random migration at any concentration.

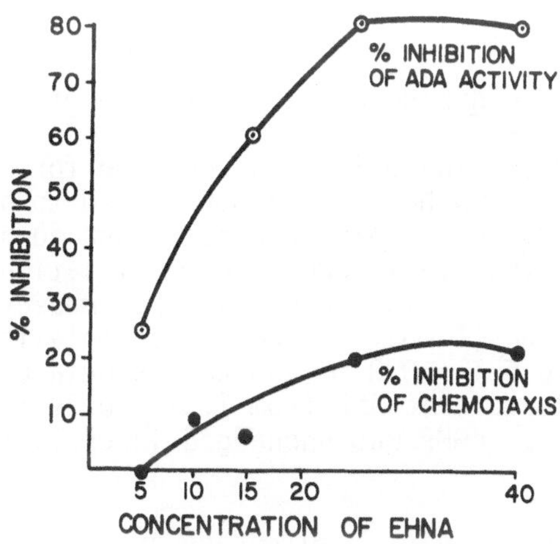

FIGURE 1. Effect of Increasing Concentration of EHNA (ug/ml) on ADA Activity and Leucocyte Chemotaxis

Adenosine in 1 mM and 5 mM concentrations (concentrations which are 1-5 thousand fold the elevations observed in ADA deficient subjects), had no effect on either random or directed leucocyte migration under agarose. Adenosine (1 mM or 5 mM) did not potentiate the decrease in leucocyte chemotaxis observed with the adenosine deaminase inhibitor, EHNA. At 10 mM concentrations, adenosine resulted in a marked decrease in leucocyte chemotaxis. This may represent direct stimulatory effect of adenosine on adenyl cyclase similar to that reported by Marone, Plant and Lichtenstein (15). At all adenosine concentrations, cell viability was not decreased as measured by trypan-blue exclusion.

Effect of ADA Inhibition on Intracellular Cyclic AMP

Since intracellular levels of cyclic AMP have been demonstrated to modulate the chemotactic responsiveness of polymorphonuclear cells, intracellular cyclic AMP was measured by radioimmunoassay in polymorphonuclear cells from five normal subjects. Mean control cyclic AMP was 0.86 ± 0.2 picomoles/10^7 cells and increased to 2.2 ± 0.46 picomoles/10^7 after pretreatment with EHNA ($p < 0.01$).

When sub-inhibitory concentrations of EHNA (20 uM) were combined with sub-inhibitory concentrations of the phosphodiesterase inhibitor, theophyllin (0.5 mM), there was an additive effect on leucocyte chemotaxis which decreased from 12.2 ± 0.2 mm to 10.0 ± 0.3 mm or 20% and on intracellular cyclic AMP which increased from 1.09 ± 0.6 picomoles/10^7 cells to 2.28 ± 0.27 picomoles/10^7 cells.

Effect of Deoxyadenosine on Leucocyte Chemotaxis

Adenosine deaminase is also required for the metabolism of deoxyadenosine which has been demonstrated to be 100-fold greater inhibitor of lymphocyte blastogenesis than adenosine (9). To test the hypothesis that deoxyadenosine accumulation might account for inhibition of chemotaxis, chemotaxis was measured in the presence of deoxyadenosine (1 mM). Leucocyte chemotaxis decreased from 8.5 ± 0.8 mm to 6.1 ± 0.9 mm and was associated with an increase in intracellular cyclic AMP from 1.38 ± 0.29 to 2.37 ± 0.39 picomoles/10^7 cells. cGMP was unchanged in cells incubated with deoxyadenosine.

Effect of L-Homocysteine Thiolactone on Leucocyte Chemotaxis

An alternative hypothesis has been proposed by Kredich and Martin (10). In the presence of ADA inhibition, the accumulation of adenosine might also stimulate the enzyme S-adenosyl homocysteine hydrolase and result in an accumulation of S-adenosyl homocysteine. L-homocysteine tiolactone is a potent inhibitor of SAM dependent methylation reactions. The effect of inhibition of SAM dependent methylation was studied by preincubating leucocytes with L-homocysteine (1 mM). L-homocysteine resulted in a marked decrease in chemotaxis from 8.5 ± 0.8 mm to 3.7 ± 0.8 mm without a significant change in either cyclic AMP or cGMP.

Effect of EHNA, Deoxyadenosine and L-Homocysteine Thiolactone on Leucocyte Phagocytosis

Since phagocytic function, as well as locomotive function, appears to be modulated by intracellular cyclic nucleotide levels, a candidal phagocytic assay (10) was performed in the presence of EHNA (100 uM), L-homocysteine (1 mM), and deoxyadenosine (1 mM). The number of leucocytes which phagocytized candida were expressed as a percentage of total cells with control cells demonstrating virtually 100% phagocytosis (Table 1). Both EHNA and deoxyadenosine resulted in a decrease in the percent of cells phagocytizing candida, with EHNA resulting in a 18.8% decrease and deoxyadenosine a 15.5% decrease. As noted before, both EHNA and deoxyadenosine decreased leucocyte chemotaxis and were associated with increases in intracellular cyclic AMP. In contrast, L-homocysteine resulted in decreased chemotaxis, with no change in phagocytosis and no rise in cAMP.

DISCUSSION

Adenosine deaminase is the major degradative enzyme for both adenosine and deoxyadenosine which are metabolized to inosine and deoxyinosine respectively. Since the initial recognition of the association of adenosine deaminase deficiency and severe combined immunodeficiency syndrome, adenosine metabolism has been extensively investigated to determine the mechanism responsible for leucopenia and depressed B and T cell function (4-11).

TABLE 1. Effect of EHNA, L-Homocysteine and Deoxyadenosine on Leucocyte Function and Intracellular cAMP

	EHNA (100 uM)	L-Homocysteine (1 mM)	Deoxyadenosine (1 mM)
Decrease in Chemotaxis	21%*	57%*	30%*
Decrease in Phagocytosis	18.8±4.9**	5.8±2.7	15.5±2.9**
cAMP Increase	156%**	12%	70%***

*$p < 0.001$
**$p < 0.01$
***$p < 0.02$

In lymphoblastoid cells and fibroblasts, various mechanisms of cellular toxicity have been proposed including pyrmidine starvation, accumulation of cytotoxic concentration of adenosine, deoxyadenosine and/or cyclic AMP and inhibition of transmethylation reactions. In addition to the marked decrease in lymphocyte function, these patients are subject to recurrent bacterial infections and inhibition of ADA with ADA inhibitors such as EHNA and deoxycoformycin have been shown to inhibit leucocyte chemotaxis (12). The studies reported here document the role of adenosine metabolism in leucocyte function. ADA inhibition was associated with a decrease in both leucocyte chemotaxis and leucocyte candidal phagocytosis. Both of these functions have been reported to be dependent upon intracellular cyclic AMP concentrations (17, 18).

In the absence of adenosine deaminase, adenosine is primarily phosphorylated to AMP by adenosine kinase, providing increasing substrate for adenyl cyclase. Adenosine is known to rapidly increase cyclic AMP concentrations in several tissues including lymphocytes (11). Deoxyadenosine might be expected to have a similar effect.

In summary, the decrease in leucocyte chemotaxis and phagocytosis observed after ADA inhibition with EHNA and deoxyadenosine appears to be modulated by a shift in adenosine metabolism to phosphorylated products which ultimately leads to an accumulation of intracellular cyclic AMP. Evidence exists that adenosine and cAMP metabolism are intrictly related, and a marked increase in cAMP levels has been shown in fibroblast extracts and lymphocytes from subjects with ADA deficiency (19, 20). Preliminary data does not suggest that deoxyadenosine accumulates within the leucocytes. L-homocysteine thiolactone has a profound effect on leucocyte chemotaxis but that may be independent of ADA inhibition and occurs by another metabolic mechanism.

REFERENCES

1. E. R. Giblett, J. E. Anderson, F. Cohen, B. Pollara, and H. J. Meuwissen, Adenosine-deaminase deficiency in two patients with severely impaired cellular immunity, Lancet. 2:1067 (1972).
2. C. R. Scott, S.-H Chen, and E. R. Giblett, Detection of the carrier state in combined immunodeficiency disease associated with adenosine deaminase deficiency. J. Clin. Invest. 53:1194 (1974).
3. J. Yount, P. Nichols, H. D. Ochs, S. P. Hammar, C. R. Scott, S.-H Chen, E. R. Giblett, and R. J. Wedgewood, Absence of erythrocyte adenosine deaminase associated with severe combined immunodeficiency, J. Pediatr. 84:173 (1974).
4. D. A. Carson and J. E. Seegmiller, Effect of adenosine deaminase inhibition upon human lymphocyte blastogenesis, J. Clin. Invest. 57:274 (1976).

5. F. F. Snyder, J. Mendelsohn, and J. E. Seegmiller, Adenosine metabolism in phytohemagglutinin-stimulated human lymphocytes, J. Clin. Invest. 58:654 (1976).
6. D. Fischer, M. B. Van der Weyden, R. Snyderman, and W. N. Kelley, A role for adenosine deaminase in human monocyte maturation, J. Clin. Invest. 58:399 (1976).
7. H. Green, and T.-S Chan, Pyrmidine starvation induced by adenosine in fibroblasts and lymphoid cells: Role of adenosine deaminase, Science. 182:836 (1973).
8. K. Ishii, and H. Green, Lethality of adenosine for cultured mammalian cells by interference with pyrmidine biosynthesis, J. Cell Sci. 13:429 (1973).
9. H. A. Simmonds, G. S. Panayi, and V. Corrigall, A role for purine metabolism in the immune response: Adenosine-deaminase activity and deoxyadenosine catabolism, Lancet. i:60 (1978).
10. N. M. Kredich, and D. W. Martin, Jr., Role of S-adenosylhomocysteine in adenosine-mediated toxicity in cultured mouse T lymphoma cells, Cell. 12:931 (1977).
11. G. Wolberg, T. P. Zimmerman, K. Hiemstra, M. Winston, and L.-C Chu, Adenosine inhibition of lymphocyte-mediated cytolysis; possible role of cyclic adenosine monophosphate, Science. 187:957 (1975).
12. A. D. Meisel, C. Natarajan, G. Sterba, and H. S. Diamond, Effect of adenosine deaminase inhibition on leucocyte function, Clin. Res. 27:331A (1979).
13. A. Boyum, Separation of leukocytes from blood and bone marrow, Scand. J. Clin. Lab. Invest. 21:Suppl. 97, 77 (1968).
14. R. D. Nelson, P. G. Quie, and R. L. Simmons, Chemotaxis under agarose: A new and simple method for measuring chemotaxis and spontaneous migration of human polymorphonuclear leukocytes and monocytes, J. Immunol. 115:1650 (1975).
15. G. Marone, M. Plaut, and L. M. Lichtenstein, Characterization of a specific adenosine receptor on human lymphocytes, J. Immunol. 11:2153 (1978).
16. R. I. Lehrer, and M. J. Cline, Interaction of candida albicans with human leucocytes and serum, J. Bacteriol. 98:996 (1969).
17. I. Rivkin, J. Rosenblatt, and E. L. Becker, The role of cyclic AMP in the chemotactic responsiveness and spontaneous motility of rabbit peritoneal neutrophils, J. Immunol. 115:1126 (1975).
18. R. Anderson, A. Glover, and A. R. Rabson, The in vitro effects of histamine and metiamide on neutrophil motility and their relationship to intracellular cyclic nucleotide levels. J. Immunol. 118:1690 (1977).
19. C. G. Mills, F. C. Schmalstieg, K. B. Trimmer, A. S. Goldman, and R. M. Goldblum, Purine metabolism in adenosine deaminase deficiency, Proc. Natl. Acad. Sci USA, 73:2867 (1976).
20. F. C. Schmalstieg, J. A. Nelson, G. C. Mills, T. M. Monahan, A. S. Goldman, and R. M. Goldblum, Increased purine nucleotides in adenosine deaminase-deficient lymphocytes, J. Pediatr. 91:48 (1977).

PURINE RIBONUCLEOSIDE AND DEOXYRIBONUCLEOSIDE METABOLISM IN THYMOCYTES

Floyd F. Snyder and Trevor Lukey

Pediatric Research Center, Alberta Children's Hospital &
Division of Pediatrics and Medical Biochemistry,
The University of Calgary, Calgary T2N 1N4, Canada

The human deficiencies of adenosine deaminase and purine nucleoside phosphorylase activities are associated with an accumulation of the purine deoxyribonucleoside triphosphates, dATP (1-4) and dGTP (4) respectively. In the absence of adenosine deaminase or purine nucleoside phosphorylase activities, deoxyadenosine or deoxyguanosine are presumed to be rephosphorylated, thereby producing increased dATP or dGTP concentrations which may in turn inhibit ribonucleotide reductase. We have therefore examined the specificity and optimal assay conditions for the principal purine ribonucleoside and deoxyribonucleoside kinase activities in mouse thymocytes. The role of these activities was evaluated by analyzing the competition for common substrates between the nucleoside kinases and adenosine deaminase or purine nucleoside phosphorylase.

Thymocytes lysates were prepared from Swiss-Webster mice in 25 mM HEPES, pH 7.2 containing 5 mM dithioerythritol. Kinase assays were started by the addition of labelled nucleoside and reactions were stopped by transferring 0.025 ml of reaction mixture to Whatman DE 81 papers for batch processing as previously reported (5). The adenosine deaminase inhibitor, EHNA (6), has been used in assays of adenosine and deoxyadenosine phosphorylation and assays were conducted in the absence of phosphate in order to minimize purine nucleoside phosphorylase activity.

A comparison of the specific activities for purine nucleoside kinases in mouse thymocytes shows adenosine kinase to be 5.6- and 7.6-fold greater than deoxyguanosine and deoxyadenosine kinase

---

Abbreviations: EHNA, erythro-9-(2-hydroxy-3-nonyl)adenine.

respectively (Table 1). The phosphorylation of both inosine and guanosine was on the lower limit of detectability for the radiochemical assays.

The pH profile for each activity was found to be quite distinct. The phosphorylation of deoxyadenosine was relatively constant over a broad range, from pH 6.5 - 8.6. Deoxyguanosine kinase activity was optimal at pH 8.4. Adenosine kinase activity was optimal at pH 6.0, having 30% greater activity in tris-maleate than in phosphate buffer. At sub-saturating concentrations of adenosine, the activity was greater at pH 7.0 than 6.0.

The deoxyribonucleoside kinases showed similar nucleoside substrate dependence, requiring 300 µM deoxyguanosine or 400 µM deoxyadenosine for optimal activity. Adenosine phosphorylation was optimal at 40 µM. Adenosine kinase exhibited optimal activity at 1.0 mM Mg^{++} with a decrease in activity at Mg^{++} concentrations greater than 1.0 mM (Fig. 1A). Deoxyadenosine and deoxyguanosine were optimal at 2 or 6 mM Mg^{++} or greater respectively. The optimal ATP concentration was 5 and 1 mM for adenosine and deoxyadenosine kinase respectively whereas deoxyguanosine kinase was optimal at 3 mM ATP and greater (Fig. 1B). The specific Mg^{++} and ATP concentration requirements found for adenosine kinase have also been observed by others (7-10). The conditions for adenosine kinase in mouse thymocytes were similar to those for human lymphoid cells (5,11). GTP has been shown to act as an alternate substrate for ATP (8) and in the present study, substitution of GTP for equal concentrations of ATP gave essentially the same activity for each of the three principal kinases. A summary of the optimal assay conditions is given in Table 2.

The phosphorylation of deoxyadenosine and deoxyguanosine have been attributed, in whole or in part, to a deoxycytidine kinase activity (12-15). We have therefore examined the effect of the pyrimidines deoxycytidine and dCTP and several purine nucleosides on the phosphorylation of adenosine, deoxyadenosine and deoxyguanosine (Table 3). The phosphorylation of adenosine was not affected

Table 1. PURINE NUCLEOSIDE KINASE ACTIVITIES IN THYMOCYTES

SUBSTRATE	ACTIVITY $nmol \cdot mg\ protein^{-1} \cdot min^{-1}$	RELATIVE
Adenosine	4.7	100
Deoxyguanosine	0.84	18
Deoxyadenosine	0.62	13
Inosine	0.06	1
Guanosine	0.03	0.6

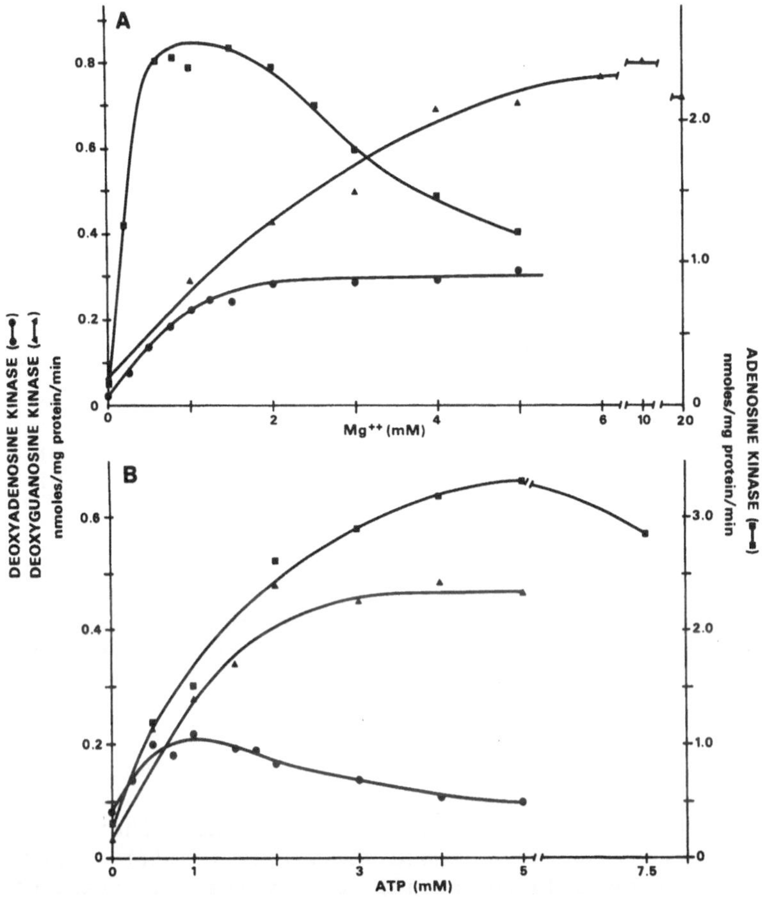

Fig. 1. Magnesium and ATP concentration dependence for the phosphorylation of adenosine, deoxyadenosine and deoxyguanosine.

Assays were conducted as described in Table 2.

Table 2. OPTIMAL ASSAY CONDITIONS FOR PURINE NUCLEOSIDE KINASES

COMPONENTS	NUCLEOSIDE SUBSTRATES Adenosine	Deoxyadenosine	Deoxyguanosine
Buffer (100 mM)	Tris-maleate	Tris-HCl	Tris-HCl
pH	6.0	7.6	8.4
ATP (mM)	5.0	1.0	3.5
Mg^{++} (mM)	1.0	5.0	10
Nucleoside (μM)	45	450	350
EHNA (μM)	5	10	-

Table 3. SPECIFICITY OF PURINE NUCLEOSIDE KINASES

ADDITIONS (200 μM)	RELATIVE ACTIVITY Adenosine	Deoxyadenosine	Deoxyguanosine
A.	45 μM	450 μM	350 μM
None	100	100	100
Deoxycytidine	99	66	7
dCTP	107	68	7
B.	45 μM	100 μM	80 μM
None	100	100	100
Adenosine	-	60	105
Deoxyadenosine	106	-	30
Deoxyinosine	105	87	110
Deoxyguanosine	106	87	-

by any of the deoxyribonucleosides tested nor dCTP. The phosphorylation of deoxyguanosine was inhibited greater than 90% by dCTP and deoxycytidine, and up to 70% by deoxyadenosine. The phosphorylation of deoxyadenosine was partially inhibited by dCTP or deoxycytidine, 35%; deoxyinosine or deoxyguanosine, 13%; and by adenosine, 40%. The present findings are consistent with there being a specific kinase for adenosine, having low affinity toward deoxyadenosine and a deoxyguanosine-deoxycytidine kinase of broad specificity which also catalyzes the phosphorylation of deoxyadenosine. There are reports of distinct deoxyguanosine kinase activities (16,17). Deoxyinosine may also be phosphorylated as indicated by its inhibition of deoxyadenosine phosphorylation which is consistent with reports of deoxyinosine phosphorylation by human and calf thymus (17,18).

A summary of the activities of purine nucleoside metabolism in thymocytes is given in Table 4. The relationship between the respective thymocyte kinase and deaminase activities was evaluated

Table 4. ACTIVITIES OF PURINE NUCLEOSIDE METABOLISM IN THYMOCYTES

ACTIVITY	$nmol \cdot mg\ protein^{-1} \cdot min^{-1}$
Adenosine kinase	4.66 ± 0.50
Deoxyadenosine kinase	0.62 ± 0.08
Deoxyguanosine kinase	0.84 ± 0.21
Adenosine deaminase	347 ± 57
Deoxyadenosine deaminase	123
Purine nucleoside phosphorylase	36.2 ± 9.8

Table 5. RELATIONSHIP OF ALTERNATE ROUTES OF ADENOSINE AND DEOXYADENOSINE METABOLISM AT VARIOUS SUBSTRATE CONCENTRATIONS

SUBSTRATE (μM)	DEAMINATION:PHOSPHORYLATION Adenosine	Deoxyadensoine
0.1	3.2	2330
1.0	5.6	2170
10	23	1320
100	64	410
1000	79	220

by substituting the maximal velocities and appropriate Michaelis constants into rate equations for two reactions competing for the same substrate (Table 5). At low substrate concentrations the ratio of deamination to phosphorylation (2000:1) shows deamination to be the predominant route of deoxyadenosine metabolism, whereas phosphorylation and deamination are approximately equal for adenosine (3:1). There is a difference of approximately 3 orders in magnitude in these ratios between deoxyadenosine and adenosine metabolism at low substrate concentrations but they approach one another at high substrate concentrations. Similar studies for deoxyguanosine gave ratios of phosphorolysis:phosphorylation of approximately 200:1 at substrate concentrations less than 10 μM. These findings imply a significant physiological role for adenosine deaminase and purine nucleoside phosphorylase activities, particularly with respect to deoxyribonucleoside metabolism at low substrate concentrations.

ACKNOWLEDGEMENTS

This work was supported by the Medical Research Council of Canada.

REFERENCES

1. M.S. Coleman, J. Donofrio, J.J. Hutton, L. Hahn, A. Daoud, B. Lampkin and J. Dyminski, J.Biol.Chem. 253:1619-1626 (1978).

2. A. Cohen, R. Hirschhorn, S.D. Horowitz, A. Rubinstein, S.H. Polmar, R. Heng and D.W. Martin, Jr, Proc.Natl.Acad.Sci.USA 75:472-476 (1978).

3. J. Donofrio, M.S. Coleman, J.J. Hutton, A. Daoud, B. Lampkin and J. Dyminski, J.Clin.Invest. 62:884-887 (1978).

4. A. Cohen, L.J. Gudas, A.J. Amman, G.E.J. Staal and D.W. Martin, Jr, J.Clin.Invest. 62:1405-1409 (1978).

5. M.S. Hershfield, F.F. Snyder and J.E. Seegmiller, Science 197: 1284-1287 (1977).

6. H.J. Schaeffer and C.F. Schwender, J.Med.Chem. 17:6-8 (1974).

7. B. Lindberg, H. Klenow and K. Hansen, J.Biol.Chem. 242: 350-356 (1967).

8. J.W. De Jong, Arch.Int.Physiol.Biochim. 85:557-569 (1977).

9. R.C. Jackson, H.P. Morris and G. Weber, Brit.J.Cancer 37:701-713 (1978).

10. R.L. Miller, D.L. Adamczyk and W.H. Miller, J.Biol.Chem. 254: 2339-2345 (1979).

11. F.F. Snyder, J. Mendelsohn and J.E. Seegmiller, J.Clin.Invest. 58:654-666 (1976).

12. J.P. Durham and D.H. Ives, J.Biol.Chem. 245:2276-2284 (1970).

13. T.A. Krenitsky, J.V. Tuttle, G.W. Koszalka, I.S. Chen, L.M. Beachem III, J.L. Rideout and G.B. Elion, J.Biol.Chem. 251: 4055-4061 (1976).

14. Y. Kozai, S. Sonoda, S. Kobayashi and Y. Sugino, J.Biochem. 71: 485-496 (1972).

15. L. J. Gudas, B. Ullman, A. Cohen and D.W. Martin, Jr, Cell 14: 531-538 (1978).

16. M. B. Meyers and W. Kreis, Arch.Biochem.Biophys. 177:10-15 (1976).

17. W.R. Gower, Jr, M.C. Carr and D.H. Ives, J.Biol.Chem. 254: 2180-2183 (1979).

18. D.A. Carson, J. Kaye and J.E. Seegmiller, Proc.Natl.Acad.Sci.USA 74:5677-5681 (1977).

MOLECULAR MECHANISM(S) OF DEOXYRIBONUCLEOSIDE TOXICITY IN T-LYMPHOBLASTS

James M. Wilson, Beverly S. Mitchell and
William N. Kelley

Departments of Internal Medicine and Biological Chemistry, Human Purine Research Center, University of Michigan Medical School, Ann Arbor, Michigan U.S.A.

The association of at least two inborn errors of purine metabolism with immunodeficiency diseases has made possible the investigation of these diseases at the molecular level. Adenosine deaminase (ADA) deficiency results in severe combined immunodeficiency disease with impairment of T- and B-lymphocyte function.[1] Purine nucleoside phosphorylase (PNP) deficiency has been associated with T-lymphocyte dysfunction in nine patients.[2] The mechanism(s) whereby these enzyme deficiency states affect lymphocyte development and/or function has (have) not been fully elucidated. However, the recent findings of markedly elevated dATP levels in ADA-deficient erythrocytes[3] and of dGTP levels in PNP-deficient erythrocytes[4] provide support for the hypothesis that deoxyribonucleosides play an important role in the pathogenesis of the immune dysfunction in these two diseases and suggest that deoxyribonucleosides or their metabolites may be toxic to lymphoid cells.

We have previously demonstrated marked cytoxicity of deoxyadenosine plus erythro-9-(2-hydroxy-3-nonyl)adenine (EHNA), an inhibitor of ADA, and of deoxyguanosine alone for human T-lymphoblasts, an effect which is significantly greater than the toxicity of these compounds for cultured human B-lymphoblasts.[5] In the present study, we have focused on the mechanism of toxicity of deoxyribonucleosides in human T-lymphoblasts.

The effects of deoxyguanosine and deoxyadenosine on deoxyribonucleoside triphosphate pools and nucleic acid synthesis in T-lymphoblasts were studied as a first step in elucidating the mechanism of their toxicity. Incubation with increasing concentrations of deoxyadenosine (0 - 50 μm) plus EHNA (5 μm) resulted in an accumulation of dATP from 66 to 318 pmol/10^6 cells. The much smaller dCTP pool was decreased from 2.5 to <0.1 pmol/10^6 cells. These changes were correlated with an inhibition of DNA synthesis, as measured by the incorporation of ^{3}H-uridine into DNA, without affecting RNA synthesis. In similar experiments, incubation with increasing concentrations of deoxyguanosine (0 - 50 μm) resulted in an accumulation of dGTP from 24 to 305 pmol/10^6 cells. The dCTP pool was depleted from 2.5 to 0.3 pmol/10^6 cells. These changes were also correlated with an inhibition of DNA synthesis but not of RNA synthesis.

One hypothesis to account for deoxyribonucleoside-mediated inhibition of DNA synthesis is an inhibition of the enzyme ribonucleotide reductase by elevated levels of dATP or dGTP. This enzyme catalyzes the reduction of ribonucleoside diphosphates to their corresponding 2'-deoxy derivatives and insures an adequate supply of deoxyribonucleotides for DNA synthesis.

Deoxyribonucleoside induced depletion of the dCTP pool and the reversal of toxicity with deoxycytidine, i.e., deoxycytidine rescue, have frequently been cited as the major evidence for ribonucleotide reductase inhibition in deoxyribonucleoside toxicity.[6,7,8] Interpretation of these data is difficult since deoxycytidine rescue and dCTP depletion could be due to a number of different mechanisms, i.e., inhibition of deoxyadenosine and deoxyguanosine phosphorylation. We have tested more directly the hypothesis that deoxyribonucleoside toxicity is mediated by an inhibition of this enzyme. In addition, we have compared the effects of deoxyguanosine with the effects of hydroxyurea, a known inhibitor of ribonucleotide reductase, on the incorporation of ribonucleosides and deoxyribonucleosides into DNA.

Ribonucleotide reductase activity is required for the incorporation of uridine, but not of deoxyuridine or thymidine, into DNA. Inhibition of this enzyme would thus be expected to result in a greater decrease in the incorporation of uridine than of either deoxyuridine or thymidine into DNA. Just such a differential effect was observed with hydroxyurea at concentrations in the mM range and with deoxyguanosine at concentrations in the μM range. Hydroxyurea at concentrations from 100 μM to 5 mM progressively inhibited incorporation of both ^{3}H-deoxyuridine and ^{14}C-thymidine into DNA (Figure 1A). Incorporation of ^{3}H-uridine, however, had fallen to 20% of control values at 100 μM hydroxyurea, a concentration which left ^{14}C-thymidine incorporation relatively unaffected (Figure 1B). Deoxyguanosine alone exhibited effects similar qualitatively to those observed with hydroxyurea (Figues 2A and B).

Since the majority of ^{3}H-uridine is incorporated into DNA via dTTP (data not shown), one may conclude that the site of the differential inhibition of ^{3}H-uridine incorporation as compared with ^{14}C-thymidine incorporation into DNA must occur between transport

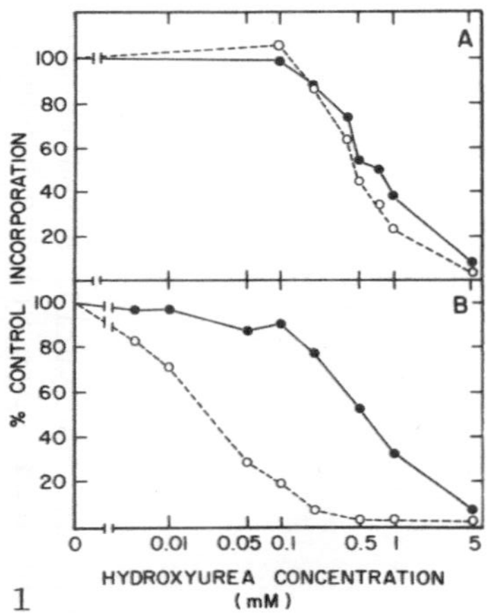

1

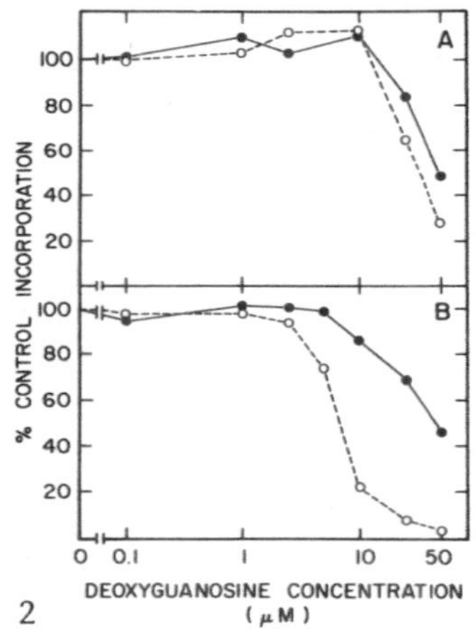

2

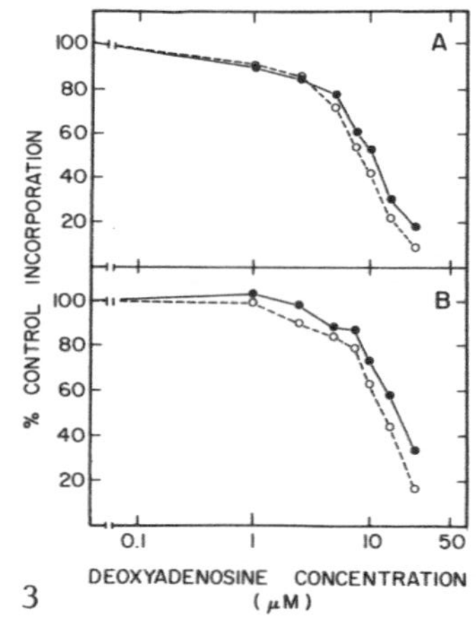

3

Figs. 1,2 &3. Differential incorporation of nucleosides into DNA: Effects of increasing concentrations of hydroxyurea (Fig. 1), deoxyguanosine (Fig. 2), or deoxyadenosine plus 5 μM EHNA (Fig. 3). Cells were incubated at a concentration of $1x10^6$/ml in RPMI + 10% heat-inactivated horse serum in the presence of the additive indicated for 1 hour. Labeled nucleosides were added for an additional 2-1/2 hr. incubation period. The RNA and DNA components were separated by KOH hydrolysis of the trichloroacetic acid precipitate. Panel A- ●——● ^{14}C-dThd ($5x10^{-8}$ M, 515 mCi/mmol); O---O ^{3}H-dUrd ($5x10^{-8}$ M, 9 Ci/mmol). Panel B- ●——● ^{14}C-dThd, as in Panel A; O---O ^{3}H-Urd ($4x10^{-8}$ M, 24.2 Ci/mmol). Values are expressed as a percentage of control values obtained in the absense of additives and are the mean of duplicate determinations of a single experiment. Similar results were obtained in two additional experiments. The mean control cpm ± 1 S.D. for 12 determinations: Panel A- ^{14}C-dThd 8730 ± 321, ^{3}H-dUrd 9072 ± 389; Panel B- ^{14}C-dThd 7442 ± 492, ^{3}H-Urd 11284 ± 551.

of uridine and phosphorylation of dTMP (Figure 4). Since uridine incorporation into RNA is not inhibited, uridine transport and phosphorylation, and UMP phosphorylation are unimpaired. We have excluded involvement of thymidylate synthetase (reaction 7, Figure 4) by demonstrating no difference in the inhibition of incorporation of deoxyuridine as compared with thymidine into DNA in cells incubated with either deoxyguanosine or hydroxyurea.

After careful consideration of these alternative explanations, we conclude that our results are due to inhibition of either the reduction of UDP by ribonucleotide reductase or the reversible conversion of dUDP to dUMP catalyzed by thymidylate kinase. Since thymidylate kinase does not appear to be regulated allosterically by nucleotides,[9] it is most probable that our results are explained by an inhibition of ribonucleotide reductase.

Further support for this conclusion can be derived from our studies on the incorporation of ^{3}H-cytidine into ribonucleotide and deoxyribonucleotide pools. ^{3}H-cytidine must be phosphorylated to CDP and reduced by ribonucletide reductase to enter the deoxyribonucleotide pool (Figure 4). A block of ribonucleotide reductase should substantially increase counts appearing in the ribonucleotide fractions while decreasing counts in the deoxyribonucleotide pool. Following incubation with hydroxyurea, there was a decrease in the incorporation of ^{3}H-cytidine into deoxyribonucleotides to a mean of 32% of control values and a corresponding increase to 144% of control in the labeled ribonucleotide pool (Table 1); almost identical results were obtained with deoxyguanosine.

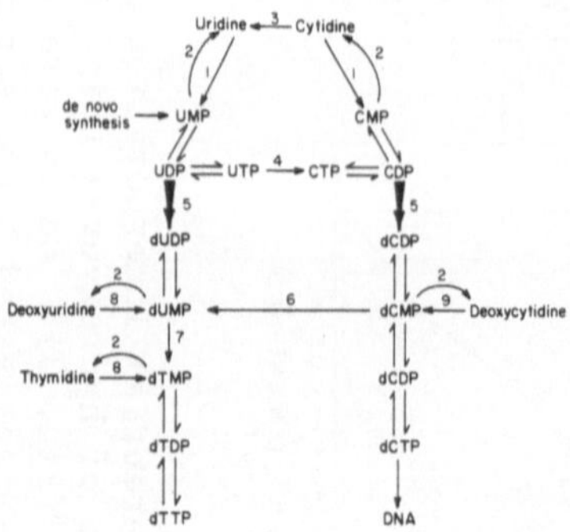

Fig. 4. Pathways of cytidine, uridine, deoxyuridine and thymidine incorporation into nucleic acids. 1 - uridine kinase; 2 - 5'-nucleotidase; 3 - cytidine deaminase; 4 - CTP synthetase; 5 - ribonucleotide reductase; 6 - dCMP deaminase; 7 - thymidylate synthetase; 8 - thymidine kinase

Table 1

^{3}H-Cytidine incorporation into ribonucleotide and deoxyribonucleotide pools in T-lymphoblasts

Additives	No.	Deoxyribonucleotides % Control	Ribonucleotides % Control
None		100	100
Hydroxyurea (2.5 mM)	5	32 ± 9	144 ± 6
Deoxyguanosine (50 μM)	3	38	142
Deoxyadenosine (50 μM) plus EHNA (5 μM)	4	70 ± 11	109 ± 1

Molt-4 lymphoblasts (10^6 cells/ml) were incubated in 2.5 ml RPMI + 10% heat-inactivated horse serum for 1 hour at 37° C in the presence of the additives indicated. ^{3}H-cytidine (28 Ci/mmole; 0.05 μM) was added for a further 75 minute incubation period. The intracellular pools were extracted in 60% menthanol at -20° C and the deoxyribonucleotides were separated from the ribonucleotides with an affinity gel (Affi-Gel 601, Bio-Rad) that selectively binds cis-diol compounds at high pH. Values for 4 or more determinations represent the mean ± 1 S.D.

Deoxyadenosine plus EHNA, in contrast to either hydroxyurea or deoxyguanosine, produced little, if any, differential effect on the incorporation of uridine and thymidine into DNA (Figure 3B). In addition, there was a far more modest increment in the ribonucleotide pool at a concentration of deoxyadenosine which effectively inhibited DNA synthesis (Table 1).

In conclusion, deoxyguanosine induced inhibition of DNA synthesis is apparently mediated through dGTP inhibition of ribonucleotide reductase. Deoxyadenosine, in the presence of an inhibitor of adenosine deaminase, inhibits DNA synthesis by mechanisms other than, or, perhaps, in addition to, inhibition of ribonucleotide reductase. One recently espoused possibility is that an increased concentration of deoxyadenosine leads to an accumulation of S-adenosylhomocysteine by suicide inactivation of S-adenosylhomocysteine hydrolase, thereby depleting the cell of S-adenosylmethionine and preventing essential methylation reactions.[10]

REFERENCES

1. E. R. Giblett, J. E. Anderson, F. Cohen, B. Pollar, and H. J. Meuwissen, Adenosine deaminase deficiency in two patients with severely impaired cellular immunity Lancet II:1067-1069 (1972).

2. E. R. Giblett, A. J. Ammann, D. W. Wara, R. Sandman, and L. K. Diamond, Nucleoside phosphorylase deficiency in a child with severely defective T-cell immunity and normal B-cell immunity Lancet I:1010-1013 (1975).

3. M. S. Coleman, J. Donofrio, J. J. Hutton, L. Hahn, A. Daoud, B. Lampkin, and J. Dyminski, Identification and quantitation of adenine deoxynucleotides in erythrocytes of a patient with adenosine deaminase deficiency and severe combined immunodeficiency J. Biol. Chem. 253:1619-1626 (1978).

4. A. Cohen, L. J. Gudas, A. J. Ammann, G. E. J. Staal, and D. W. Martin, Jr., Deoxyguanosine triphosphate as a possible toxic metabolite in the immunodeficiency associated with purine nucleoside phosphorylase deficiency J. Clin. Invest. 61:1405-1409 (1978).

5. B. S. Mitchell, E. Mejias, P. E. Daddona, and W. N. Kelley, Purinogenic immunodeficiency diseases: selective toxicity of deoxyribonucleosides for T cells Proc. Natl. Acad. Sci. U.S.A. 75:5011-5014 (1978).

6. B. Ullman, L. J. Gudas, A. Cohen, and D. W. Martin, Jr., Deoxyadenosine metabolism and cytoxicity in cultured mouse T lymphoma cells: a model for immunodeficiency disease Cell. 14:365-375 (1978).

7. D. A. Carson, J. Kaye, and J. E. Seegmiller, Lymphospecific toxicity in adenosine deaminase deficiency and purine nucleoside phosphorylase deficiency: possible role of nucleoside kinase(s) Proc. Natl. Acad. Sci. U.S.A. 74:5677-5681 (1977).

8. L. J. Gudas, B. Ullman, A. Cohen, and D. W. Martin, Jr., Deoxyguanosine toxicity in a mouse T lymphoma: relationship to purine nucleoside phosphorylase-associated immune dysfunction Cell. 14:531-538 (1978).

9. J. F. Henderson and A. R. P. Paterson, "Nucleotide Metabolism: An Introduction," Academic Press, New York (1973).

10. M. S. Hershfield, Apparent suicide inactivation of human lymphoblast S-adenosylhomocysteine hydrolase by 2'-deoxyadenosine and adenine arabinoside: a basis for direct toxic effects of analogs of adenosine J. Biol. Chem. 254:22-25 (1979).

INHIBITION OF IMMUNE CELL FUNCTION BY ADENOSINE: BIOCHEMICAL STUDIES

Thomas P. Zimmerman, Gerald Wolberg, Gail S. Duncan, Robert D. Deeprose and Robert J. Harvey

Wellcome Research Laboratories
Research Triangle Park, N.C., USA, 27709

INTRODUCTION

Adenosine (Ado) has been shown to inhibit the cytolysis of tumor cells by specifically-sensitized mouse lymphocytes[1]. The mechanism of this immunosuppressive effect of Ado is of interest in view of the apparent causal relationship between Ado deaminase deficiency and severe combined immunodeficiency disease[2]. Thus far, Ado has been shown to cause two distinct biochemical effects in the cytolytic lymphocytes: (i) an elevation of adenosine 3':5'-monophosphate (cAMP)[1]; and (ii) an elevation of S-adenosylhomocysteine (AdoHcy)[3]. Ado has little or no effect on the pool sizes of CTP, UTP, ATP or GTP in these cells[1,4]. Studies with other agents have shown that a selective elevation of <u>either</u> cAMP, as caused by prostaglandin E_1[5], <u>or</u> AdoHcy, as caused by 3-deazaadenosine[3], is sufficient to inhibit the cytolytic activity of these lymphocytes. The present studies were therefore undertaken in an attempt to discern the relative importance of cAMP and AdoHcy to this immunosuppressive action of Ado.

MATERIALS AND METHODS

The materials and methods used in the present work have been described in detail elsewhere[3,4,6].

RESULTS AND DISCUSSION

Ado alone is a poor inhibitor of lymphocyte-mediated cytolysis (LMC), exhibiting a 50% inhibitory concentration (IC_{50}) of approximately 150 μM[1]. However, in the presence of 7.9 μM <u>erythro</u>-9-(2-hydroxy-3-nonyl)adenine (EHNA), a potent inhibitor

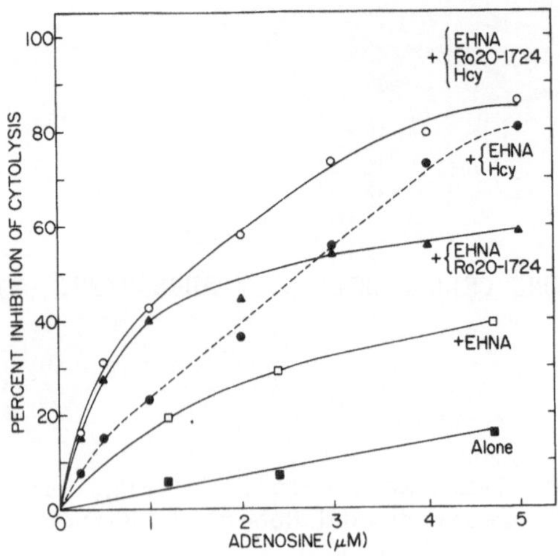

Fig. 1. Effects of EHNA, Ro 20-1724 and L-homocysteine on the inhibition of LMC by adenosine. Adenosine alone (■-■), adenosine + 7.9 μM EHNA (□-□), adenosine + 7.9 μM EHNA + 50 μM Ro 20-1724 (▲-▲), adenosine + 7.9 μM EHNA + 200 μM L-homocysteine (Hcy) (●-●), adenosine + 7.9 μM EHNA + 200 μM L-homocysteine + 50 μM Ro 20-1724 (O-O). Assay time: 70 min. Each point represents the mean of triplicate assays.

of adenosine deaminase[7], the IC_{50} for Ado is reduced 20-fold to approximately 7 μM[4]. 4-(3-Butoxy-4-methoxybenzyl)-2-imidazolidinone (Ro 20-1724), an inhibitor of cAMP phosphodiesterase[8], acts to potentiate further this inhibitory activity of Ado (Fig. 1). In addition, L-homocysteine (Hcy), which together with Ado can serve as a co-substrate in the cellular formation of AdoHcy via S-adenosylhomocysteinase[9], also potentiates this immunosuppressive activity of Ado (Fig. 1).

In the hope of better defining the biochemical mechanism whereby Ado inhibits LMC, theophylline (an antagonist of Ado-stimulated elevation of cellular cAMP[10]) and uridine (Urd, an inhibitor of Ado transport[11]) have been examined for their ability to antagonize the inhibition of LMC by Ado (Table 1). At concentrations of 75 μM and 20 mM, respectively, neither theophylline nor Urd is inhibitory to LMC. Theophylline (75 μM) is able to antagonize by 42-57% the inhibition of LMC caused by 2.0 μM Ado, regardless of whether Ado was tested alone or in combination with Ro 20-1724 or Hcy. By contrast, Urd (20 mM) antagonizes partially (57%) only the inhibition of LMC caused by the combination of Ado + Hcy and actually enhances the inhibition caused by Ado alone or Ado + Ro 20-1724. The combination of theophylline + Urd

TABLE 1. EFFECTS OF THEOPHYLLINE AND URIDINE ON THE INHIBITION OF LMC BY ADENOSINE IN THE ABSENCE AND PRESENCE OF Ro 20-1724 OR L-HOMOCYSTEINE*

Additive(s)	Reversal agent(s) added			
	None	Theo	Urd	Theo + Urd
	percent inhibition of cytolysis			
Ado	21	9	42	31
Ado + Ro 20-1724	52	30	58	42
Ado + Hcy	61	30	26	15

*Each assay was performed in duplicate in the presence of 7.9 μM EHNA. At the concentrations employed, Ro 20-1724 (25 μM), L-homocysteine (Hcy, 200 μM), theophylline (Theo, 75 μM) and uridine (Urd, 20 mM), either alone or in the specified combinations, are not inhibitory to LMC. The concentration of adenosine (Ado) was 2.0 μM.

is less effective than theophylline alone in reversing the inhibition of LMC due to Ado alone or to Ado + Ro 20-1724 but is more effective than either agent alone in antagonizing the inhibition of LMC caused by Ado + Hcy.

Ado has been shown to cause an elevation of cAMP in these cytolytic lymphocytes[1] and to be potentiated in this activity both by Ro 20-1724[4] and by Hcy[12]. As shown in Table 2, 75 μM theophylline is able to antagonize by 77-87% the elevation of lymphocyte cAMP caused by 2.0 μM Ado, regardless of whether Ro 20-1724 or Hcy is present. Urd (20 mM) appears to have little or no effect on the magnitude of the cAMP elevation caused by Ado ± Ro 20-1724 but does partially (55%) antagonize the increase in cAMP due to Ado + Hcy. The combination of theophylline + Urd is similar to, or slightly less effective than, theophylline alone in preventing the Ado-stimulated elevation of cAMP.

Direct measurement of AdoHcy in extracts of lymphocytes by reversed-phase HPLC was not possible, due both to the small lymphocyte pool (<20 pmoles/10^6 cells) of AdoHcy and to the small amount of tissue available. However, the ability of Ado to cause a buildup of [^{3}H]AdoHcy in lymphocytes prelabeled with L-[2-^{3}H]-methionine is documented in Table 3. Ro 20-1724 appears to enhance slightly this buildup of [^{3}H]AdoHcy due to Ado. The apparent reversal of this effect of Ado by Hcy is attributed to S-adenosylhomocysteinase-catalyzed exchange of the metabolically generated [^{3}H]Hcy present in [^{3}H]AdoHcy with exogenous, non-radioactive Hcy and consequent isotopic dilution of the conjugated amino acid. Theophylline has little or no effect on this cellular buildup of [^{3}H]AdoHcy caused by Ado. However, Urd completely prevents this effect of Ado on [^{3}H]AdoHcy levels (Table 3). Lymphocytes treated with both Ado and Hcy accumulate large

TABLE 2. EFFECTS OF THEOPHYLLINE AND URIDINE ON THE ADENOSINE-STIMULATED ELEVATION OF LYMPHOCYTE CYCLIC AMP*

Additive(s)	Reversal agent(s) added			
	None	Theo	Urd	Theo + Urd
	picomoles of cAMP per 10^7 cells			
None	0.59±0.03	0.78±0.09	0.73±0.08	0.84±0.05
Ado	2.92±0.20	1.12±0.14	2.91±0.06	1.18±0.11
Ado + Ro 20-1724	8.40±0.28	2.15±0.12	9.94±0.48	2.77±0.09
Ado + Hcy	4.89±0.27	1.16±0.02	2.53±0.11	1.45±0.11

*Cytolytic lymphocytes (1.0 x 10^7 cells/5.0 ml of medium) were incubated in the presence of 7.9 μM EHNA with the specified additives at 37° for 30 min, after which the cells were acid-extracted. Concentrations of reagents were the same as in Table 1. All cellular incubations and cAMP radioimmunoassays were performed in duplicate. Each value represents the mean ± the standard error of the mean for four determinations.

TABLE 3. EFFECTS OF THEOPHYLLINE AND URIDINE ON THE ADENOSINE-STIMULATED BUILDUP OF S-[^{3}H]ADENOSYLHOMOCYSTEINE IN LYMPHOCYTES PRELABELED WITH L-[2-^{3}H]METHIONINE*

Additive(s)	Reversal agent(s) added			
	None	Theo	Urd	Theo + Urd
	dpm of [^{3}H]AdoHcy per 10^6 cells			
None	145 ± 1	129 ± 23	142 ± 31	99 ± 21
Ado	539 ± 75	640 ± 5	138 ± 5	139 ± 7
Ado + Ro 20-1724	759 ± 15	734 ± 64	127 ± 10	121 ± 21
Ado + Hcy	167 ± 11	107 ± 10	51 ± 4	60 ± 6

*Cellular levels of [^{3}H]AdoHcy were determined after 30-min incubations of L-[2-^{3}H]methionine-prelabeled lymphocytes with the specified additives in the presence of 7.9 μM EHNA. Concentrations of reagents were the same as in Table 1. This experiment was performed in duplicate and the results are expressed as the mean ± the standard error of the mean for the two analyses.

amounts of AdoHcy which can be quantitated by ultraviolet monitoring of the effluent from the liquid chromatograph. Urd (20 mM) inhibits completely the cellular buildup of non-radioactive AdoHcy from 5.0 µM Ado + 200 µM Hcy, while theophylline (75 µM) has no effect on AdoHcy formation (data not shown).

S-Adenosylmethionine-dependent methyltransferases in general exhibit strong inhibition by AdoHcy, the common product of all such methylation reactions. As a further criterion for monitoring the effect of Ado on AdoHcy levels and cellular transmethylation reactions, protein carboxymethylation was assayed in intact lymphocytes treated with the various drugs. Ado (2.0 µM) inhibits this cellular methylation reaction by 30%. Although theophylline (75 µM) has no effect on this activity of Ado, Urd (20 mM) antagonizes completely the inhibition of protein carboxymethylation caused by Ado (data not shown).

The present results suggest that low concentrations (≦2 µM) of Ado inhibit the cytolytic activity of sensitized mouse lymphocytes primarily due to an elevation of cellular cAMP and not to an elevation of AdoHcy. This conclusion is based upon: (i) the ability of theophylline to antagonize both the inhibition of LMC and the elevation of cAMP caused by Ado (±Ro 20-1724 or Hcy) without affecting the associated cellular buildup of AdoHcy or the inhibition of protein carboxymethylation caused by Ado; (ii) the inability of Urd to antagonize the inhibition of LMC caused by Ado ± Ro 20-1724 even though Urd could completely prevent the effects of Ado on AdoHcy accumulation and protein carboxymethylation; and (iii) the parallel antagonistic effects of Urd on the inhibition of LMC and the elevation of cAMP caused by Ado + Hcy. Since concentrations of theophylline >75 µM are inhibitory to LMC, it was not feasible to antagonize more fully the effects of Ado on cAMP and LMC by using higher concentrations of theophylline.

Nevertheless, the present results do suggest that AdoHcy may play an important role in the inhibition of LMC caused by higher concentrations of Ado in combination with Hcy. As shown in Fig. 1, higher concentrations (>3 µM) of Ado consistently inhibit LMC more strongly in the presence of Hcy than in the presence of Ro 20-1724, even though Ro 20-1724 is superior to Hcy in potentiating the elevation of lymphocyte cAMP by these higher concentrations of Ado (data not shown). Hill plot analysis[13] of the LMC inhibition data in Fig. 1 yields slopes of 0.78 ± 0.15, 0.60 ± 0.02, 0.64 ± 0.04, 1.30 ± 0.08 and 1.12 ± 0.05, respectively, for the data obtained with Ado alone, Ado + EHNA, Ado + EHNA + Ro 20-1724, Ado + EHNA + Hcy, and Ado + EHNA + Hcy + Ro 20-1724. The fact that Hcy, when added to the LMC assays together with Ado ± Ro 20-1724, causes a doubling of the Hill plot slopes to values greater than unity indicates that Hcy can potentiate the inhibition of LMC by Ado by two different mechanisms - possibly one involving cAMP and one involving AdoHcy - and further indicates that these two sites of Hcy activity interact with each other in a partially cooperative manner. Moreover, a comparison of the fractional inhibition observed at each concentration of Ado in the presence of either Hcy or Ro 20-1724 with that observed in the presence of both Hcy and Ro 20-1724 indicates that Hcy and Ro 20-1724 are mutually antagonistic, as defined by Webb[14], in their ability to potentiate the inhibition of LMC by Ado. In other words, Ro 20-1724 appears to exert its effect upon these cells via one of the same two mechanisms by which Hcy produces its effect; presumably, this common mechanism of action represents the effects of these two agents on cAMP levels.

ACKNOWLEDGEMENTS

The excellent assistance of Mrs. Marvin S. Winston and Robert L. Veasey is acknowledged. We express our appreciation to Dr. Gertrude B. Elion for her interest in and support of this work.

REFERENCES

1. G. Wolberg, T.P. Zimmerman, K. Hiemstra, M. Winston, and L.-C. Chu (1975) Science 187, 957.
2. E.R. Giblett, J.E. Anderson, F. Cohen, B. Pollara, and H.J. Meuwissen (1972) Lancet 2, 1067.
3. T.P. Zimmerman, G. Wolberg, and G.S. Duncan (1978) Proc. Nat. Acad. Sci. 75, 6220.
4. G. Wolberg, T.P. Zimmerman, G.S. Duncan, K.H. Singer, and G.B. Elion (1978) Biochem. Pharmacol. 27, 1487.
5. C.S. Henney, H.E. Bourne, and L.M. Lichtenstein (1972) J. Immunol. 108, 1526.
6. T.P. Zimmerman, G. Wolberg, C.R. Stopford, and G.S. Duncan (1979) in Transmethylation, eds. E. Usdin, R.T. Borchardt, and C.R. Creveling, Elsevier North Holland, Inc., New York, p. 187.
7. H.J. Schaeffer and C.F. Schwender (1974) J. Med. Chem. 17, 6.
8. H. Sheppard and G. Wiggan (1971) Mol. Pharmacol. 7, 111.
9. N.M. Kredich and D.W. Martin, Jr. (1977) Cell 12, 931.
10. A. Sattin and T.W. Rall (1970) Mol. Pharmacol. 6, 13.
11. A. Cohen, B. Ullman, and D.W. Martin, Jr. (1979) J. Biol. Chem. 254, 112.
12. T.P. Zimmerman, R.D. Deeprose, G. Wolberg, and G.S. Duncan (in press) Biochem. Pharmacol.
13. J. Monod, J.-P. Changeux, and F. Jacob (1963) J. Mol. Biol. 6, 306.
14. J.L. Webb (1965) in Enzyme and Metabolic Inhibitors, Vol. 1, Academic Press, New York, p. 507.

INTERACTIONS BETWEEN ENERGY METABOLISM AND ADENINE NUCLEOTIDE METABOLISM IN HUMAN LYMPHOBLASTS.

S. S. Matsumoto, K. O. Raivio*, R. C. Willis, and J. E. Seegmiller.
University of California San Diego, La Jolla, Calif. 92093 USA, and *Children's Hospital, Univ. of Helsinki, SF-00290, Helsinki 29, Finland.

Adenine nucleotides play an important role in the transfer of chemical energy for metabolic processes. In addition, adenine nucleotide degradation can be a major source of purine bases formed during certain kinds of metabolic stress. These two properties of adenine nucleotides may be related. Fructose and 2-deoxyglucose can produce elevated levels of purines *in vivo* (1), in perfused organs (2), and in cultured cells (3,4). The mechanism of nucleotide degradation caused by fructose or 2-deoxyglucose involves the utilization of ATP to form a slowly metabolized hexose phosphate which accumulates and decreases the intracellular concentration of inorganic phosphate (1-4). AMP deaminase, which is normally inhibited by phosphate, becomes more active when the phosphate concentration decreases and adenine nucleotides are broken down by the reactions: AMP → IMP + NH_3; IMP → inosine + PO_4^{2-}. There also exists another pathway for adenine nucleotide degradation catalyzed by the enzymes purine 5'-nucleotidase and adenosine deaminase: AMP → adenosine + PO_4^{2-}; adenosine → inosine + NH_3.

It is possible to determine the relative amount of adenine nucleotide degradation proceeding by either the AMP deaminase or 5'-nucleotidase initiated pathway in the following system (cf. 3). A human lymphoblast cell line, possessing less than one percent of normal adenosine kinase activity (5), was incubated with [8-^{14}C] adenine to radioactively label the intracellular purine nucleotide pools. The cells were then placed in medium containing the adenosine deaminase inhibitor, deoxycoformycin, and incubated under various conditions which caused nucleotide breakdown. Any nucleotide broken down through the 5'-nucleotidase pathway is trapped as adenosine. On the other hand, nucleotide catabolism via the AMP

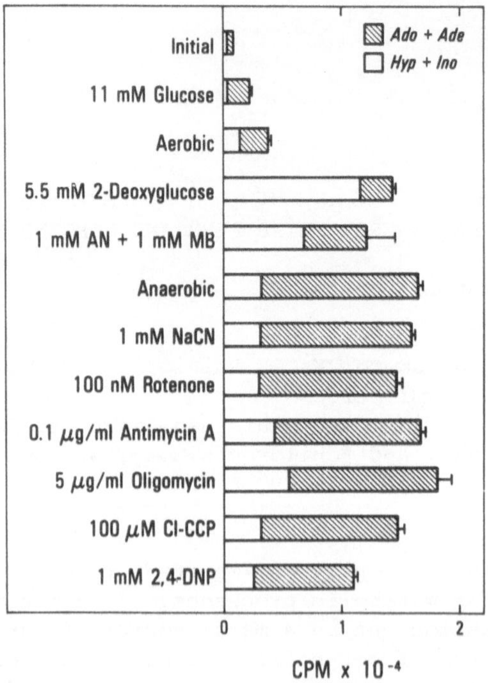

Figure 1. Accumulation of nucleosides and bases during stress of energy metabolism. Adenosine kinase-deficient lymphoblasts were prelabelled with [^{14}C]-adenine. The cells were resuspended and incubated for 4 hr in medium without glucose and containing deoxycoformycin plus the indicated additions. In the anaerobic case the cells were allowed to settle into a pellet. In all the other cases the cell suspensions were shaken in an atmosphere of air. AN = 6-aminonicotinamide; MB = methylene blue; Cl-CCP - carbonylcyanide-m-chlorophenylhydrazone; 2,4-DNP - 2,4-dinitrophenol.

deaminase pathway gives rise to hypoxanthine. The resulting proportions of adenosine and hypoxanthine produced by various incubation conditions is shown in Figure 1. When the cells were incubated with glucose only a small amount of nucleotide degradation occurred. Even when glucose was absent, nucleotide degradation was low if the cells were able to oxidatively metabolize glutamine in the medium. When cells were incubated with 2-deoxyglucose, degradation was greatly increased and hypoxanthine was the major breakdown product. Therefore, 2-deoxyglucose induced nucleotide breakdown via the AMP deaminase pathway, which is completely consistent with previous reports. A large amount of nucleotide degradation also occurred when respiration and oxidative phosphorylation were inhibited at several sites in the absence of glucose. For example, anaerobic conditions which developed as the cells

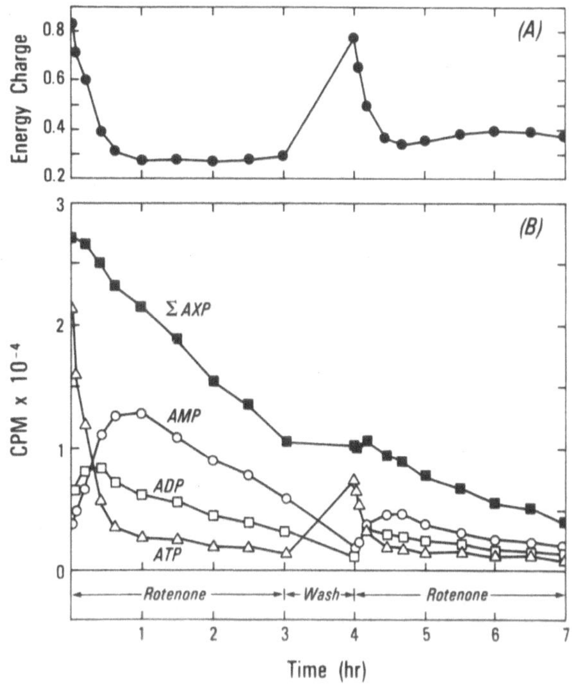

Figure 2. Variation of nucleotide pools during energy depletion. WI-L2 lymphoblasts were incubated in medium with 100 nM rotenone and without glucose. After 3 hr the cells were washed free of rotenone and resuspended in medium containing 11 mM glucose. After 1 hr the cells were washed free of glucose and resuspended in medium containing 100 nM rotenone. ΣAXP = ATP + ADP + AMP.

settled into a pellet, or the addition of the electron transfer inhibitor, rotenone, or the addition of the uncoupler of oxidative phosphorylation, 2,4-dinitrophenol, all had a similar effect. In these cases however, adenosine was the main product of nucleotide breakdown in the adenosine kinase-deficient mutant cell plus the adenosine deaminase inhibitor. Therefore, when the cell can use neither oxidative phosphorylation nor glycolysis to regenerate ATP, adenine nucleotides are degraded via the pathway involving the initial action of 5'-nucleotidase.

The presence of two separate pathways for adenine nucleotide degradation may allow the cell to respond with flexibility to different stimuli. A decrease in inorganic phosphate concentration is the principal factor in causing nucleotide degradation via the AMP deaminase pathway. When oxidative phosphorylation is inhibited, cellular inorganic phosphate increases rather than decreases. Some other stimulus must exist for adenine nucleotide degradation via

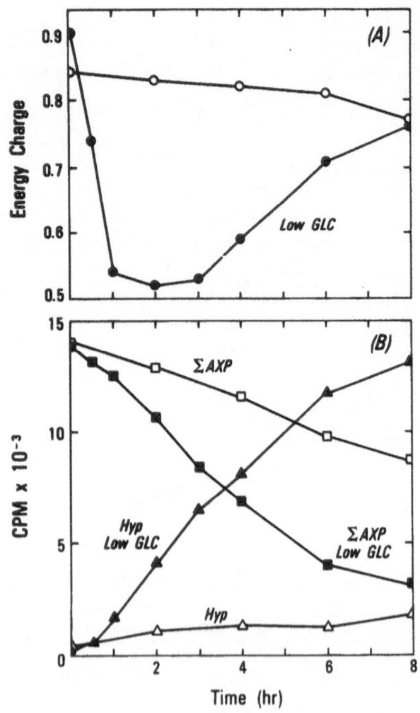

Figure 3. Adenine nucleotide degradation and adenylate energy charge recovery during anaerobic stress. WI-L2 lymphoblasts were suspended in medium containing 11 mM glucose (open symbols) or 1 mM glucose (solid sympols, "Low GLC"). The cells were allowed to settle into a pellet and were incubated at 37° without shaking.

the nucleotidase pathway. Nucleotidase activity possibly could be increased by an increase in the concentration of the substrate,AMP, or by decreases in the concentrations of the nucleotidase inhibitors, ADP and ATP. To test these possibilities, the variations in the individual adenine nucleotide concentrations were examined during the inhibition of oxidative phosphorylation. In the experiment shown in Figure 2, a lymphoblast culture was incubated with rotenone in the absence of glucose to cause nucleotide degradation. (This lymphoblast line had normal adenosine kinase activity and deoxycoformycin was not present.) During the first phase of stress induced by rotenone, the ATP concentration rapidly fell. Both AMP and ADP concentrations transiently rose and then, both declined. As a result, the sum, ATP + ADP + AMP, and the adenylate energy charge (ATP + 1/2 ADP)/(ATP + ADP + AMP), also decreased. After the first phase of stress, the cells were washed free of rotenone and purine breakdown products, and were incubated with glucose to allow energy regeneration. During this recovery period, nucleotide degradation ceased and the ATP concentration rose to about one-third of its original value. The adenylate energy charge returned to its initial value. When glucose was washed away and rotenone was added back, degradation began again. In the second phase of stress the AMP concentration never rose above its original value at the start of the experiment and nucleotide degradation continued even after AMP fell below its original concentration. Thus, an elevated concentration of substrate, AMP, could not have been the sole stimulus for nucleotide degradation by the nucleotidase pathway. During the recovery period when no degradation occurred, the ATP concentration remained in the low range that accompanied degradation during the first phase of stress. Thus, a lower concentration of the nucleotidase inhibitor, ATP, could not have been the sole stimulus for degradation. The ADP concentration actually rose when degradation first began. Thus, a lower concentration of the nucleotidase inhibitor, ADP, could not have been the sole stimulus for degradation. However, nucleotide degradation always occurred when the adenylate energy charge dropped to low values and no degradation occurred when the energy charge was re-established at high values. Thus, the adenylate energy charge may summarize the combined regulatory effects of ATP, ADP and AMP upon purine 5'-nucleotidase activity.

The adenylate energy charge is a calculated parameter which reflects the energy state of the cell. In many biological systems, the energy charge has been found to coordinate the activities of many metabolic reactions and it has been proposed to be of importance in maintaining cellular homeostasis (6). AMP degradation may lead in some cases to an increase in the energy charge value as shown in Figure 3. This is because AMP appears only in the denominator of the energy charge expression. In other words, removal of AMP from the cell increases the amount of ATP relative to ADP and AMP. Taken together, the experiments shown in Figures 2 and 3 imply that a decrease in the adenylate energy charge may stimulate adenine nucleotide degradation, while nucleotide degradation may cause the energy charge to rise to normal values. Such a reciprocal relationship is a characteristic of closely regulated metabolic systems.

In conclusion, the pathway including purine 5'-nucleotidase and adenosine deaminase is used for adenine nucleotide degradation during conditions of energy depletion. This phenomenon may be related to the occurrence *in vivo* of adenosine production in heart and brain during hypoxia and uric acid production during hypoglycemia in glycogen storage disease. Adenine nucleotide degradation may have important regulatory consequences in cellular energy metabolism.

1. K. O. Raivio, M. P. Kekomaki and P. H. Maenpaa, Depletion of liver adenine nucleotides induced by D-fructose, *Biochem. Pharmacol.* 18: 2615 (1969).

2. G. Van Den Berghe, M. Bronfman, R. Vanneste, and H-G Hers, The mechanism of adenosine triphosphate depletion in the liver after a load of fructose. Biochem. J. 162: 601, (1977).
3. C. A. Lomax and J. F. Henderson, Adenosine formation and metabolism during adenosine triphosphate catabolism in Ehrlich ascites tumor cells, Cancer Res. 33: 2825 (1973).
4. L. A. Sauer, Control of adenosine monophosphate catabolism in mouse ascites tumor cells. Cancer Res. 38: 1057 (1978).
5. M. S. Hershfield, F. F. Snyder, and J. E. Seegmiller, Adenine and adenosine are toxic to human lymphoblast mutants defective in purine salvage enzymes. Science 197: 1284 (1977).
6. D. E. Atkinson, "Cellular Energy Metabolism and Its Regulation", Academic Press, New York (1977).

This work was supported in part by NIH Grants GM17702, AM13622, AM07318 and the Kroc Foundation.

ENZYMES OF PURINE INTERCONVERSIONS IN SUBFRACTIONS OF LYMPHOCYTES

J.P.R.M. van Laarhoven, G.Th. Spierenburg,
C.H.M.M. de Bruyn and E.D.A.M. Schretlen[o].
Dept. of Human Genetics and [o]Dept. of Pediatrics,
Faculty of Medicine, University of Nijmegen, Nijmegen,
The Netherlands.

INTRODUCTION

Previously described micromethods for the determination of purine interconversion enzyme activities in lymphocytes (1) enable us to analyse purine metabolism systematically in lymphocyte subfractions using a relatively small number of cells (500-5000). A relation between purine interconversion defects and immune dysfunctions has been established (2-4). The mechanism by which adenosine deaminase (ADA) deficiency leads to impairment of the B and T cell and purine nucleoside phosphorylase (PNP) deficiency leads to T cell dysfunction is not yet completely understood. A better understanding of purine interconversions in B and T cell subfractions might help to obtain a better view on B or T cell specificity in these immune diseases. One of the possibilities to achieve this might be a systematic enzymological analysis of purine metabolism in T and non-T lymphocytes. Nine purine enzyme activities were measured in T and non-T lymphocyte subpopulations using 500-5000 cells per assay.

METHODS AND MATERIALS

Isolation of Lymphocyte Subfractions

Lymphocytes were isolated from peripheral blood by nylon wool filtration and Ficoll-Isopaque density centrifugation (1,5) and diluted to 2×10^6 cells/ml. Sheep erythrocytes were treated with neuraminidase (Behringwerke) and diluted in Earle's BSS to 1×10^8 cells/ml. Equal volumes of lymphocytes and neuraminidase treated sheep erythrocytes were mixed and incubated at 37 °C; after 15 min.

the cells were spun down and E-rosettes were allowed to form (1 hr., 0 °C). After a second Ficoll-Isopaque density centrifugation the non-rosetted non-T cells were collected from the interphase (6). The rosetted T cells were obtained from the pellet, after lysing the sheep erythrocytes with a solution containing 155 mM NH_4Cl, 10 mM $KHCO_3$ and 0.1 mM disodium EDTA (pH 7.4).

"Micro" Assays with lymphocytes

Enzyme incubations were carried out according to van Laarhoven et al. (1). The following enzymes were assayed: ADA with adenosine and deoxyadenosine as a substrate, PNP with inosine and hypoxanthine as a substrate, hypoxanthine-guanine phosphoribosyltransferase (HG-PRT) with both hypoxanthine and guanine as a substrate, adenine phosphoribosyltransferase (APRT), adenosine kinase (AK) and purine-5'-nucleotidase (5'N) with AMP as a substrate.

Mitogenic Stimulation of Unfractionated Lymphocyte Preparations

The isolated lymphocytes were resuspended in Tris buffered minimal essential medium(MEM, Gibco F-14; pH 7.4) containing 20 % human A^+ serum. After dilution of the cells to a concentration of 3×10^5 cells/ml, portions of 1 ml each were divided into sterile tubes (Nunc nr. 1090). To test PHA stimulation an equal number of tubes was cultured for 3 days at 37 °C with and without addition of 0.5 µl phytohaemagglutinin P (Difco 3110-57). In similar experiments 25 µl/ml poke weed mitogen (Gibco) was added to a final concentration of 25 µgr/ml. Cells were cultured for 7 days at 37 °C. 24 hours before harvesting 0.5 µCi/ml ^{3}H-thymidine (spec. act. 24 Ci/mmol; Radiochemical Centre Amersham, UK) was added to the cultures. The cells were harvested by filtration on glass fiber filters (Millipore AP 2002500). After incubation of the filters (½ hr., 20 °C) in liquid scintillation counting vials with 0.5 ml NCS tissue solubilizer (Radiochemical Centre Amersham, UK) diluted (1:3) with counting fluid (MI 92, Packard), 10 ml MI 92 was added, containing 1 % acetic acid (v/v). For the enzymatic assays the cells were collected after 3 days (PHA) and 7 days (PWM) by centrifugation (600 g, 10 min.), resuspended in 0.9 % NaCl (w/v) and lyophilised as described above.

RESULTS AND DISCUSSION

The activities of HG-PRT and PNP in T and non-T subfractions were in the same range (table 1). Mean activities of APRT, AK and 5'N were higher in T cells as compared to non-T cells. One enzyme was more active in non-T than in T cells: ADA. This was found with both adenosine and deoxyadenosine as a substrate (table 1). These

Table 1. Purine Interconversion Enzymes in Lymphocyte Subfractions.

Enzyme	Substrate	T fraction	non-T fraction	n
HG-PRT	hypoxanthine	3.24 ± 1.01	3.14 ± 1.68	10
	guanine	5.88 ± 2.28	5.66 ± 2.54	10
APRT	adenine	9.82 ± 5.88	4.48 ± 2.54	10
ADA	adenosine	61.67 ± 27.73	91.62 ± 49.01	10
	deoxyadenosine	35.53 ± 13.18	63.72 ± 38.65	9
PNP	hypoxanthine	74.45 ± 43.44	66.89 ± 34.21	10
	inosine	15.56 ± 4.60	12.59 ± 4.24	10
AK	adenosine	0.88 ± 0.63	0.47 ± 0.31	8
5'N	AMP	12.31 ± 7.08	3.29 ± 2.46	7

Activities ± standard deviation, calculated from a group of healthy individuals and expressed as 10^{-9} mol product formed/ 10^{6} lymphocytes.hour.

latter findings are different from those reported by Huang et al. (7) who found that ADA activities of complement-receptor negative (T) cells were approximately 10 times higher than those of complement-receptor positive (B) cells. Differences in ADA activity in B and T cells have also been reported by several other groups (8,9). Purine nucleoside phosphorylase was shown to display similar activities in T and non-T cells. A number of investigators, however, has reported that PNP activity is higher in T cells. PNP has even been suggested as a T cell marker on the basis of histochemical findings (10). These inconsistencies might be due to the differences of isolation procedures used. Our isolation procedure is based on the E-rosette forming capacity. Other workers made use of other immunological markers such as complement receptors (7). Moreover, our subfractions are hardly contaminated ($<$3%) with monocytes or granulocytes. Such data are not available from other reports (7,10). Tritsch and Minowada (8) compared leukemic T cell lines with normal B-cell lines and reported higher ADA activities in malignant T cell-lines.
To our knowledge there are no reports on HG-PRT, APRT, AK and 5'N activities in lymphocyte subfractions.

As can be seen in table 1 there exists a rather wide range of enzyme activities in healthy controls. This is due to individual variation and not to methodological errors (1). In addition, in a given individual enzyme activity ratios are constant, e.g. PNP activities are consistently higher than APRT activities (table 1).

Table 2. Effect of Culture Time on Purine Interconversion Enzymes in Unfractionated Lymphocytes in the Absence of Mitogens.

Enzyme	Culture Time					
	0 days		3 days		7 days	
	P[1]	C[2]	P	C	P	C
H-PRT	100% (4.5)[3]	100% (5.7)	68%	83%	39%	50%
ADA	100% (99)	100% (88)	70%	74%	33%	33%
PNP[4]	100% (12.8)	100% (15.7)	95%	76%	77%	67%
5'N	100% (19.7)	100% (20.0)	54%	77%	91%	112%

1) "pure" (P) fractions were isolated as described in materials and methods (contamination with non-lymphoid leukocytes less than 3%)
2) "contaminated" (C) fractions are isolated in the same way as "pure" fractions but the nylon wool filtration step is omitted (contamination: 2-3% monocytes, 20-30% granulocytes)
3) values in parentheses represent the absolute enzyme activities in 10^{-9} mol/10^{6} cells.hr.
4) ^{14}C-inosine was used as a substrate for the PNP assay.

Mitogenic Stimulation of Unfractionated Lymphocytes

In order to test the effect of the mitogens PHA and PWM on unfractionated lymphocytes, two different lymphocyte preparations were employed. A "pure" fraction was obtained by the complete procedure as described under methods and materials. A "contaminated" fraction was obtained by following the same procedure, but omitting the nylon wool filtration step. The "pure" preparation contained more than 97% lymphocytes, 1% monocytes and 1% granulocytes (according to Hemalog D determinations). The "contaminated" fraction contained 72% lymphocytes, 2% monocytes and 25% granulocytes.

In order to be able to assess the real stimulation index of purine enzyme activities an experiment was performed in which, in the absence of mitogens, the activity of HG-PRT, ADA, PNP and 5'N was tested after 3 days (culture time for PHA stimulation test) and after 7 days (culture time for PWM stimulation test). Three out of four enzymes tested (HG-PRT, ADA and PNP) showed a decrease in activity during culturing, whereas one enzyme (5'N) displayed a lower activity after 3 days, but after 7 days the activities were again around the original level.
It was concluded that the results of the PHA stimulation experiments should be related to the control values obtained after 3 days of culturing in the absence of this mitogen. The same was done in the PWM experiments (culturing time 7 days).

Table 3. Effect of Mitogens on ^{3}H-thymidine Incorporation and on Purine Interconversion Enzymes in Unfractionated Lymphocytes.

	PHA (3 days)		PWM (7 days)	
	P[1]	C[1]	P	C
^{3}H-thymidine incorporation index[2]	10	120	25	25
H-PRT	1.3	2.5	3.6	4.6
ADA	1.0	1.0	2.4	2.3
PNP[1]	1.1	1.3	1.5	1.6
5'N	1.2	1.0	1.1	3.4

1) see legend from table 2
2) the values given represent the stimulation index, i.e. the ratio between the activity measured with and without mitogen; for PHA stimulation the values after 3 days of culturing were used, and for PWM stimulation the values after 7 days.

The "pure" lymphocytes were hardly stimulated by PHA as judged from the ^{3}H-thymidine incorporation. The response of the "contaminated" lymphocytes to PHA was considerably higher (table 3). PWM stimulated both the "pure" and the "contaminated" cells to the same extent. The activities of the purine enzymes tested remained unchanged after PHA stimulation in both lymphocyte preparations with the possible exception of HG-PRT in the "contaminated" lymphocytes. The HG-PRT activities of "pure" and "contaminated" lymphocytes after 7 days of PWM stimulation were clearly increased. A less pronounced rise was observed for ADA, whereas PNP activity did not show significant stimulation. The marked rise after PWM stimulation of 5'N activity in "contaminated" lymphocytes as compared to "pure" cells was ascribed to contaminating cells, especially granulocytes. Our results concerning the effect of culture time on lymphocytic HG-PRT, PNP, ADA and 5'N activities in the absence of the mitogens are essentially in agreement with data reported by Raivio and Hovi (11) on T cell enriched lymphocyte preparations.
On the other hand, our data on ADA after PHA stimulation do not show the increase in activity as described by the latter workers.
It should be emphasized, however, that the present study was carried out under other experimental conditions. This makes comparison with other reports (7,8,9) difficult.
On the basis of findings with PHA stimulated intact lymphocytes (11) increased levels of purine-PRT activities might be expected.
This is indeed the case with H-PRT in the "contaminated" cells, but not in "pure" cells. As judged from the ^{3}H-thymidine incorpo-

ration the "contaminated" cell population was synthetising DNA more actively than the "pure" lymphocytes after PHA stimulation. Therefore, it seems that the rate of purine reutilisation is associated with the rate of DNA synthesis. This phenomenon is also observed in PWM stimulated "pure" and "contaminated" cells. Although satisfying explanations for our findings can not yet be given the differential responses to PHA and PWM seen with our lymphocyte preparations seem to point to differences in purine metabolism in PHA and PWM responsive cells.

ACKNOWLEDGEMENTS

This study was supported by the Koningin Wilhelmina Fonds (Foundation for Cancer Research in the Netherlands).
The authors thank Dr. J. Bakkeren, Dr. R. de Abreu and Dr. G. de Vaan (Department of Pediatrics, Faculty of Medicine, University of Nijmegen), Dr. T.L. Oei and Prof. Dr. S.J. Geerts (Department of Human Genetics, Faculty of Medicine, University of Nijmegen) for stimulating discussions.

REFERENCES

1. J.P.R.M. van Laarhoven, G.Th. Spierenburg, F.T.J.J. Oerlemans and C.H.M.M. de Bruyn, Adv. Exp. Med. Biol., this volume
2. E.R. Giblett, J.E. Anderson, F. Cohen, B. Pollara and H.J. Meuwissen, Lancet i:1067 (1972).
3. E.R. Giblett, A.J. Ammann, D.W. Ward, R. Sandman and L.K. Diamond, Lancet ii:1010 (1975).
4. N.L. Edwards, D.B. Magilavy, J.T. Cassidy and I.H. Fox, Science 201:628 (1978).
5. B.E.J. de Pauw, J.M.C. Wessels, E.J.M. Geestman, J.B.J.M. Smeulders, D.J.Th. Wagener and C. Haanen, J. Imm. Meth. 25:291 (1979).
6. M.H.J. van Oers, W.P. Zeylemaker and P.T.A. Schellekens, Eur. J. Imm. 7:143 (1977).
7. A.T. Huang, G.L. Logue and H.L. Engelbrecht, Brit. J. Haematol. 34:631 (1976).
8. G.L. Tritsch and J.Minowada, J. Natl. Cancer Inst. 60:1301 (1978).
9. R.Tung, R.Silber, F. Quagliata, M. Conklyn, J. Gottesman, R. Hirschhorn, J. Clin. Invest. 57:756 (1976).
10. M. Borgers, H. Verhaegen, M. de Brabander, F. Thoné, J. van Reempts and G. Geuens, J. Imm. Meth. 16:101 (1977).
11. K.O. Raivio and T. Hovi, Exp. Cell Res. 116:75 (1978).

MEASUREMENT OF THE RATES OF SYNTHESIS AND DEGRADATION OF HYPOXANTHINE-GUANINE PHOSPHORIBOSYLTRANSFERASE IN HUMAN LYMPHOBLASTS

Pamela Moore Mattes and William N. Kelley

Departments of Internal Medicine and Biological Chemistry, Human Purine Research Center, University of Michigan Medical School, Ann Arbor, Michigan U.S.A.

INTRODUCTION

The steady state amount of cellular proteins is determined by the net sum of both their rates of synthesis and degradation. Individual proteins in mammalian cells each have a characteristic rate of degradation as well as synthesis. In recent years, protein catabolism has gained increased recognition as a fundamental intracellular process: 1) during nutritional starvation, 2) in the regulation of enzyme levels, and 3) in the elimination of abnormal proteins.[1,2,3]

We have developed a method for determining the apparent rates of synthesis and degradation of the purine salvage pathway enzyme hypoxanthine-guanine phosphoribosyltransferase (HGPRT) in a normal human lymphoblast cell line. Nearly complete absence of this enzyme's activity is displayed by patients with the X-linked genetic disorder Lesch-Nyhan syndrome.[4] Most of these patients have been found to lack not only HGPRT enzyme activity but also to lack HGPRT protein in their red blood cells as determined by immunochemical methods.[5,6] We are interested in determining whether HGPRT deficiency in some patients is caused by selective proteolysis of abnormal enzyme protein.

MEASUREMENT OF THE SYNTHESIS OF HGPRT

The rate of synthesis of human HGPRT is measured by a procedure in which we radioactively label the protein in lymphoblast cultures for different periods of time and isolate and quantitate labeled HGPRT by a combination of immunochemical and electrophoretic techniques.

In order to measure the rate of HGPRT synthesis, lymphoblast cultures (1x10^6 cells) are washed with leucine-free growth medium (RPMI) containing 10% dialyzed fetal calf serum and then resuspended in 1 ml. of medium containing 40 to 70 μCi of ^{3}H-leucine at a final concentration of 72 μM. Leucine is an essential amino acid; supplementing the leucine-free medium with 72 μM leucine permits the cells to grow at a normal rate over the period of cell labeling. Following incubation at 37°, the cells are washed with phosphate-buffered saline containing 2mM leucine and the cell pellets are extracted with 10 mM Tris-HCl pH 7.4 containing 0.5% Triton X-100. The cell extracts are centrifuged for 30 minutes at 100,000 x g. Ninety-eight percent of the cellular HGPRT is found in the supernatant following centrifugation. The supernatants are then incubated at 85° in the presence of 1mM Mg-PRPP (5-phosphoribosyl-1-pyrophosphate) which has been shown to stabilize HGPRT to heat denaturation.[7] The extracts are then filtered and the filtrates are incubated overnight at 4° with rabbit anti-HGPRT antibodies which have been covalently bound to cyanogen-bromide activated Sepharose.[8] The following day the Sepharose beads are washed with a high salt buffer containing 1% Triton X-100 to remove non-specifically bound protein from the beads. The remaining bead-bound protein is then eluted with 1.6% sodium dodecyl sulfate (SDS) and 5.4 M urea at 85°. This material and the beads are loaded upon 12-1/2% polyacrylamide-SDS disc gels and electrophoresed. The gels are cut into 1 mM slices, digested, and the radioactivity in each slice counted in a toluene-based scintillator.

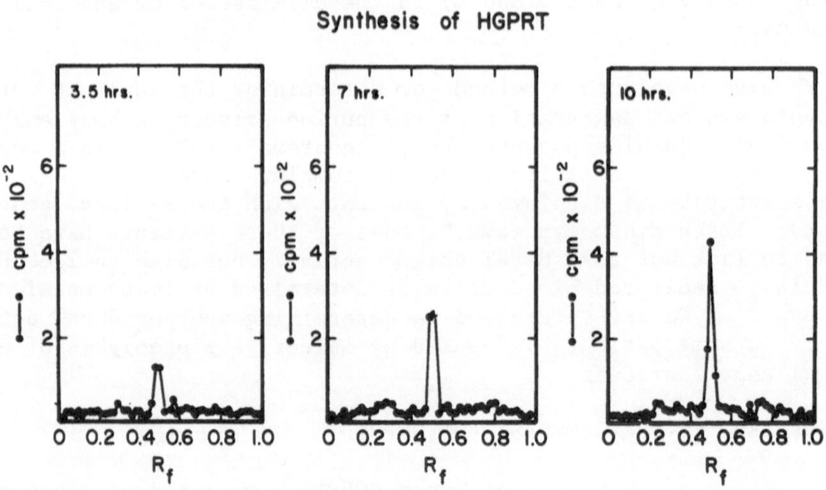

Fig. 1. SDS-polyacrylamide gel profiles of HGPRT synthesized in human lymphoblasts after 3.5, 7, and 10 hours of growth in medium containing ^{3}H-leucine.

Figure 1 shows the SDS-polyacrylamide gel profiles obtained from labeling lymphoblasts with ^{3}H-leucine for 3.5, 7, and 10 hours. The gels display a major peak of radioactivity with an R_f value of approximately 0.52. This value corresponds well to a molecular weight of 24,000 daltons on our gels. The subunit molecular weight of the HGPRT tetramer purified to homogeneity is 24,000 daltons, as reported by our laboratory.[9] Furthermore, incorporation of radioactivity into the major peak on the gels is linear with time over the duration of the period of cell labeling.

CONFIRMATION OF THE IDENTITY OF THE MAJOR PEAK AS HGPRT

To show that the major radioactively labeled peak on SDS-polyacrylamide gels is HGPRT, we carried out two control experiments. For the first experiment, we grew duplicate lymphoblast cultures in radioactive growth medium as described above. We then prepared extracts and carried the extracts through the high speed centrifugation and Mg-PRPP heat treatment steps. An excess of purified human HGPRT (10 μg) was added to one of the preparations, and both samples were incubated with the anti-HGPRT Sepharose beads. Figure 2 shows the gel profiles obtained from each preparation after the protein was eluted from the Sepharose beads and electrophoresed on polyacrylamide gels in the presence of SDS. Figure 2(b) is the gel profile of the extract incubated with the excess of purified HGPRT and anti-HGPRT Sepharose beads and Figure 2(a) is the gel profile from the control extract. In Figure 2(b) the large excess of unlabeled purified HGPRT displaces the labeled lymphoblast HGPRT from binding to the anti-HGPRT antibody bound to the Sepharose beads.

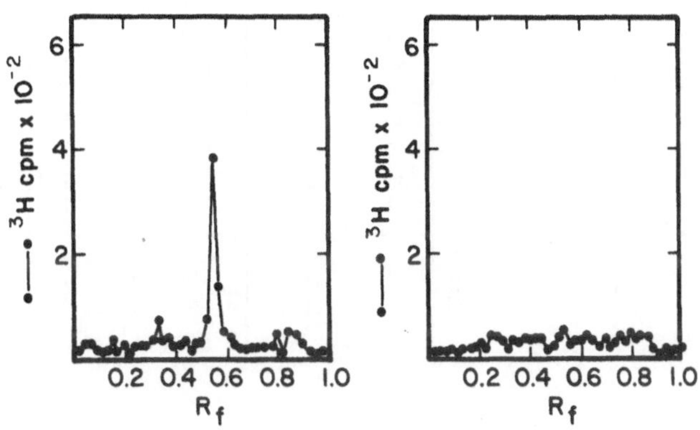

Fig. 2. Displacement of bound lymphoblast tritium-labeled HGPRT from anti-HGPRT Sepharose beads by an excess of unlabeled human HGPRT. (a) control (b) control with addition of 10 μg purified HGPRT.

To further demonstrate the identity of the gel peak as HGPRT, we did a second experiment. Identical labeled lymphoblast extracts were centrifuged, heat-treated and then incubated with either anti-HGPRT Sepharose beads or with Sepharose beads to which pre-immune rabbit serum antibodies were conjugated. No 24,000 molecular weight peak was detected when pre-immune serum Sepharose was incubated with a labeled cell extract rather than Sepharose to which antiserum specific for HGPRT was bound. Thus, the 24,000 dalton species detected on gels binds specifically to Sepharose-linked anti-HGPRT antibodies.

MEASUREMENT OF HGPRT DEGRADATION

We have also measured the relative rate of degradation of HGPRT in cultured human lymphoblasts. To do so, we pulse-label lymphoblast cultures with ^{3}H-leucine for 12 hours and then wash the cultures with growth medium containing 3 mM leucine. The cells are resuspended in the same high leucine medium and grown at 37° for varying lengths of time. The cultures are harvested and the extracts are prepared as usual.

Figure 3 shows the resulting gel profiles from an experiment done to determine the apparent half life of human lymphoblast HGPRT in cells grown under normal growth conditions. The lymphoblasts were labeled for 12 hours in growth medium containing ^{3}H-leucine, washed, and then grown in medium containing 3 mM leucine as a cold chase for intervals of up to 27 hours. As can be seen in Figure 3,

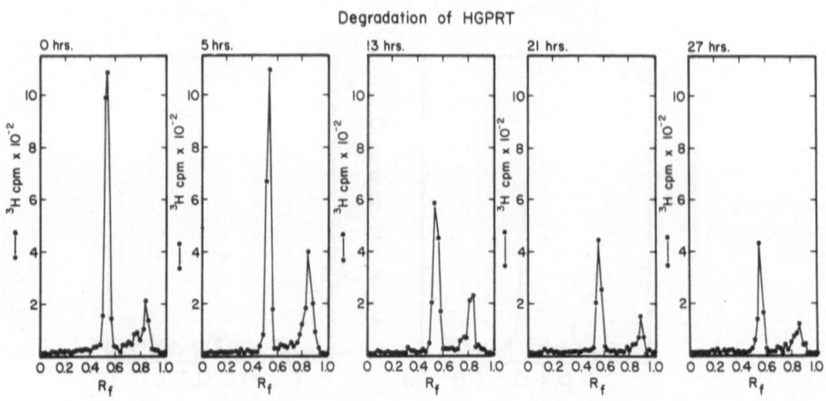

Fig. 3. SDS-polyacrylamide gel profiles of the degradation of HGPRT in human lymphoblasts. Lymphoblast cultures were pulse-labeled for 12 hours, washed, and grown for 0, 5, 13, 21 and 27 hours in growth medium containing 3 mM leucine. The extracts were prepared as described in the text.

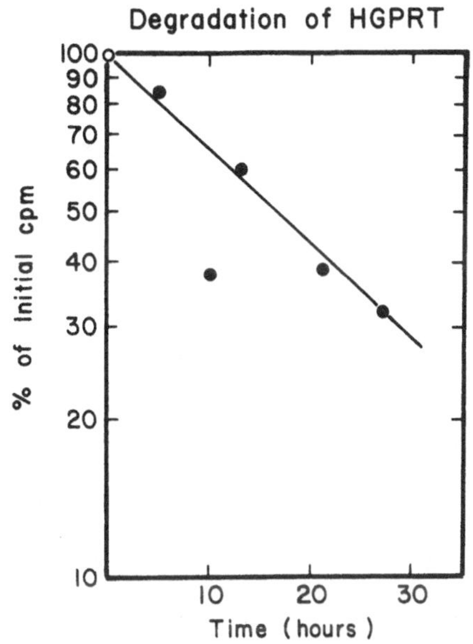

Fig. 4. Semi-logarithmic plot of the degradation of tritium-labeled HGPRT in human lymphoblasts.

increasing the time of incubation of labeled cell cultures with growth medium containing 3 mM leucine clearly decreases the radioactivity in the 24,000 molecular weight peak of the corresponding SDS-polyacrylamide gels. The identity of the smaller molecular weight species with an R_f value of approximately 0.85 is presently unknown, although the possibility that it is a degradative product of HGPRT is being investigated.

Figure 4 is a semi-logarithmic plot of the disappearance of the radioactivity in the 24,000 dalton peak with time of incubation of the cell cultures in medium containing 3 mM leucine. Since protein degradation is a first order process, we are able to calculate the apparent half life of an HGPRT molecule as approximately 16 hours in lymphoblasts grown under normal growth conditions.

In conclusion, with the preceding methodology we will be able to determine and compare the mechanism of possible hormonal, nutritional, or drug effects on the level of HGPRT activity in human cells. Finally, we shall also be able to determine the rate of degradation of HGPRT in lymphoblasts from HGPRT-deficient patients in order to see whether the reduction in HGPRT enzyme protein in these patients is the result of selectively accelerated degradation of abnormal enzyme protein.

REFERENCES

1. A. L. Goldberg and J. F. Dice, Intracellular protein degradation in mammalian and bacterial cells, Ann. Rev. Biochem. 43:835 (1975).

2. A. L. Goldberg and A. C. St. John, Intracellular protein degradation in mammalian and bacterial cells: Part 2, Ann. Rev. Biochem. 45:747 (1976).

3. R. T. Schimke and N. Katanuma, eds., "Intracellular Protein Turnover," Academic Press, New York (1975).

4. M. Lesch and W. L. Nyhan, A familial disorder of uric acid metabolism and central nervous system function, Am. J. Med. 36:561 (1964).

5. S. Upchurch, A. Leyva, W. J. Arnold, E. A. Holmes, and W. N. Kelley, Hypoxanthine phosphoribosyl transferase deficiency: association of reduced catalytic enzyme activity with reduced levels of immunologically detectable enzyme protein, Proc. Natl. Acad. Sci. U.S.A. 72:4142 (1975).

6. G. Ghangas and G. Milman, Radioimmune determination of hypoxanthine phosphoribosyl transferase crossreactive material in erythrocytes of Lesch-Nyhan patients, Proc. Natl. Acad. Sci. U.S.A. 72:4147 (1975).

7. A. S. Olsen and G. Milman, Chinese hamster hypoxanthine-guanine phosphoribosyl transferase, J. Biol. Chem. 249:4030 (1974).

8. P. Cuatrecasas, Protein purification by affinity chromatography, J. Biol. Chem. 245:3059 (1970).

9. J. A. Holden and W. N. Kelley, Human hypoxanthine-guanine phosphoribosyl transferase, J. Biol. Chem. 253:4459 (1978).

HUMAN 5´-NUCLEOTIDASE. PROPERTIES AND CHARACTERIZATION OF THE ENZYME FROM PLACENTA, LYMPHOCYTES AND LYMPHOBLASTOID CELLS IN CULTURE.

Wolf Gutensohn

Institut für Anthropologie und Humangenetik der Universität, Arbeitsgruppe Biochemische Humangenetik, Schillerstr. 42, D 8000 München 2, Fed. Rep. Germany

INTRODUCTION

Significantly decreased activities of the ectoenzyme 5´-nucleotidase (EC 3.1.3.5) in peripheral lymphocytes of patients with primary hypogammaglobulinemia have been described. There is conflicting evidence in the literature whether this concerns the X-linked type of the disease or rather the socalled "adult onset" hypogammaglobulinemia[1, 2, 3]. Nevertheless, it seemed interesting to study the properties of human 5´-nucleotidase (5´-N) more closely. For reasons of availability of starting material the following approach was chosen: First the placental 5´-N was purified and characterized followed by an extensive comparison of the properties of this enzyme with 5´-N from human lymphocytes and lymphoblastoid cells

METHODS

The marker enzyme activities of 5´-N, alkaline phosphatase (AP) and alkaline phosphodiesterase (PD) are enriched in microsomal and plasmamembrane fractions from placenta obtained by gradient centrifugation. Solubilization of these enzymes by various detergents and purification of 5´-N by a three step procedure including DEAE-cellulose chromatography and affinity chromatography on lentil-lectin-Sepharose and 5´-AMP-Sepharose have been described elsewhere[4].

Enzyme assays: 5´-N was tested by a radiochemical assay using ^{3}H-AMP. Activity is expressed as that part of total AMPase activity inhibitable by 200 µM α,β-methylene-adenosine-diphosphonate (AOPCP). Ecto-5´-N activity on intact cells was determined by the same procedure using isotonic incubation con ditions. AP and PD were tested by continuous or discontinuous optical assays using p-nitrophenyl-substrates at pH 10.0. On intact cells phosphatase was measured under isotonic conditions and pH 7.4.

Cell counts were obtained either in a hemacytometer or in a Coulter counter.

RESULTS AND DISCUSSION

The purified placental 5´-N is free of unspecific phosphatase activity and on SDS-gel-electrophoresis (Weber Osborn system) a homogeneous band corresponding to a molecular weight of about 78 000 is seen at 3 different gel concentrations. The enzyme was also characterized by preparative isoelectric focu sing. The patterns obtained were not absolutely identical for individual placentas, but enzyme peaks were consistently found in the ranges of pI 5.75 - 5.85; 6.1 - 6.2; 6.3 - 6.4 and 6.5. After neuraminidase treatment of the enzyme, activity was not decreased, however, a definite shift to a single peak at pI 7.2 in isoelectric focusing as well as a lowered electrophoretic mobility on PAGE in a nondenaturing system were observed[4].

In a study with lectins both 5´-N and AP from placenta were shown to be receptors for Concanavalin A and Lens culinaris lectin; however, enzyme activity was only blocked with the 5´-N. There was a marked difference in susceptibility to lectin inhibition between membrane bound and isolated 5´-N[5].

Organspecific isoenzymes or subtypes of AP have been extensively studied and characterized. In order to show or exclude a similar situation for 5´-N an extensive comparison of the placental enzyme with that of two lymphoblastoid cell lines ("IDU" and Just B2; see Table) was carried out. Great similarities were observed in the following properties: Affinity to lentil-lectin-Sepharose and 5´-AMP-Sepharose; K_M-values for AMP; reversible inhibition by a number of purine- and pyrimidine-nucleotides; inactivation by EDTA and reactivation by various metal ions; Arrhenius plots; heat inactivation; crossreactivity with an antiserum against placental 5´-N (see below). There was a slight but consistent difference in the extractability by detergents of 5´-N from placental and lymphoid membranes. This would suggest a different anchorage of the enzyme in the membranes.

5´-N activity in lymphocytes or lymphoblasts is usually much lower than in placenta, however, here the "ectoenzyme" nature can be clearly shown. Activity on intact cells tested under isotonic conditions is usually at least as high as in detergent extracts. Using this technique we have measured 5'-N on peripheral lymphocytes of healthy blood donors and patients with lymphoproliferative disorders and confirmed results from Zucker-Franklin et al.[6] Most patients with CLL, morbus Hodgkin, ALL, CML and AML have low to undetectable levels of 5'-N activity on their lymphocytes, however, occasionally a patient with extremely high activity is found (for example one patient with a T-blast crisis in CML).

The 5'-N activities in various established lymphoblastoid cell lines are summarized in Tab. 1. It can be seen that enzyme activity varies in a considerable range and that there is no relationship between the T- or B-cell origin or character of a cell line and the presence or absence of 5'-N.

Table 1. Ectoenzyme activity in human lymphoblastoid cell lines.

Cell Line	Origin	Type	Ecto-5'-N (nMoles/h/10^6cells)	Ecto-Phosphatase (nMoles/h/10^6cells)
JUST-B2	EBV-infected PBL	B	9.2	51
RAJI	Burkitt-L.	B	0	15
DG 75	Burkitt-L	B	0	46
MOLT-4	T-Leukemia	T	3.5	13
CCRF-CEM	T-Leukemia	T	0	9.5
"IDU"	?	?	26.8	78

Two Burkitt lymphoma lines (Ramos and BJAB) which have lost the Epstein Barr virus nuclear antigen (EBNA) and a number of lines derived from these by superinfection with EBV were studied. All were 5'-N-negative, except for one superinfected line which had regained 5'-N activity. In a preliminary experiment this enzyme seemed to be more resistant to the anti-5'-N serum.

It is clear that cell lines lacking 5'-N do not show any impairment in growth. Nevertheless it seemed interesting to look whether 5'-N is growth related in a line with high 5'-N activity. Ecto-5'-N activity in the "IDU"-line related to cell number showed a distinct peak at the onset and logarithmic phase of growth and a decline in stationary phase. This observation is in contrast to results obtained with HeLa cells and it is not known, whether this also holds for other lines. On the other hand growth of the "IDU"-line in culture is not inhibited by 200 μM AOPCP in the medium.

An antiserum against purified human placental 5´-N was raised in rabbits and is now tested for its specificity. The antibody is precipitating as shown by the Ouchterlony diffusion assay It inhibits isolated as well as plasmamembrane bound placental 5´-N, but not AP from the same membranes. It crossreacts with microsomal 5´-N from human liver, kidney, spleen, brain and peripheral lymphocytes. Results on crossreactivity with human serum-5´-N are still ambiguous. Crossreactivity was also shown for 5´-N from liver, kidney, spleen and brain of rat and mouse. Studies towards using this antibody as a cell surface label are underway.

ACKNOWLEDGMENT

This work was supported by the Deutsche Forschungsgemeinschaft.

REFERENCES

1. S. M. Johnson, G. L. Asherson, R. W. E. Watts, M. E. North, J. Allsop, and A. D. B. Webster, Lymphocyte purine 5´-nucleotidase deficiency in primary hypogammaglobulinaemia, Lancet 1977, i:168 - 170.
2. A. D. B. Webster, M. North, J. Allsop, G.L. Asherson, and R. W E. Watts, Purine metabolism in lymphocytes from patients with primary hypogammaglobulinaemia, Clin. exp. Immunol. 31:456 - 463 (1978).
3. N. L. Edwards, D. B. Magilavy, J. T. Cassidy, and I. H Fox, Lymphocyte ecto-5´-nucleotidase deficiency in agammaglobulinemia, Science 201:628 - 630 (1978).
4. W. Gutensohn, Fractionation of human plasmamembrane ectoenzymes by ionexchange-, lectin-specific and substrate-specific affinity-chromatography, in: "Proceedings of the 27th Colloquium: Protides of the Biological Fluids", (Brussels, 1979), Pergamon Press, Oxford (1979), in press.
5. W. Gutensohn, Human placental 5´-nucleotidase and alkaline phosphatase. Receptors for Concanavalin A and Lens culinaris lectin, Hoppe Seyler´s Z. Physiol Chem. 359:1599 - 1602 (1978).
6. R. Silber, M. Conklyn, G. Grusky, and D. Zucker-Franklin, Human lymphocytes: 5´-Nucleotidase-positive and -negative subpopulations, J. Clin. Invest. 56:1324 - 1327 (1975).

METABOLISM AND TOXICITY OF 9-β-D-ARABINOFURANOSYLADENINE IN HUMAN MALIGNANT T CELLS AND B CELLS IN TISSUE CULTURE

Dennis A. Carson, Jonathan Kaye, and J. E. Seegmiller

Department of Clinical Research, Research Institute of Scripps Clinic, La Jolla, California 92037 and Department of Medicine, University of California, San Diego, La Jolla, California 92093

INTRODUCTION

In humans, a genetic deficiency of the enzyme adenosine deaminase (ADA) leads to the specific impairment of the development of the lymphoid system, with resulting immunodeficiency disease[1]. Recent studies have suggested that the remarkable lymphospecific toxicity seen in ADA deficiency may result from the selective accumulation of deoxyadenosine nucleotides in lymphoid tissues, particularly the thymus, which when compared to other tissues have high deoxyadenosine phosphorylating activity, and low deoxyribonucleotide dephosphorylating activity[2-5]. The dATP thereby produced inhibits DNA synthesis, probably by inhibiting ribonucleotide reductase and perhaps via other as yet undescribed mechanisms.

While some ADA deficient children suffer from combined immunodeficiency disease, in many cases it is the T cell related functions that are most severely impaired[1]. In previous studies, we noted a marked *in vitro* hypersensitivity of T leukemia cells to deoxyadenosine toxicity which appeared to faithfully reproduce the *in vivo* suppression of T cell development and function seen in ADA deficiency[6]. We therefore reasoned that one might be able to use human T leukemia cell lines for the *in vitro* analysis of potential T cell specific cytotoxic agents, as suggested by Ohnuma and coworkers[7].

The purine nucleoside 9-β-D-arabinofuranosyl adenine (ara-A) is an ADA substrate with anti-viral and cytotoxic properties, remarkably free of undesirable side effects at the usual therapeutic doses[8]. The structural similarities between ara-A and deoxyadenosine, however, prompted us to study the toxic effects of ara-A toward human T cell leukemia and B cell lines grown in tissue culture in the presence of ADA inhibitors. We found that ara-A, like deoxyadenosine, was particularly cytotoxic toward human T leukemia cells, which phosphorylated the nucleoside via deoxycytidine kinase, and sequestered the toxic metabolite ara-ATP. The growth inhibitory effects of ara-A toward T cells, however, could be prevented by the addition of deoxycytidine to the culture medium.

MATERIALS AND METHODS

Cell Lines

The human leukemia cell line CCRF-CEM was obtained from the American Type Culture Collection (Rockville, MD) and maintained in suspension culture in RPMI-1640 medium supplemented with 10% heat inactivated dialyzed fetal calf serum (Flow Laboratories, Rockville, MD), 2 mM glutamine, penicillin 100 U/ml, and streptomycin 100 μg/ml. The human B cell line Wil-2 and an adenosine kinase deficient variant (Wil-2 (AK-)) were obtained and selected by Dr. M. Hershfield as previously described[9], and were grown as described for CCRF-CEM.

Growth Curves

To 5 ml CCRF-CEM cells at a concentration of 2×10^5 cells/ml or Wil-2 cells at 1×10^5 cells/ml in complete medium was added either the ADA inhibitor erythro-9-(2-hydroxy-3 nonyl) adenine hydrochloride (EHNA) (Burroughs-Wellcome, Research Triangle Park, NC), or deoxycoformycin (Pentostatin, Parke-Davis, Detroit, MI)[10,11], to final concentrations of 5 μM and also, where indicated, varying concentrations of ara-A and deoxycytidine. After three days the number of live cells was determined in a hemocytometer using trypan blue. The percent growth equals:

$$\frac{\text{live cells with nucleoside + ADA inhibitor}}{\text{live cells with ADA inhibitor alone}} \times 100.$$

At a concentration of 5 μM, both EHNA and deoxycoformycin inhibited growth by ∿15-20%. Growth curves were plotted on semi-log paper and the concentration of ara-A required to inhibit growth by 50% was determined.

Enzyme Assays

The deoxyadenosine, adenosine, ara-A and deoxycytidine phosphorylating activities of cell extracts were determined radiochemically by a modification of the method of Ives, exactly as performed previously[2,12]. The final concentration of reactants was ATP 5 mM, magnesium chloride 2.5 mM, sodium fluoride 15 mM, 50 mM Tris (final pH 7.4), 5 μM EHNA, substrate 300 μM, and protein 2 mg/ml in a final volume of 100 μl. Reactions were terminated in a boiling water bath and after cooling products were separated from reactants by thin layer chromatography on PEI cellulose in methanol-water (1:1)[2]. Activities are expressed as nanomoles product/min/mg protein. Protein was determined by Lowry's method using bovine serum albumin as a standard[13].

Measurement of ara-ATP Levels

To 10 ml cells at a concentration of 10^6 cells/ml in complete medium was added 30 μM ara-A, and 5 μM deoxycoformycin, either with or without 10 μM deoxycytidine. Control cultures contained cells and medium alone. After 24 hours at 37°C, the cultures were cooled on ice, the medium was removed, and the perchloric acid soluble nucleotides were extracted and neutralized as previously described [5,6]. Ara-ATP was separated from ATP and dATP by gradient high pressure liquid chromatography on a Whatman Partisil SAX column (Whatman, Clifton, NJ) using the same elution conditions we have previously used to separate ATP and dATP[5,6]. Each nucleotide triphosphate was identified by comparison of retention volumes with known standards.

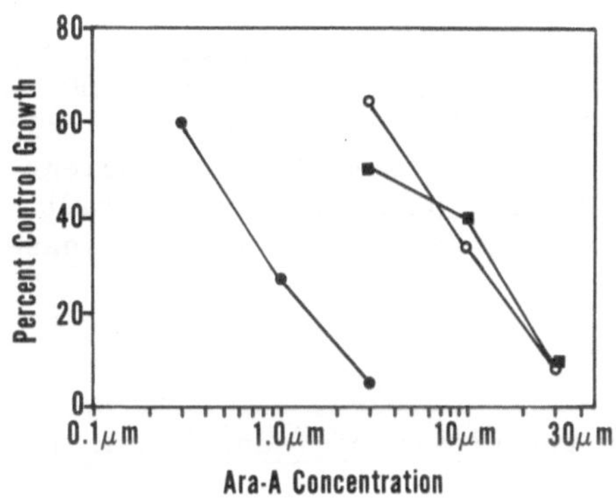

Fig. 1. Effect of ara-A on the growth of T and B cell lines. To the human T cell line CCRF-CEM (●——●), the B cell lines Wil-2 (o——o) or an adenosine kinase deficient variant (■——■), was added varying concentrations of ara-A in complete medium containing the ADA inhibitor EHNA at a concentration of 5 μM. After 72 hours in tissue culture, the percent control growth was determined as described in the Methods.

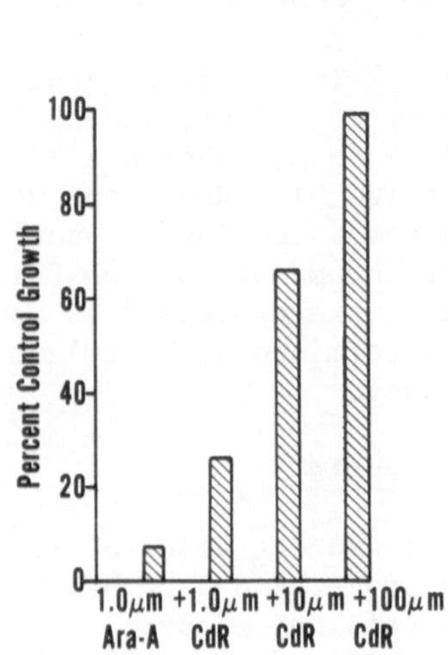

Fig. 2. Deoxycytidine reversal of ara-A toxicity in the CCRF-CEM human T leukemia cell line. This experiment was performed as described in Figure 1 except that the varying concentrations of deoxycytidine were included in the growth medium.

Ion Exchange Chromatography

Human tonsillar tissue obtained at surgery was minced with a scalpel, homogenized in 10 mM Tris, pH 8.0, in a Waring blender and centrifuged at 27,000 x g for 40 minutes. Twenty milligrams of the supernatant fraction was dialyzed against 0.005 M sodium phosphate, pH 7.9, and applied to a 1 ml column of DE52 cellulose (Whatman) and chromatographed with a 20 ml linear gradient of 0.005 M phosphate, pH 7.9, to 0.005 M phosphate, 0.5 M NaCl. Individual column fractions were assayed radiochemically for kinase activity as described above.

RESULTS

Effect of ara-A on the Growth of T and B Cell Lines

In the presence of 5 μM EHNA, the growth of the human T cell line CCRF-CEM was inhibited 50% by 0.5 μM ara-A; the human B cell line Wil-2 was ten fold less sensitive with 50% inhibition at 5 μM ara-A (Fig. 1). Similar results were obtained when deoxycoformycin was substituted for EHNA. In our hands, the adenosine kinase deficient Wil-2 did not differ significantly from the wild type in its sensitivity to ara-A inhibition of growth. The addition of as little as 1 μM deoxycytidine to the culture medium reversed the inhibitory effects of ara-A and EHNA on the growth of the CCRF-CEM cell line (Fig. 2).

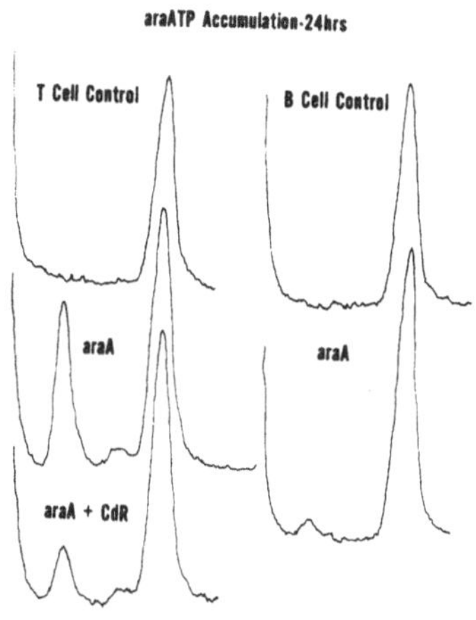

Fig. 3. Accumulation of ara-ATP in T and B cell lines. To the T cell line CCRF-CEM or the B cell line Wil-2 was added 5 μM EHNA and either 30 μM ara-A or ara-A + 10 μM deoxycytidine. After 24 hours the acid soluble nucleotides were extracted and ara-ATP was detected by high pressure liquid chromatography. The changing peak in the middle of the chromatogram is ara-ATP. The constant peak to the right is GTP. The tracings on the left from top to bottom are: T cell control, T cells + ara-A, T cells + ara-A + deoxycytidine. The tracings on the right are B cell control, B cells + ara-A.

Accumulation of ara-ATP in T and B Cells

When T and B cell lines were incubated with 30 μM ara-A and 5 μM deoxycoformycin, the T cells accumulated more than ten times as much ara-ATP as the B cells over a 24 hour period (Fig. 3). The addition of 10 μM deoxycytidine to the culture medium inhibited by 70% the accumulation of ara-ATP by T leukemia cells incubated with 30 μM ara-A.

Enzymes Responsible for ara-A Phosphorylation

Table 1 shows the phosphorylating activities for adenosine, deoxyadenosine and ara-A in CCRF-CEM, Wil-2 and Wil-2(AK-) cell lines determined at a substrate concentration of 300 μM. A true V_{max} for ara-A phosphorylating activity could not be determined because of difficulties in solubilizing the nucleoside. Of note, extracts of the adenosine kinase deficient Wil-2 had similar ara-A phosphorylating activity as wild type cells.

Table 1. Nucleoside Phosphorylating Activity in T and B Cell Lines

Cell Line	Type	Adenosine Phosphorylating Activity	Deoxyadenosine Phosphorylating Activity	Ara-A Phosphorylating Activity
CCRF-CEM	T cell	2.65	0.85	0.10
Wil-2	B cell	1.70	0.68	0.05
Wil-2 (AK-)	B cell	<0.01	0.51	0.05

Enzyme activities were determined radiochemically at a substrate concentration of 300 μM as described in the Methods. Activities are expressed as nanomoles product/min/mg protein.

When a human tonsillar extract was fractionated on DE52 cellulose, most of the ara-A, deoxyadenosine, and deoxycytidine phosphorylating activities co-chromatographed (Fig. 4). Additionally a minor fraction of ara-A phosphorylating activity was found in the adenosine kinase peak. In dialyzed tonsillar extracts, phosphorylation of 1 μM ara-A was inhibited 86% by 20 μM deoxycytidine, 60% by 100 μM deoxyguanosine or deoxyadenosine, but not by up to 100 μM adenosine.

In intact cells, adenosine kinase under certain conditions could phosphorylate ara-A. Thus, human red cells which have nearly undetectable levels of both deoxycytidine kinase and 5'-nucleotidase could effectively accumulate ara-ATP when incubated with ara-A in the presence of an ADA inhibitor[14]. Furthermore, while intact Wil-2 (AK-) cells phosphorylated ara-A as well as wild type cells at ara-A concentrations of up to 20 μM, at nucleoside concentrations of 200 μM, the AK- cells were deficient. These results indicate that at either high ara-A concentrations or in cells with low levels of 5'-nucleotidase, ara-A could be phosphorylated by adenosine kinase.

DISCUSSION

The results presented herein show that when ADA is inhibited, ara-A is much more toxic to human malignant T cell lines than to B cell lines. The increased sensitivity of the malignant T cells to growth inhibition by ara-A parallels the increased ability of these same cells to phosphorylate the nucleoside and accumulate the toxic metabolite ara-ATP. However, there is no correlation between the ara-A phosphorylating activity in cell extracts, and the ability of the intact cells to sequester ara-ATP. Rather, the increased sensitivity of T cells to ara-A, as well as deoxyribonucleosides in general, reflects both a high ara-A phosphorylating activity, and a low deoxyribonucleotide dephosphorylating activity[5,6].

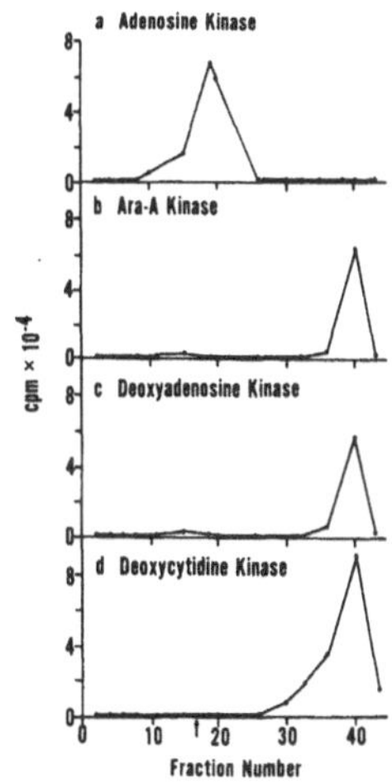

Fig. 4. DE52 chromatography of ara-A kinase in a human tonsillar extract. A human tonsillar extract was fractionated on a DE52 column using a sodium chloride gradient in 0.005 M phosphate pH 7.9 as described in the Methods. Each fraction was assayed radiochemically for a) adenosine phosphorylating activity, b) ara-A phosphorylating activity, c) deoxyadenosine phosphorylating activity, and d) deoxycytidine phosphorylating activity. As can be seen, the deoxyribonucleoside and ara-A phosphorylating activities co-chromatograph.

While both deoxycytidine kinase and adenosine kinase potentially can mediate ara-A phosphorylation, it is deoxycytidine kinase which is most important at low but toxic substrate concentrations. Thus it is not surprising that ara-A toxicity toward malignant T lymphocytes can be prevented by the addition of deoxycytidine to the culture medium. As shown here, deoxycytidine probably acts by inhibiting ara-A phosphorylation, thereby lowering the intracellular concentration of the active metabolite, ara-ATP. This point is of clinical importance, in that animals have substantial levels of deoxycytidine in the circulation; in rat blood the concentration of 40 μM has been reported[15]. Indeed, one would expect that agents which decreased deoxycytidylate synthesis might potentiate ara-A toxicity as is the case with 9-β-D-arabinofuranosylcytosine[16]. Smith et al have also recently reported that deoxyadenosine could antagonize the anti-viral effects of ara-A[17]. As with deoxycytidine, this effect may result from deoxyadenosine mediated inhibition of ara-A phosphorylation.

The addition of ADA inhibitors to in vitro systems potentiates both the anti-viral and cytotoxic effects of ara-A by preventing its deamination[19]. To what extent ADA inhibitors favorably alter the *in vivo* therapeutic index of ara-A is uncertain. However, the similarities between the toxicity of deoxyadenosine and ara-A, combined with the known effects of deoxyadenosine in producing immunodeficiency disease in ADA deficient patients, suggest that the combination of ara-A and an ADA inhibitor might have immunosuppressive effects which would counteract its anti-viral actions. On the other hand, the combination of ara-A plus an ADA inhibitor might be valuable chemotherapeutically in the treatment of T cell leukemias and in other situations where selective T cell killing is desirable. In particular, the ability to rescue cells from ara-A toxicity through the administration of deoxycytidine might have clinical applications.

REFERENCES

1. E. R. Giblett, J. E. Anderson, F. Cohen, B. Pollara, and H. J. Meuwissen, Adenosine-deaminase deficiency in two patients with impaired cellular immunity, Lancet 2:1067 (1972).
2. D. A. Carson, J. Kaye, and J. E. Seegmiller, Lymphospecific toxicity in adenosine deaminase deficiency and purine nucleoside phosphorylase deficiency: Possible role of nucleoside kinase(s), Proc. Natl. Acad. Sci. USA 74:5677 (1977).
3. A. Cohen, R. Hirschhorn, D. Horowitz, A. Rubinstein, S. H. Polmar, R. Hong, and D. W. Martin, Jr., Deoxyadenosine triphosphate as a potentially toxic metabolite in adenosine deaminase deficiency, Proc. Natl. Acad. Sci. USA 75:472 (1978).
4. M. S. Coleman, J. Donofrio, J. J. Hutton, and L. Hahn, Identification and quantitation of adenine nucleotides in erythrocytes of a patient with adenosine deaminase deficiency and severe combined immunodeficiency, J. Biol. Chem. 253:1619 (1978).
5. D. A. Carson, J. Kaye, S. Matsumoto, J. E. Seegmiller, and L. Thompson, Biochemical basis for the enhanced toxicity of deoxyribonucleosides toward malignant human T cell lines, Proc. Natl. Acad. Sci. USA 76:2430 (1979).
6. D. A. Carson, J. Kaye, and J. E. Seegmiller, Differential sensitivity of human leukemic T cell lines and B cell lines to growth inhibition by deoxyadenosine, J. Immunol. 121:1726 (1978).
7. T. Ohnuma, H. Arkin, J. Minowada, and J. F. Holland, Differential chemotherapeutic susceptibility of human T-lymphocytes and B-lymphocytes in culture, J. Natl. Cancer Inst. 60:749 (1978).
8. R. J. Whitely, S.-J. Soong, R. Dolin, G. J. Galasso, L. T. Ch'ien, C. A. Alford and the Collaborative Study Group, Adenine arabinoside therapy of biopsy-proved herpes simplex encephalitis, National Institute of Allergy and Infectious Diseases Collaborative Antiviral Study, New Engl. J. Med. 297:289 (1977).

9. M. S. Hershfield, F. F. Snyder, and J. E. Seegmiller, Adenine and adenosine are toxic to human lymphoblast lines deficient in purine salvage enzymes, Science 197:1284 (1977).
10. H. J. Schaeffer, and D. F. Schwender, Enzyme inhibitors. 26. Bridging hydrophobic and hydrophilic regions on adenosine deaminase with some 9-(2-hydroxy-3-alkyl) adenines, J. Med. Chem. 17:6 (1974).
11. R. W. K. Woo, H. W. Dion, S. M. Large, L. T. Dahl, and L. J. Durham, A novel adenosine and ara-A deaminase inhibitor, (R)-3-(2-deoxy-β-D-erythropentofuranosyl)-3,6,7,8,-tetrahydro-imidazo 4,5-d 1,3 diozepin-8-ol, J. Heterocyclic. Chem. 11:641 (1974).
12. D. H. Ives, J. P. Durham, and V. S. Tucker, Rapid determination of nucleoside kinase and nucleotidase activities with tritium-labeled substrates, Anal. Biochem. 28:192 (1969).
13. O. H. Lowry, N. J. Rosebrough, A. Farr, and R. J. Randall, Protein measurement with the folin phenol reagent, J. Biol. Chem. 193:265 (1951).
14. T. Chang, and A. J. Glazko, Effect of an adenosine deaminase inhibitor on the uptake and metabolism of arabinosyl adenine (vidarabine) by intact human erythrocytes, Res. Comm. Chem. Path. Pharm. 14:127 (1976).
15. J. Rotherham, and W. C. Schneider, Deoxyribosyl compounds in animal tissues, J. Biol. Chem. 252:853 (1958).
16. C. Mills-Yamamoto, G. J. Lauzon, and A. R. P. Paterson, Toxicity of combinations of arabinosylcytosine and 3'-deazauridine toward neoplastic cells in culture, Biochem. Pharm. 27:181 (1978).
17. S. H. Smith, C. Shipman, Jr., and J. C. Drach, Deoxyadenosine antagonism of the antiviral activity of 9-β-D-arabinofuranosyl adenine and 9-β-D-arabinofuranosyl hypoxanthine, Cancer Res. 38:1916 (1978).
18. S. H. Lee, N. Caron, and A. P. Kimball, Therapeutic effects of 9-β-D-arabinofuranosyladenine and 2'-deoxycoformycin combinations in intracerebral leukemia, Cancer Res. 37:1953 (1977).
19. W. Plunkett, and S. S. Cohen, Two approaches that increase the activity of analogs of adenine nucleosides in animal cells, Cancer Res. 35:1547 (1975).

Supported by National Institutes of Health Grants AM 23200, RR 05514, AM 13622, and GM 17702 and by grants from the Kroc Foundation and the Arthritis Foundation.

SUPPRESSION OF CELLULAR IMMUNITY DUE TO INHIBITION OF PURINE NUCLEOSIDE PHOSPHORYLASE BY ALLOPURINOL-RIBOSIDE

Y. Nishida, N. Kamatani, K. Tanimoto and I. Akaoka

Department of Medicine and Physical Therapy
Faculty of Medicine, University of Tokyo
Bunkyo-ku, Tokyo, Japan 113.

Recently, children with purine nucleoside phosphorylase (PNP) deficiency have been reported to show defective T cell immunity but intact B cell immunity (1). These findings suggest that inhibition of PNP may result in suppression of T cell functions and an inhibitor of PNP may also be a useful immunosuppressive agent.

In the present report the effects of allopurinol-riboside (A-R) on PNP activity were studied in vitro and on the lymphocyte proliferation induced by mitogens. The effects of A-R on humoral and cellular immunity to sheep red blood cells (SRBC) were also studied in mice.

MATERIALS AND METHODS

A-R was donated from Kyowa Hakko Corp. Ltd.

Enzyme assay

The inhibition of A-R on PNP activity was determined by a spectrophotometric method (2). After addition of xanthine oxidase (PL Biochem.) and PNP (Boehringer Manheim) to the sample cuvette which contained inosine as substrate and or plus A-R in 0.05 M PBS ph 7.4, the increase in optical density at 293 n m was recorded.

Conversion of A-R

A-R was incubated with PNP for 30 min at 37°C. The reaction mixture was spotted on TLC plates and developed with butanol: mehtanol:water:25%NH_4OH (60:20:20:1 v/v). It was also applied to the high pressure liquid chromatogram using Hitachi Castam gel 2618 and eluted with 0.4N ammonium acetate pH4.0 at a pressure of $100kg/mm^3$.

Cell cultures

Lymphocytes were separated by Ficoll-hypaque solution as previously described (3). The washed lymphocytes were suspended in medium RPMI 1640 containing 10% foetal calf serum, 200v/ml penicillin G and 100ug/ml streptomycin. $1x10^5$ lymphocytes containing the various concentrations of sterilized A-R were put into each well of sterile microplates. 100ul of mitogens dissolved in culture medium (4ug/ml:PHA, 50ug/ml:Con A, 1:100diluted PWM, 100ug/ml LPS) were added. They were cultured for 5 days in triplicate. 1uCi of ^{3}H-thymidine was added to each well for a further 24 hour incubation. Cells were harvested on glass fiber filter paper and the radioactivity of thymidine incorporated into lymphocytes was measured in a liquid-scintillation counter.

Humoral and Cellular immunity

$1x10^8$ SRBC were injected into the peritoneal cavity of 20 mice. 10 mice were injected with A-R in a dose of 1 mg/g wt/day intramuscularly for successive 7 days. Seven days after injection, the blood haemagglutinin titres were measured in microtitration trays using 0.1 ml of serum diluted serially in gelatin-veronal buffer to which an equal volume of 1% SRBC was added.

$5x10^5$ SRBC were injected into peritoneal cavity of 20 mice. A-R (1 mg/g wt/day) was given in 10 mice for successive 5 days, and $1x10^8$ SRBC were injected into the left foot pad. 24 hours later, the increase in foot pad weight was measured.

RESULTS AND DISCUSSION

A-R suppressed PNP activity competitively with inosine. 50% inhibition was achieved at $277x10^{-6}$ M of A-R (Fig I).

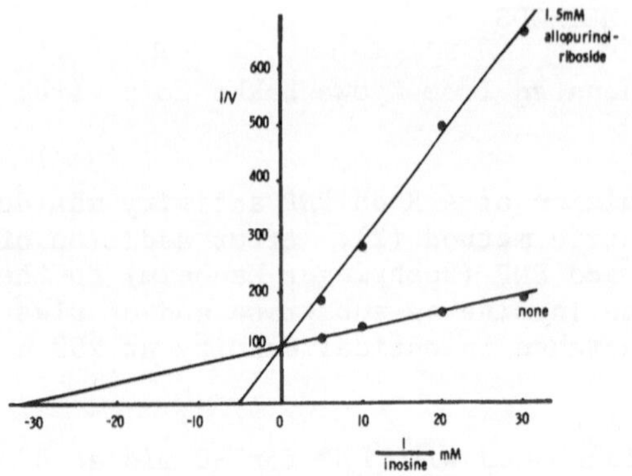

Fig. 1 Lineweaver-Burk plot. Inhibition study with allopurinol-riboside showing the competitive manner having ki of $277x10^{-6}$M.

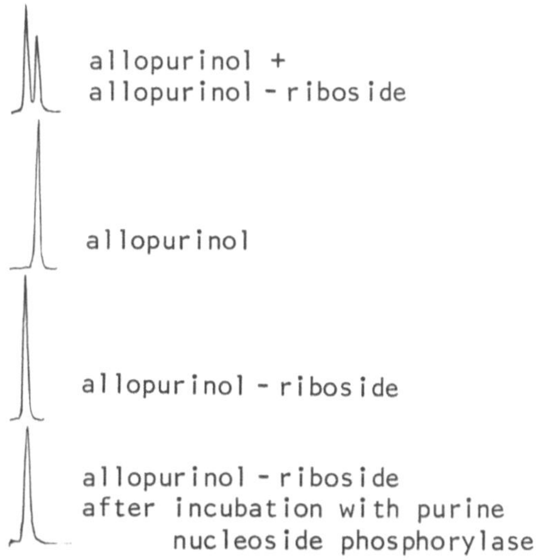

Fig II Conversion of allopurinol-riboside with purine nucleoside phosphorylase. Reaction mixtures incubated with allopurinol-riboside and purine nucleoside phosphorylase were applied to the high liquid chromatogram using Hitachi castam gel 2618 and eluted with 0.4 N-ammonium acetate PH 4.0.

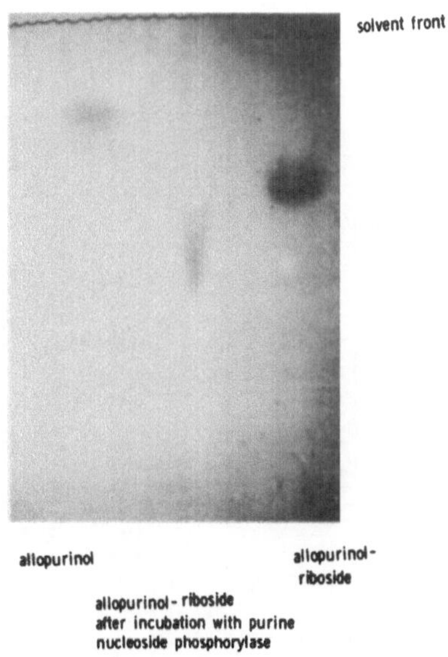

Fig III Conversion of allopurinol-riboside with purine nucleoside phosphorylase. Reaction mixtures incubated with allopurinol-riboside and purine nucleoside phosphorylase were spotted on the thin layer chromatogram, and developed with butanol: methanol:water:25%NH_4OH (60:20:20:1 v/v).

Allopurinol was not formed after incubation of A-R with PNP in vitro which was confirmed by high pressure liquid chromatography and/or T.L.C. (Fig II,III). These results may indicate that A-R binds to active site of PNP, and accordingly is useful as an inhibitor of PNP.

The results of ^{3}H-thymidine incorporation in lymphocytes stimulated with various mitogens are shown in Fig IV. The mitogen-induced proliferation of human lymphocytes was in general suppressed by A-R. Blastogenesis induced by selective T cell stimulant PHA and Con A, was markedly inhibited by A-R, while that induced by PWM and LPS was less inhibited. A-R seemed to suppress mainly the T-cell function.

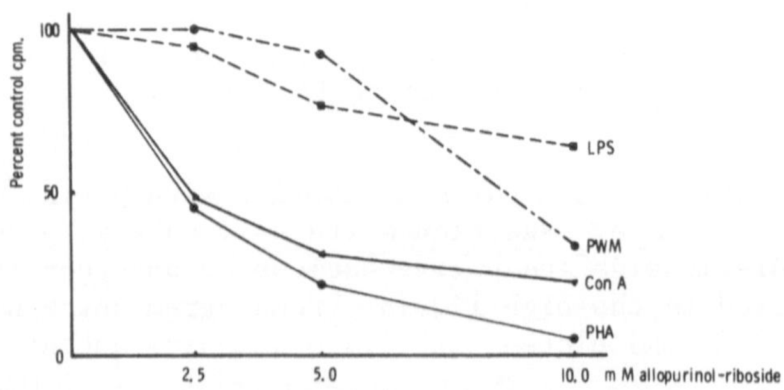

Fig IV The effect of allopurinol-riboside on the proliferative response of lymphocytes.Control (100%) cpm were 31.516:PHA, 3.221:Con A, 3.620 PWM, 2.518:LPS.

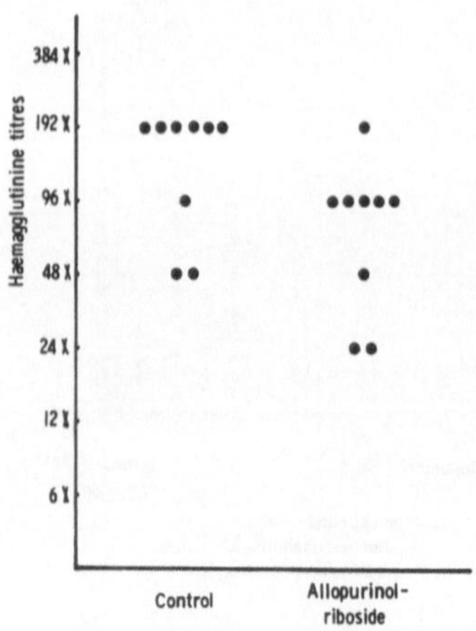

Fig V The effects of allopurinol-riboside on humoral immunity to SRBC in mice.

Haemagglutinin titres are shown in Fig V. No significant difference was seen between the mice control group and the group treated with A-R. In contrast, the increase of foot pad weight in mice treated with A-R was significantly lower, 4.3 ± 1.4 mg, than the average increase of 17.9 ± 3.3 mg in the controls (Fig VI).

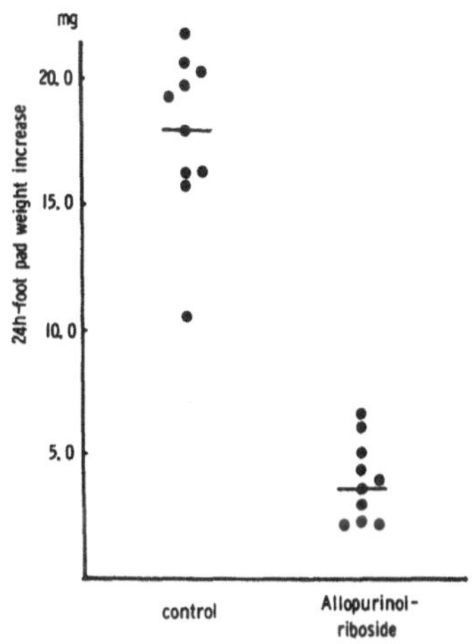

Fig VI The effects of allopurinol-riboside on cellular immunity to SRBC in mice.

Humoral immunity was not suppressed by A-R, but delayed type hypersensivity was suppressed significantly by A-R in vivo experiments.

A-R in an adequate dose may produce animal effects experimentally similar to PNP deficiency and may also be a useful and powerful inhibitor of cellular immunity.

REFERENCES

1. Giblett, E.R. et al
Lancet 1: 1010, 1975

2. Kalcker H.M.
J. Biol.Chem. 167: 429, 1947

3. Boyum, A.
Scand. J. Clin. Lab. Invest 21: 77, 1968.

LYMPHOCYTE 5'-NUCLEOTIDASE DEFICIENCY: CLINICAL AND METABOLIC CHARACTERISTICS OF THE ASSOCIATED HYPOGAMMAGLOBULINEMIA

N. Lawrence Edwards, James T. Cassidy, and Irving H. Fox

Human Purine Research Center
Departments of Internal Medicine and Biological Chemistry, University of Michigan, Ann Arbor, Michigan, 48109

Decreased activity of 5'-nucleotidase on the surface of peripheral blood mononuclear cells (PBM's) has recently been observed in young males with congenital agammaglobulinemia[1], primary non-familial hypogammaglobulinemia[1-3], and selective IgA deficiency[3,4]. A partial deficiency of the ecto-enzyme occurred in 12 of 12 patients with congenital agammaglobulinemia, in 4 of 15 with acquired (common variable) immunodeficiency, and in 2 of 10 with selective IgA deficiency. The enzyme values and clinical classification of the 18 enzyme deficient subjects with primary hypogammaglobulinemia are shown in Table 1.

The clinical and immunologic criteria used to classify our patients with primary hypogammaglobulinemia are listed in Table 2. No single variable used in disease classification uniquely correlated with the lymphocyte 5'-nucleotidase deficiency. In general, patients diagnosed as either congenital agammaglobulinemia or common variable immunodeficiency with low 5'-nucleotidase had lower percentage of circulating B-cells and earlier age of onset of symptoms that suggested immune dysfunction.

Systemic purine metabolism was evaluated in the patients with low lymphocyte 5'-nucleotidase activity by plasma urate levels, urinary uric acid excretion, and urinary excretion of radioactivity following purine nucleotide pool labeling with (8-^{14}C)adenine. Plasma and urinary uric acids were normal in the enzyme-deficient subjects. The amount of radioactivity excreted as urinary purines following (8-^{14}C)adenine infusion in 5 males with congenital agammaglobulinemia and ecto-5'-nucleotidase deficiency is 1.6 ± 0.8

TABLE 1. CLINICAL CLASSIFICATION OF SUBJECTS WITH HYPOGAMMAGLOBULINEMIA ASSOCIATED WITH 5'-NUCLEOTIDASE DEFICIENCY

Patient	5'-Nucleotidase (nmol/hr/10^6)	Clinical Diagnosis[a]
	(normal range: 12.3-41.1)	
1	3.7	cAγII
2	4.0	cAγIII
3	4.8	cAγII
4	5.2	cAγIII
5	5.5	cAγII
6	5.6	cAγI
7	6.1	cAγIII
8	6.3	cAγII
9	6.5	cAγII
10	6.5	cAγIII
11	6.9	cAγII
12	7.2	CVID
13	7.7	cAγII
14	8.6	CVID
15	9.8	sIqAD
16	10.4	CVID
17	10.4	sI gAD
18	11.3	CVID

a. cAγ - Congenital agammaglobulinemia (I-classic X-linked; II-familial male affected without lateral transmission; III-infantile onset with no known affected relatives); CVID-common variable immunodeficiency; sIgAD-selective IgA deficiency.

TABLE 2. THE CLINICAL AND IMMUNOLOGIC CHARACTERISTICS OF PRIMARY HYPOGAMMAGLOBULINEMIA

	Positive Familial Inheritance	Age of Onset (Mo.)	Serum Immunoglobulins (mg/dl)			B-Cells (%)	Cell-Mediated Immunity
			IgG	IgA	IgM		
			(>750)	(>80)	(>70)		
Congenital Agammaglobulinemia	8/12	4-24	0-220	0	0-50	0-3	Normal
Common Variable Hypogammaglobulinemia	1/15	4-192	0-320	0-30	0-2080	3-20	Variable
Selective IgA Deficiency	1/10	7-360	Normal	0	Normal	5-13	Variable

TABLE 3. PERIPHERAL MONONUCLEAR CELL 5'-NUCLEOTIDASE IN CONGENITAL AGAMMAGLOBULINEMIA

	5'-Nucleotidase[a] (nmol/hr/10^6 cells)	
	Normal Subjects (9)[b]	Congenital Agammaglobulinemia (5)[b]
Unfractionated Cells	21.9 (12.8-47.1)	5.9 (3.3-11.8)
E-Rosette Forming Cells	21.2 (11.9-33.6)	7.1 (3.3-11.8)
Non-E-Rosette Forming Cells	24.0 (12.0-43.2)	6.6 (2.8-8.4)

a. Enzyme levels are mean values for multiple determinations with the range of values in parentheses.
b. Number of individual determinations performed in duplicate.

percent of the administered radioactivity per day. This compares to baseline excretion of 1.4 and 2.1 percent per day in two males with common variable immunodeficiency and no ecto-enzyme deficiency.

The enzyme deficiency involves both E-rosette forming cells (T-cells) and non-E-rosette forming cells (non-T-cells) from patients with congenital agammaglobulinemia (Table 3) suggesting that the reduced levels of lymphocyte 5'-nucleotidase in this disease is not secondary to an altered T:B cell ratio[5].

The effect of monocyte contamination on peripheral mononuclear cell 5'-nucleotidase activity was investigated on intact cells prior to and following monocyte depletion by either the carbonyl iron or adherence techniques. Monocyte depletion of peripheral blood mononuclear cells increases the 5'-nucleotidase activity in normal cells from a mean value of 19.6 to 28.8 nmoles/hr/10^6 cells and the cells deficient in 5'-nucleotidase from a mean value of 6.2 to 7.7 nmoles/hr/10^6 cells (Figure 1).

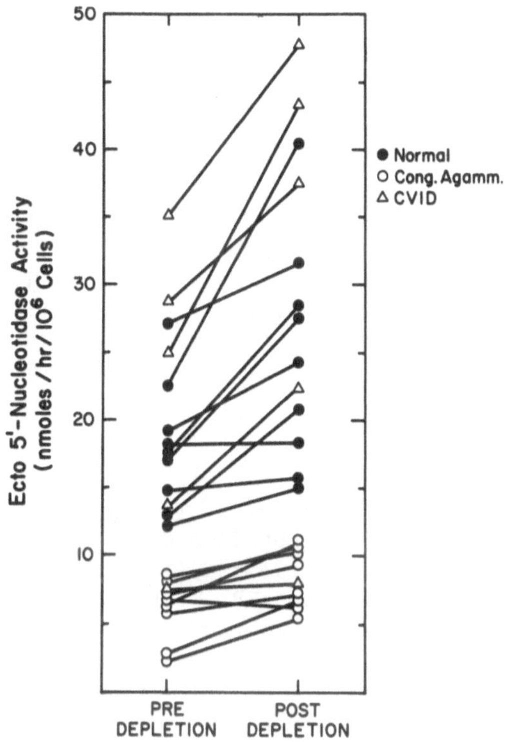

Fig. 1. Effect of monocyte depletion on peripheral mononuclear cell 5'-nucleotidase activity. The Ficoll-Hypaque separated cells prior to monocyte depletion had 22 ± 6 percent monocytes in normals and 29 ± 11 percent in congenital agammaglobulinemia. Post depletion samples had only 7 ± 2 percent monocytes in normals and 9 ± 3 percent monocytes in congenital agammaglobulinemia.

The effect of 5'-nucleotidase deficiency on gammaglobulin synthesis was tested in vitro by adding adenosine-α, β-methylene diphosphonate (AOPCP) to antigen stimulated peripheral blood lymphocyte cultures. Ten to 100 μM AOPCP, a specific inhibitor of 5'-nucleotidase, did not decrease IgG synthesis in the culture model.

The data presented here suggest that 5'-nucleotidase deficiency (a) may be associated with certain predominant clinical features, (b) may indicate a T-lymphocyte abnormality, (c) is not accounted for by the relative monocytosis of congenital agammaglobulinemia, (d) has no detectable systemic disturbance of purine metabolism, and (e) is not directly correlated with immunoglobulin secretion in vitro. Whether the enzyme deficiency causes the immune dysfunction or is a marker for an intrinsic lymphocyte abnormality cannot be distinguished at present.

ACKNOWLEDGMENTS

The authors wish to thank Jumana Judeh for the typing of this manuscript. These studies were supported by USPHS grants AM19674 and 5M01RR42 and a grant from the American Heart Foundation (77-849) and the Michigan Heart Association. N.L.E. is the recipient of a Clinical Associate Physician Award from the National Institutes of Health for the General Clinical Research Centers Branch.

REFERENCES

1. N. L. Edwards, D. B. Magilavy, J. T. Cassidy, and I. H. Fox, Lymphocyte ecto-5'-nucleotidase deficiency in agammaglobulinemia, Science 201:628 (1978).
2. S. M. Johson, G. L. Asherson, R. W. E. Watts, M. E. North, J. Allsop, and A. D. B. Webster, Lymphocyte purine 5'-nucleotidase deficiency in primary hypogammaglobulinemia, Lancet 1:168 (1977).
3. N. L. Edwards, J. T. Cassidy, and I. H. Fox, Lymphocyte 5'-nucleotidase deficiency: Clinical characteristics of the associated hypogammaglobulinemia. Clin. Res. 27:324A (1979).
4. A. D. B. Webster, M. North, J. Allsop, G. L. Asherson, and R. W. E. Watts, Purine metabolism in lymphocytes from patients with primary hypogammaglobulinemia. Clin. Exp. Immunol. 31:456 (1978).
5. N. L. Edwards, E. W. Gelfand, L. Burk, H. M. Dosch, and I. H. Fox, Distribution of 5'-nucleotidase in human lymphoid tissues. Proc. Natl. Acad. Sci. USA 76:(In Press, July) (1979).

IMMUNOLOGICAL STUDIES ON LESCH-NYHAN PATIENTS

C. de Bruyn[*], Ph. Gausset[o], J. Duchateau[o], E. Vamos[o], S. Kulakowski[**] and G. Delespesse[o].
[*]Department of Human Genetics, Faculty of Medicine, University of Nijmegen, Nijmegen, The Netherlands.
[o]Laboratory of Immunology and Bloodtransfusion, University Hospital "St. Pierre", Brussels, Belgium.
[**]Institute "Les Petites Abeilles", Vlezenbeek, Brussels, Belgium.

INTRODUCTION

Minor abnormalities of B-lymphocyte function have been reported in association with the Lesch-Nyhan (LN) syndrome (1). Other authors have not been able to confirm this (2,3). LN patients are not more susceptible to infections than healthy control individuals and no clinical evidence for impaired immune competence is available. Therefore, the immunological abnormalities, if present at all, might be rather marginal. In studies on phytohaemagglutinin (PHA) stimulated lymphocytes from one LN patient we observed a severely depressed lymphocytic proliferative response under certain in vitro conditions. This prompted us to further investigate several immunological parameters in three LN patients.

METHODS AND MATERIALS

Patients

All three patents showed the classical picture of the LN syndrome (4). Patient 1 (aged 7) and patient 2 (aged 16) are siblings. They have been admitted to two different institutions: the former receives allopurinol, the latter does not. Patient 3 (aged 17) is not institutionalised; he does not receive allopurinol. Details on these patients are presented elsewhere (5).

Immunological Methods

Lymphocytes separated on Ficoll-Hypaque gradient (6), were characterised by the E-rosette technique (7) and by membrane fluorescence (8) for their content of T and B lymphocytes. Proliferative responses to T cell mitogens (PHA, concanavalin A, (con A), T and B cell mitogens (pokeweed mitogen, protein A from staphylococcus) and soluble antigens (streptokinase, streptodornase, SKD) were tested as previously described (9). ^{3}H-thymidine, ^{3}H-uridine and ^{3}H-leucine incorporation was measured by adding 5 µCi of each precursor 18 hours before harvesting the cells. IgG secretion was measured by coprecipitation of ^{3}H-leucine labeled IgG (10).

Biochemical Methods

Hypoxanthine-guanine phosphoribosyltransferase (HG-PRT) and adenine phosphoribosyltransferase (APRT) activities in erythrocyte and lymphocyte lysates were measured as described elsewhere (11,12).

RESULTS AND DISCUSSION

Severe HG-PRT deficiency was demonstrated in the erythrocyte lysate of all three patients (table 1). The APRT activity was increased as compared to normal. In the lymphocytes the deficiency was much less pronounced. The residual activities of the H-PRT reaction were always lower as compared to those of the G-PRT reaction. Lymphocytic APRT activities seemed within the normal range (table 1).

Table 1. Purine Phosphoribosyl Transferase Activities in Erythrocyte and Lymphocyte Lysates of Patients with the Lesch-Nyhan Syndrome and controls.

	erythrocytes*			lymphocytes°		
	H-PRT	G-PRT	A-PRT	H-PRT	G-PRT	A-PRT
LN patient 1	0.1	0.7	49.6	0.4	1.8	8.5
LN patient 2	0.1	0.2	51.7	0.3	1.2	6.2
LN patient 3	0.3	0.05	52.5	1.0	2.2	13.2
mean controls	71.0 (n=9)	117 (n=5)	15.2 (n=9)	2.6	4.0	9.8

*Erythrocyte values are given in 10^{-9} moles/mg protein.hr.

°Lymphocyte values are given in 10^{-9} moles/10^{6} cells.hr.

Table 2. Effect of Pokeweed Mitogen and Staphylococcal Protein A on Total ^{3}H-Leucine Incorporation by Lymphocytes from Lesch-Nyhan Patients and Control Individuals. The Data are Given in cpm.

	unstimulated	PWM 1:100	PWM 1:400	prot. A 10 μgr/ml
LN patient 1	12,268	53,949	57,911	210,305
LN patient 2	8,105	16,631	15,320	151,504
LN patient 3	17,361	75,555	72,322	276,104
mean	12,578	48,712	48,517	212,637
(S.D.)	(4,636)	(29,809)	(29,639)	(62,333)
control 1	9,496	74,589	62,374	88,051
control 2	55,098	79,203	100,519	60,192
control 3	8,703	54,085	40,533	196,024
mean	24,432	69,292	67,808	114,756
(S.D.)	(26,560)	(13,370)	(30,360)	(71,745)

Table 3. Effect of Pokeweed Mitogen and Staphylococcal Protein A on ^{3}H-labeled Immunoglobulin Secretion by Lymphocytes from Lesch-Nyhan Patients and Control Individuals. The Data are Given in cpm/250,000 cells.

	unstimulated	PWM 1:100	PWM 1:400	prot. A 10 μgr/ml
LN patient 1	865	21,630	20,465	82,220
LN patient 2	1,345	15,210	3,420	55,475
LN patient 3	1,175	12,470	15,220	57,960
mean	1,128	16,443	13,035	65,218
(S.D.)	(243)	(4,639)	(8,730)	(14,776)
Control 1	1,385	14,995	15,830	25,705
Control 2	6,605	7,565	12,980	16,610
Control 3	760	11,870	10,685	65,925
mean	2,916	11,616	13,165	36,080
(S.D.)	(3,209)	(3,523)	(2,577)	(26,243)

The percentage of surface immunoglobulin bearing circulating lymphocytes in the three LN patients was 13.2, 8.2 and 9.2, respectively. In three healthy controls these values were 12.8, 19.8 and 7.7, respectively. On this basis it seems justified to conclude that the proportions of circulating B lymphocytes in our LN patients are in the normal range. Immunoglobulin levels in the serum were also comparable with normal (5). B lymphocyte function in vitro was tested by measuring both the incorporation into protein of ^{3}H-leucine and secretion of radioactive immunoglobulin (Ig) in response to pokeweed mitogen (PWM) and staphylococcal protein A (Sp-A). Patients and controls reacted similarly to PWM as judged from the ^{3}H-leucine incorporation (table 2) and the secretion of IgG (table 3). When stimulated with Sp-A, the patients' lymphocytes even incorporated slightly higher amounts of ^{3}H-leucine as compared to normal (table 2). The same was found with IgG secretion (table 3). These findings do not suggest impaired B lymphocyte function in vitro.

As parameters for T lymphocyte function in vitro the responses to PHA, concanavalin A (con A) and a mixture of streptokinase and streptodornase (SKD) were studied. Results of a PHA stimulation experiment are summarised in table 4. Under "optimal" culture conditions (humified atmosphere containing 5% CO_2; HEPES-buffered medium) the incorporation of ^{3}H-thymidine by LN lymphocytes was in the same range as the control lymphocytes. However, when no CO_2 was supplied during cultivation the proliferative response of LN cells was much more reduced than that of normal cells. This was not the only difference noted between normal and LN cells: in a 5% CO_2 atmosphere, but with additional bicarbonate (2 µg/ml) in the medium, a depressed proliferative respons of control cells was observed, whereas LN lymphocytes even showed an increased ^{3}H-thymidine incorporation (table 4). An explanation for these observations might be that different culture conditions result in different kinetics of the proliferative respons in LN and control lymphocytes. Omission of CO_2 might lead to a slower rate of proliferation in LN

Table 4. Effect of Culture Conditions on the Incorporation of ^{3}H-thymidine by Lymphocytes Stimulated with PHA (1 µgr/ml).

	^{3}H-thymidine incorporation (cpm)	
	LN patient 1	control
5% CO_2; no bicarbonate ("optimal" system)	219,590 ± 7,760	221,150 ± 1,510
no CO_2; no bicarbonate	77,290 ± 2,920	167,850 ± 7,030
5% CO_2; 2 µgr/ml bicarbonate	272,450 ± 6,900	134,380 ± 3,950

cells, wheras in the presence of both CO_2 and bicarbonate a reversed situation might arise. Because in all experiments the cells have been harvested after 3 days, we might have been dealing with different stages of the proliferative respons (e.g. ascending phase, plateau, descending phase). Further studies on ^{3}H-thymidine incorporation are consistent with this hypothesis (5).

Apart from the incorporation of ^{3}H-thymidine, the incorporation of ^{3}H-uridine and of ^{3}H-leucine has also been investigated after stimulation with PHA, con A and SKD. These experiments were performed in the absence of 5% CO_2 in the atmosphere. The results indicated reduced incorporation of all three metabolites by LN lymphocytes (data not shown).

LN patients do not show increased susceptibility towards immunological challenges. This is confirmed in our patients. Either the in vivo conditions allow the purine de novo pathway to compensate for the deficient salvage pathway or the relatively high residual HG-PRT activity in the lymphocytes of our patients is sufficient to maintain an apparently normal immunological status. It might be hypothesised that under certain severely disturbed conditions LN patients might become unable to cope adequately with immunological challenges.

The present findings illustrate that minor modifications of culture conditions may affect the lymphocyte proliferative responses not only of normal cells but especially those of LN cells. We conclude that these observations should be taken into account when interpreting results of stimulation experiments. In addition, subclinical immunodeficiency states might be revealed when testing both at optimal and suboptimal culture conditions.

ACKNOLEDGEMENTS

The authors express their gratitude to Dr. H. Janseune and Dr. A. Lambrechts (Institute "St. Josef", Antwerp, Belgium) for their kind cooperation in the study of two of the LN patients and to Dr. J.A.J. Bakkeren (Department of Pediatrics, University Hospital, Nijmegen) for his interest and suggestions.

REFERENCES

1. A.C. Allison, T. Hovi, R.W.E. Watts and A.D.B. Webster, Lancet ii:1179 (1975).
2. J.E. Seegmiller, T. Watanabe, M.H. Schreier and T.A. Waldmann, Adv. Exp. Med. Biol. 76A:412 (1977).
3. E.W. Gelfland, I.H. Fox, M. Stuckey and H.M. Dosch, Clin. Exp. Immunol. 31:205 (1978).

4. M. Lesch and W.L. Nyhan, Ann. J. Med. 36:561 (1964).
5. Ph. Gausset, E.Vamos, S. Kulakowski, J. Duchateau, C. de Bruyn and G. Delespesse, ms. in preparation.
6. A. Bøyum, Scand. J. Clin. Lab. Imm. 21 (suppl. 97):77 (1968).
7. M. Jondal, J. Exp. Med. 136:207 (1972).
8. J.L. Preud'homme and G. Flandrin, J. Immunol. 113:1650 (1974).
9. G. Delespesse, J. Duchateau, Ph. Gausset et al. J. Immunol. 116:437 (1976).
10. Ph. Gausset, J. Duchateau, H. Collit and G. Delespesse, ms. in preparation.
11. M.P. Uitendaal, C.H.M.M. de Bruyn, T.L. Oei and P. Hösli, Biochem. Genet. 16:1187 (1978).
12. J.P.R.M. van Laarhoven, G.Th. Spierenburg, F.T.J.J. Oerlemans and C.H.M.M. de Bruyn, Adv. Exp. Med. Biol., this volume.

ACTIVITY OF ECTO-5'-NUCLEOTIDASE IN LYMPHOBLASTOID CELL LINES DERIVED FROM CARRIERS OF CONGENITAL X-LINKED AGAMMAGLOBULINEMIA

Linda F. Thompson, Gerry R. Boss, Annie Bianchino, and J. Edwin Seegmiller
Department of Medicine, University of California, San Diego, La Jolla, California 92093

Patients with congenital X-linked agammaglobulinemia (CAG)* have 1/2 to 1/3 the normal activity of ecto-5'-nucleotidase (ecto-5'-NT) in their peripheral blood mononuclear cells (1,2). Since peripheral B cells have at least 3 times more ecto-5'-NT activity than peripheral T cells (2,3) this deficiency can be largely explained by the absence of circulating B cells in these patients (4). In an attempt to develop a biochemical test for detection of carriers for CAG, ecto-5'-NT was measured in peripheral B cells and in lymphoblastoid (B) cell lines established from mothers and sisters of patients with CAG.

METHODS

Peripheral blood mononuclear cells (PBMs) were separated from fresh whole blood (with ACD anticoagulant) by Ficoll-Hypaque gradient centrifugation (5). Monocytes were largely removed by adherence to plastic petri dishes. T and B lymphocytes were separated by rosetting with neuraminidase-treated sheep erythrocytes (6) followed by a second Ficoll-Hypaque gradient. The sheep erythrocyte rosetting fraction contained an average of 88% T cells as determined by a second rosetting with neuraminidase-treated sheep erythrocytes, and less than 2% monocytes. The non-rosetting

*Abbreviations: CAG, congenital X-linked agammaglobulinemia; ecto-5'-NT, ecto-5'-nucleotidase; PBMs, peripheral blood mononuclear cells; sIg, surface immunoglobulin; EBV, Epstein-Barr virus; Hepes, N-2-hydroxyethylpiperazine-N'-2-ethanesulfonic acid; IMP, inosine 5'-monophosphate; AOPCP, α,β-methylene adenosine 5'-diphosphate.

cell fraction contained an average of 17% monocytes, 3% T cells, and 41% surface immunoglobulin bearing (sIg^+) B cells. In most cases, ecto-5'-NT assays were performed within 12 hours of venipuncture. Lymphoblastoid cell lines were established by the addition of Epstein-Barr virus (EBV) to PBMs according to the method of Sly (7). The separated lymphocytes or cultured lymphoblasts were washed twice in 40 mM sodium Hepes, pH 7.4, 130 mM NaCl, and 0.4% bovine serum albumin and resuspended in the same buffer at a final density of 2.5 to 10 x 10^6 cell/ml. Each assay contained (in a final volume of 35 µl): 8.6 mM $MgCl_2$, 50 mM Tris HCl, pH 6.9, 240 µM [8-^{14}C]-IMP (Amersham, diluted to 6 µCi/µmole with carrier IMP), and 2.5 to 20 x 10^4 cells. In duplicate assays, α,β-methylene adenosine 5'-diphosphate (AOPCP) (Sigma) was added at 2.85 mM. All results are expressed as AOPCP-inhibitable ecto-5'-nucleotidase activity. The reactions were carried out at 37°C for 1 hr and were terminated with 5 µl 8 M HCOOH. A portion of each reaction mixture was separated by thin layer chromatography on Eastman Cellulose sheets with fluorescent indicator in methanol: water (1:1). The reaction products (inosine and hypoxanthine) and substrate (IMP) were located with non-radioactive markers; the spots were cut out and quantitated by liquid scintillation counting.

RESULTS AND DISCUSSION

The activity of ecto-5'-NT in PBMs, rosetting cells, and non-rosetting cells isolated from control subjects, patients with CAG, and female relatives of patients with CAG are shown in Fig. 1. The low activity of ecto-5'-NT in the PBMs of the CAG patients is largely explained by the absence of a cell population high in ecto-5'-NT activity; i.e., by an absence of sIg^+ B cells. Female relatives of CAG patients have normal activities of ecto-5'-NT in their PBMs and T cells. Although the average activity in their non-rosetting cells is somewhat lower than that in the control subjects, all the individual values fall within the normal range, making it impossible to identify carriers for CAG by the measurement of this enzyme activity.

Ecto-5'-NT activity was also measured in lymphoblastoid cell lines established from control subjects and female relatives of patients with CAG. Since these cell lines were established by EBV infection, they are solely of B-cell origin. The activity was measured as soon as enough cells were available for assay (49-88 days after the addition of EBV) and at 5 to 7-day intervals thereafter. Figure 2 shows the ecto-5'-NT activity of both groups at the time of the initial assay and after an additional five weeks of culture. The female members of four families have been studied. At the time of initial assay, ecto-5'-NT activity was markedly deficient (<10% of normal) in the lymphoblastoid cell lines established from 3 of the 4 mothers and two of the three sisters of the CAG patients. Three of the four families have more

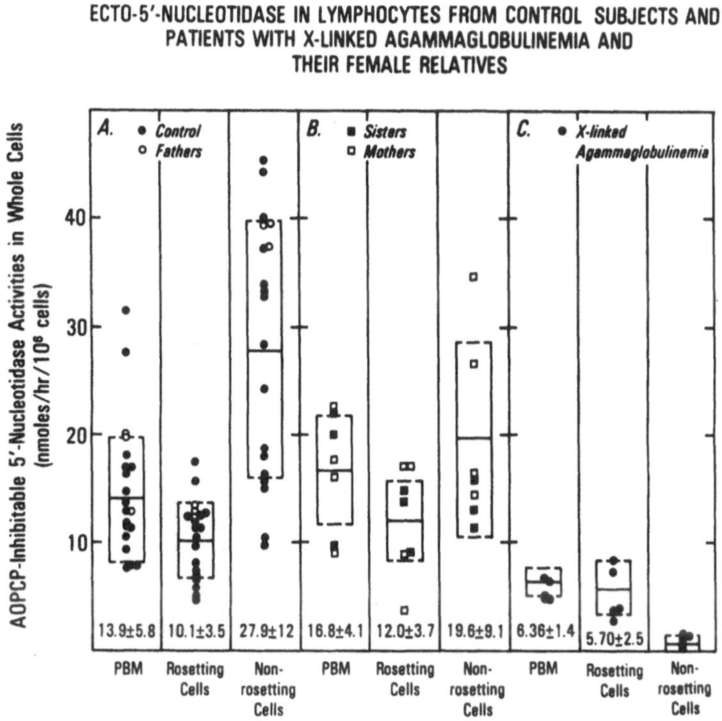

Fig. 1. Lymphocyte subpopulations were separated from peripheral blood samples and assayed for ecto-5'-nucleotidase activity as described in Materials and Methods.

than one affected child, the only exception being the family whose mother's lymphoblast line had normal ecto-5'-NT activity. We propose that the affected male in this family represents a new mutation for CAG and that the female relatives whose lymphoblastoid cell lines show low ecto-5'-NT activity when intially established are heterozygous carriers for CAG.

The following model has been proposed to explain the gradual increase in ecto-5'-NT activity in the lymphoblast cell lines derived from the presumed CAG heterozygotes. Ecto-5'-NT appears to be a marker for B-cell maturation (8). Both fetal spleen** and cord blood B lymphocytes (8), which are sIg^+, are markedly deficient in ecto-5'-NT activity and newborns do not gain the ability to produce their own immunoglobulins until several months after birth (9). In CAG patients, B-lymphocyte maturation is presumably blocked at some early stage before the acquisition of sIg (4) and consequently also before the acquisition of ecto-5'-NT

**Data in preparation

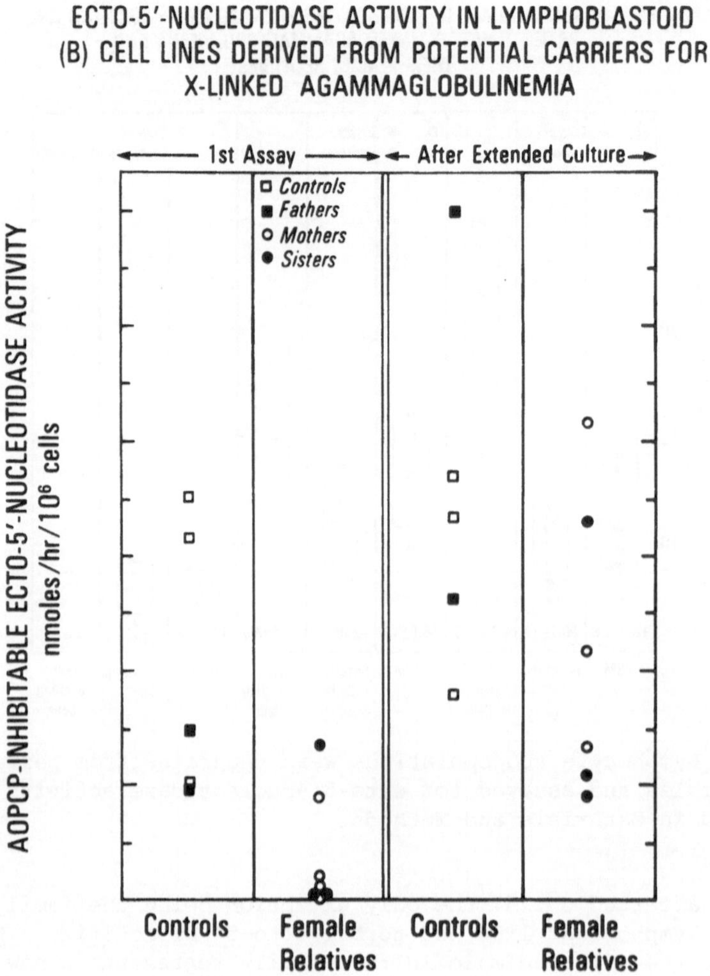

Fig. 2. Cultured cells from permanent lymphoblastoid cell lines were assayed for ecto-5'-nucleotidase activity as described in Materials and Methods.

activity and the ability to produce immunoglobulins. According to Lyon's Hypothesis (10) carriers for CAG should have two populations of B lymphocytes: one with sIg and normal ecto-5'-NT activity and one which is blocked in maturation and which lacks sIg and ecto-5'-NT activity. Low activity of ecto-5'-NT in newly initiated lymphoblast lines derived from carriers for CAG implies EBV transformation of both types of B lymphocytes. However, the cells lacking ecto-5'-NT activity appear to be at a selective disadvantage since over a period of 3 to 5 weeks, the growth rate of the culture increases and the activity of ecto-5'-NT increases until it reaches the normal range, as if cells with ecto-5'-NT activity gradually overgrew cells with little or no activity. Certainly, lymphoblastoid cell lines from additional carriers for CAG must be established and assayed for ecto-5'-NT activity in order to assess the validity of this heterozygote detection test. Experiments are also in progress to identify ecto-5'-NT positive and negative cells in the cultures derived from the presumed carriers.

Thus, the partial deficiency of ecto-5'-NT in PBMs from patients with CAG does not appear to be caused by a mutation in the structural gene for the enzyme. As further evidence for this hypothesis, ecto-5'-NT activities from lymphoblast lines derived from 2 sisters of a CAG patient were characterized with respect to pH optimum, K_M and V_{MAX}. (Table 1.) Even though the catalytic activity of ecto-5'-NT from line #1016 was 65 times greater than that in #1015, the enzyme from both cell lines exhibited essentially the same pH optimum and K_M for IMP as substrate. Although not conclusive proof, these data are consistent with low levels of structurally normal ecto-5'-NT in cell line #1015.

Although assessment of ecto-5'-NT activity in newly initiated lymphoblastoid cell lines appears to provide a means for the detection of heterozygous carriers for CAG, the data are perplexing in at least two respects. First, why is the ecto-5'-NT activity so low in the lymphoblast lines derived from the carriers (i.e. <10% of normal, rather than 50% as expected)? This implies either an excess of ecto-5'-NT negative lymphocytes in the carriers or preferential transformation of the defective cells by EBV. In either case, it is difficult to explain the inability to obtain lymphoblast lines from the CAG patients, or previous reports of an absence of EBV receptors on CAG lymphocytes (11).

TABLE 1

Characterization of Lymphoblastoid Cell Line Ecto-5'-Nucleotidase Activity

Cell Line	pH Optimum	V_{MAX}	K_M
Angela G. #1016	6.9-7.0	65 nmoles/hr/ 10^6 cells	28 μM
Sherry G. #1015	6.9-7.0	1.0 nmoles/hr/ 10^6 cells	11 μM

REFERENCES

1. N. L. Edwards, D. B. Magilavy, J. T. Cassidy, and I. H. Fox, Lymphocyte ecto-5'-nucleotidase deficiency in agammaglobulinemia. Science 201: 628 (1978).
2. G. R. Boss, L. F. Thompson, I. V. Jansen, R. D. O'Connor, H. L. Spiegelberg, T. A. Waldmann, and R. N. Hamburger, 5'-Nucleotidase activity in T and B lymphocytes of normal subjects and patients with congenital X-linked agammaglobulinemia. Fed. Proc. 38: 496 (1979).
3. M. Rowe, C. G. DeGase, T. A. E. Platts-Mills, G. L. Asherson, A. D. B. Webster, and S. M. Johnson, 5'-Nucleotidase of B and T lymphocytes isolated from human peripheral blood. Clin. Exp. Immunol. 36: 97 (1979).
4. R. S. Geha, F. S. Rosen, and E. Merler, Identification and characterization of subpopulations in human peripheral blood after fractionation on discontinuous gradients of albumin. The cellular defect in X-linked agammaglobulinemia. J. Clin. Invest. 52: 1726 (1974).
5. A. Böyum, A separation of leukocytes from blood and bone marrow. Scan. J. Clin. Lab. Invest. 21 (Suppl. 97): 77 (1968).
6. M. S. Weiner, C. Bianco, and V. Nusseuzweig, Enhanced binding of neuraminidase-treated sheep erythrocytes to human T lymphocytes. Blood 42: 939 (1973).
7. W. S. Sly, G. S. Sekhon, R. Kennett, W. F. Bodmer, and J. Bodmer, Permanent lymphoid lines from genetically marked lymphocytes: Success with lymphocytes recovered from frozen storage. Tissue Antigens 7: 165 (1976).
8. G. R. Boss, L. F. Thompson, H. L. Spiegelberg, T. A. Waldmann, R. D. O'Connor, R. N. Hamburger, and J. E. Seegmiller, Lymphocyte ecto-5'-nucleotidase activity as a marker of B-cell maturation. Trans. Assoc. Am. Phys., in press (1979).
9. E. R. Stiehm, Fetal defense mechanisms. Am. J. Dis. Child 129: 438 (1975).
10. R. C. Davidson, H. M. Nitowsky, and B. Childs, Demonstration of two populations of cells in the human female heterozygous for glucose-6-phosphate dehydrogenase variants. Proc. Natl. Acad. Sci. USA 50: 481 (1963).
11. A. R. Hayward and M. F. Greaves, Central failure of B-lymphocyte induction in pan-hypogammaglobulinemia. Clin. Immunol. and Immunopath. 3: 461 (1975).

ACKNOWLEDGEMENTS

This work was supported in part by NIH Grants GM17702 and AM13622 and the Kroc Foundation. Linda Thompson is the recipient of a postdoctoral fellowship from the Arthritis Foundation. The authors wish to acknowledge Dr. Richard O'Connor for collecting blood samples, Dr. Hans Spiegelberg for B cell sIg determinations, and the excellent technical assistance of Inga Jansen and Greg Schmunk.

ADENOSINE DEAMINASE AND PURINE NUCLEOSIDE PHOSPHORYLASE IN ACUTE AND CHRONIC LYMPHATIC LEUKEMIA

Heinz Ludwig, Heide Winterleitner, Rudolf Kuzmits, and Mathias M. Müller
Department of Internal Medicine II and Institute of Medical Chemistry, University of Vienna and St. Anna-Children-Hospital Vienna

INTRODUCTION

Significant progress in the classification of leukemia has been achieved by the introduction of methods which make detailled characterisation of cell membrane antigens or receptors possible (1). These methods have proved to be of considerable clinical relevance as tools for precise classification of the underlying leukemic process, thus providing more concise evaluation of prognosis and often serving as valuable guide for therapy. However, by these techniques the biochemical properties inherent in the individual leukemic cell clone cannot be analysed. These biochemical characteristics of neoplastic cells are the prerequisits for better understanding of the leukemic process and may furthermore be used in the development of new cytostatic agents and the establishment of new chemotherapeutical schemes (2). In addition, the analysis of the metabolism of leukemic cells might possibly allow prediction of sensitivity or resistance of the individual leukemic clone to certain cytostatic agents.

The purine interconversion system is one of the metabolic pathways most important for lymphocyte reactivity, since normal lymphocytes lack the capacity for purine de novo-synthesis (3). Thus, it was of interest to study activities of adenosine deaminase (ADA) and purine nucleoside phosphorylase (PNP), two enzymes of the purine interconversion system, in patients with immunologically well characterized forms of acute and chronic lymphatic leukemia. Furthermore, as the acute form of lymphatic leukemia is no longer thought of representing a single disease entity (4) it should be elucidated whether determination of biochemical characteristics allows an even more precise classification of those acute lymphatic leukemias which are at present unclassifiable by immunological techniques.

MATERIAL AND METHODS:

25 patients with chronic lymphatic leukemia (CLL), 17 patients with acute lymphatic leukemia (ALL) and 23 healthy controls were studied. All CLL-patients presented with B-lymphocyte leukemia and did not receive cytostatic treatment for a minimum of 4 weeks before testing. The ALL-patients were studied shortly after diagnosis and before any initiation of chemotherapy. 10 presented common acute lymphatic leukemia, 4 were unclassifiable and 3 presented T-cell lymphatic leukemia. The sheep erythrocyte rosette technique was used for identification of T-lymphocytes, a direct immunofluorescence technique using Fab_2 anti-human immunoglobulin conjugates for determination of B-lymphocytes and an anti-common ALL-antiserum (kindly provided by Dr. Greaves) for characterisation of common ALL-determinants. Leukemic lymphocytes, which could not be classified by one of the above mentioned techniques were designated as unclassifiable.

Enzyme activities of adenosine deaminase and of purine nucleoside phosphorylase were measured according to Uitendaal et al., (5) and Uitendaal et al., (6), respectively. In short: Homogenates of 3×10^3 (for PNP) or 3×10^4 lymphocytes (for ADA) were incubated for 20 minutes at optimal conditions with specific ^{14}-C radiolabelled substrate. (The Radio Chemical Center, Amersham, UK). Subsequently, the amount of radioactive conversion products were determined in a liquid scintillation counter (Nuclear Chicago, USA) after separation from the non-converted ^{14} C-labelled substrate. Separation was achieved by either high voltage electrophoresis for ADA or paper chromatography for PNP. Enzyme activities were calculated in nMol conversion products/10^6/cells/hour. In the CLL-patients the Student's-t-test was used for comparison of log transformed values; in the ALL-patients Pukey's test was used.

RESULTS:

In the 25 CLL and 17 ALL-patients a statistically significant decrease of PNP-activities ($p < 0{,}05$, $p < 0{,}01$, respectively) was observed (Table 1). Likewise ADA-activities were significantly reduced in the total group of ALL-patients ($p < 0,05$) and decreased at levels of borderline significancy in patients with CLL ($p < 0,05$). The median of PNP-activities was 345,0 in the CLL-patients, 138,0 in the entire group of children with ALL and 444,0 in the controls. The respective medians for ADA-activities were 192,0, 135,0 and 324,0.

Table 1. Adenosine deaminase and purine nucleoside phosphorylase activity in patients with chronic or acute lymphatic leukemia and in controls

PATIENTS	number tested	Adenosine Deaminase (median)	Purine Nucleoside Phosphorylase (median)
Chronic lymphatic leukemia	25	192,0 [a]	345,0 [d]
Acute lymphatic leukemia	17	135,0 [b]	138,0 [e]
Controls	23	324,0 [c]	444,0 [f]

a : c = $p < 0,05$　　d : f = $p < 0,05$
b : c = $p < 0,05$　　e : f = $p < 0,01$

In 3 patients with T-ALL and 1 patient with unclassifiable acute lymphatic leukemia adenosine deaminase was markedly increased whereas in 10 patients with common ALL and in the 4 patients with unclassifiable ALL ADA-levels were found within normal or below normal range (Table 2). By contrast, purine nucleosid-phosphorylase activities did not differ within the different groups of ALL-patients.

Table 2. Adenosine deaminase in different subtypes of acute lymphatic leukemia

	Acute lymphatic leukemia: T-lymphocyte n=3	common n=10	unclassifiable n=4
Adenosine Deaminase	696.0 [a]	113,5 [b]	132,5 [c]

a : b = $p < 0,001$
a : c = ns.

DISCUSSION:

The data obtained in the present study show a significantly decreased PNP-activity in the entire group of ALL-patients and in patients with CLL. In addition,they confirm the reduced ADA levels observed in patients with CLL (7,8) and also in those with ALL as has been recently described (9). However, they do not accord with data observed in studies by Smith and Harrap (10) and Smith et al. (11), where ADA activity was found to be significantly increased in two groups of ALL patients. These discrepancies might be explained by a predominance of children with T-ALL in the latter studies. In the present investigation a marked heterogeneity of ADA activities was observed in the entire group of ALL patients. ADA was found to be significantly higher in the patients with T-ALL compared to those with common or unclassifiable acute lymphatic leukemia. These findings are in agreement with observations obtained in different lymphoid cells which have shown that normal T-lymphocytes and T-lymphoblastoid cell lines have significantly higher ADA activities than other lymphocyte subpopulations (8,12).

The finding of one patient with immunologically unclassifiable ALL and high ADA activities indicates that lymphatic cell clones often retain their enzymatic potential even when neoplastic transformation does occur. This patient seems to suffer from a malignant proliferation of lymphoid cells with biochemical characteristics of T-lymphocytes which lack the sheep erythrocyte receptor. However, on behalf of enzymatic studies he probably should be classified as T-ALL. Interestingly, for PNP no such heterogeneity in the different ALL subgroups and no association between high PNP activities and T-lymphocyte cell clones was observed.

In conclusion, the present investigation indicates that determination of ADA-levels may allow more precise characterisation of some patients with hitherto unclassifiable acute lymphatic leukemia, which in case of increased ADA-levels probably can be regarded as leukemias of the T-lymphocyte subtype.

LITERATUR:

1. Koziner B., Filippa DA, Mertelamann R., Gupta S., Clarkson B., Good RA, Siegal FP: Characterization of the malignant lymphomas in leukemic phase by multiple differentiation markers of mononuclear cells. Am.J.Med. 63, 556, 1977.

2. Scholar EM, Calabresi P.: Identification of the enzymatic pathways of nucleotide metabolism in human lymphocytes and leukemia cells. Cancer Res. 33, 94, 1973.

3. Allison AC., Hovi T., Watts RWE., Webster ADB.: The role of de novo-purine synthesis in lymphocyte transformation. In: Purine and Pyrimidine Metabolism. Ciba Foundation Symposium 48 pp 207, Elsevier, Exerpta Medica-North Holland. Amsterdam (1977).

4. Tsukimoto I., Wong KY., Lampkin BC.: Surface markers and prognostic factors in acute lymphoblastic leukemia. N.Engl.J.Med. 294, 245 (1976)

5. Uitendaal MP., De Bruyn CHMM., Oei TL., Hösli P.: Ultramicrochemical studies on enzyme kinetics. Adv.Exp.Med.Biol. 76A, 597 (1977)

6. Uitendaal MP., De Bruyn, CHMM., Oei TL., Hösli P.: Characterisation of purine nucleoside phosphorylase from fibroblast using ultra-micro-chemical methods. Human Heredity 28, 151 (1978)

7. Ramot B., Brok-Simoni F., Barnea N., Bank I., Holtzmann F.: Adenosine deaminase (ADA) activity in lymphocytes of normal individuals and patients with chronic lymphatic leukemia. British Journal of Haematology 36, 66 (1977).

8. Tung R., Silber R., Quagliata R., Conklyn M., Gottesman J., Hirschhorn R.: Adenosine deaminase activity in chronic lymphatic leukemia. Journal of Clinical Investigation 57, 756 (1976).

9. Zimmer J., Khalifa AS., Lightbody JJ.: Decreased adenosine deaminase activity in acute lymphocytic leukemic children and their parents. Cancer Res.35, 68 (1975)

1o.Smyth JF., Poplack DG., Holiman BJ., Leventhal BG.: Correlation of adenosine deaminase activity with cell surface markers in acute lymphoblastic leukemia. J.Clin.Invest. 71O, 1978

11.Symth JF., Harrap KR: Adenosine deaminase activity in leukemia. Br.J.Cancer 31, 544 (1975)

12. Sullivan JL., Osborne WRA., Wedgwood RJ.: Adenosine activity in lymphocytes. Brit.J.Haematol. 38, 15 (1978)

PURINE SALVAGE ENZYMES IN LYMPHOCYTES AND GRANULOCYTES FROM PATIENTS WITH SMALL-CELL CARCINOMA OF THE LUNG

Per Nygaard and Johannes Mejer,

University Institute of Biological Chemistry B
Sølvgade 83,
Department of Medicine C and Bloodbank,
Bispebjerg Hospital, Copenhagen, Denmark

An impaired immune function is frequently seen in patients with malignant diseases. This suggests that the cell-mediated immune response plays a role in the defence against the development of such diseases. Disordered immune function has been associated with specific defects in some purine enzymes: adenosine deaminase (ADA), purine nucleoside phosphorylase (PNP), and 5'-nucleotidase (5'N)(1). The association between these enzyme deficiencies and the immune disorder has been explained to be due to an accumulation of purine nucleosides and nucleotides, which exert toxic effects especially on lymphoid cells. Deficiencies in other purine enzymes such as hypoxanthine phosphoribosyltransferase (HGPRT) and adenine phosphoribosyltransferase (APRT) have not been associated with immunological abnormalities.

Reports from numerous laboratories have shown that levels of purine enzymes especially ADA are altered in lymphocytes from patients with various diseases. Thus both reduced ADA activity (2) and increased activity (3) have been reported in lymphocytes of individuals with various types of solid tumors. Studies on human leukemia have revealed great variations in the level of ADA (4). It appears likely that regulation of purine salvage and degradation (excretion) is exerted, in part at least, through a control of the activity of certain key enzymes. Some of these enzymes are essential for normal lymphocyte function. These ideas prompted the present investigation of six purine enzymes in lymphocytes and granulocytes of patients with small-cell carcinoma of the lung.

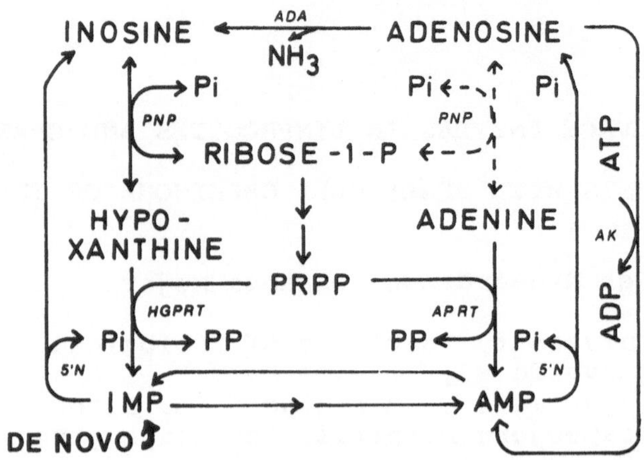

Figure I. Pathways and Enzymes of Purine Metabolism. Adenine phosphoribosyltransferase (EC 2.4.2.7) APRT; adenosine kinase (EC 2.7.1.20) AK; 5'-nucleotidase (EC 3.1.3.5) 5'N; adenosine deaminase (EC 3.5.4.4) ADA; purine nucleoside phosphorylase (EC 2.4.2.1) PNP; and hypoxanthine phosphoribosyltransferase (EC 2.4.2.8) HGPRT.

MATERIALS AND METHODS

Patients: Newly diagnosed patients were studied, all had microscopically-proven small-cell (oat-cell) carcinoma of the lung (5). The patients were admitted to the Department of Internal Medicine C, Bispebjerg Hospital, Copenhagen. None of the patients had received cytostatic treatment before collection of their leukocytes. Eighteen donors with normal leukocyte, differential counts, and hemoglobin age 27-60 years were used as controls.

Cell Collection: Mononuclear cells (lymphocytes and monocytes) and granulocytes were harvested from heparinized peripheral blood after separation on a Ficoll-Isopaque gradient (6). The proportion of lymphocytes in the separated monuclear layer from both controls and patients exceeded 95 %, the rest being monocytes, some samples contained a small proportion <2 % of granulocytes.

The granulocyte fraction contained more than 94 % granulocytes.

Enzyme Analysis: The separated cells were frozen and thawed by resuspension in 0.1 M Tris-Cl pH 8.1. The cells were disrupted by sonification and were centrifuged at 8000xg for 5 min, the supernatant was used for enzymatic analysis. Purine nucleoside phosphorylase and adenosine deaminase were determined spectrophotometrically (7). Adenine phosphoribosyltransferase, hypoxanthine phosphoribosyltransferase, adenosine kinase and 5'-nucleotidase (1) were measured in radiochemical assays (7). Activities are expressed as nmole substrate converted per hour per mg protein at 37°. Protein was determined according to Lowry (8).

Statistics: Mann-Whitney rank sum test and Spearman test.

TABLE 1

Clinical Data from 11 Patients with Small-Cell Bronchogenic Carcinoma

pt	age (years) and sex	leukocytes (cells/μl)	% lymph./gran.	length of survival (days)
1	44 F	7850	15/76	479
2	75 M	7150	19/75	222
3	65 F	12300	15/80	487
4	71 M	8100	23/68	445
5	56 M	15600	13/81	510
6	56 M	12300	13/76	401
7	65 M	9550	26/58	243
8	45 M	13850	13/82	297
9	64 M	5850	25/69	369
10	61 M	12650	25/69	40
11	55 F	5250	22/68	355

RESULTS

Six purine enzymes (Fig. I) were determined in both lymphocytes and granulocytes of patients with small-cell carcinoma of the lung and in controls. Some clinical data are shown in Table 1. The leukocyte counts of most of the patients revealed numbers which were above the normal range.

The activities of the purine enzymes in the patients and the control group are shown in Fig. II. A comparison of the enzyme data with results from the control group is illustrated in Table 2. The most salient feature is that PNP activity in lymphocytes of the patients is significantly higher than the activity found in the control group. Furthermore the activity of APRT and 5'N in granulocytes from the patients differs from the levels in the control group. No correlation was found between enzyme levels and survival time.

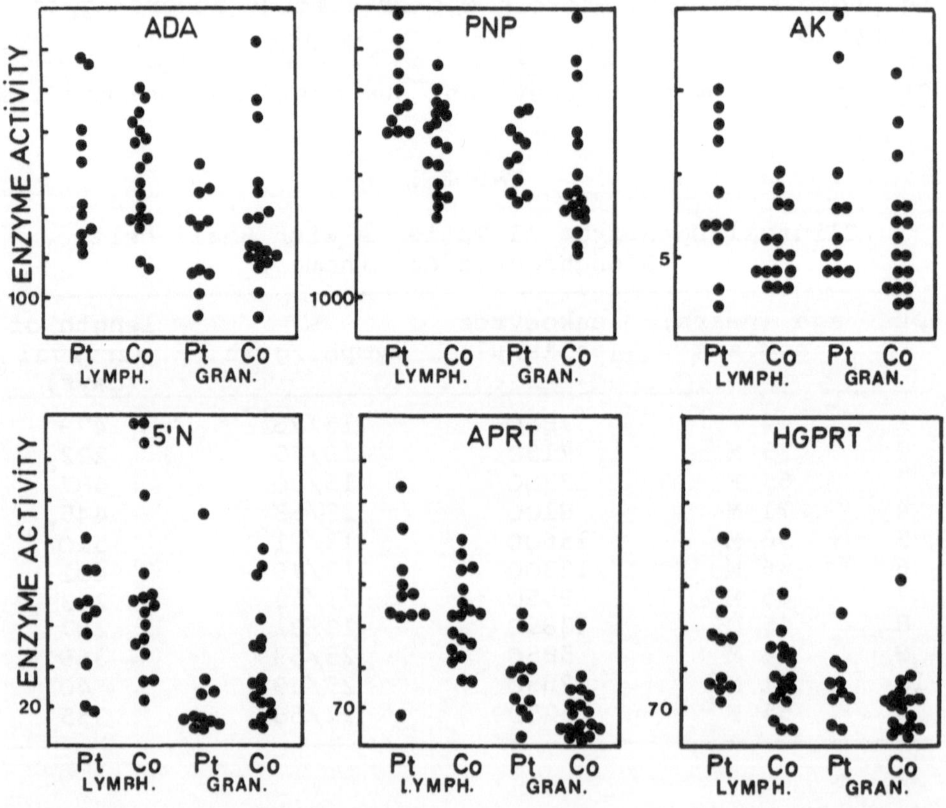

Figure II. Purine enzyme activities of lymphocytes and granulocytes from patients (Pt) with small-cell bronchogenic carcinoma and from control persons (Co).

TABLE 2

Comparison of Purine Enzyme Values between Normals and Patients with Small-Cell Carcinoma of the Lung. Mean Values ± S.D. (nmoles/mg protein per hour)

Enzyme	Cell type	Patients	Controls	P
ADA	lymphocytes	395 ± 169	407 ± 139	
	granulocytes	239 ± 117	291 ± 170	
PNP	lymphocytes	6064 ± 926	4700 ± 1069	<0.01
	granulocytes	4331 ± 814	3957 ± 1622	
APRT	lymphocytes	262 ± 98	222 ± 71	
	granulocytes	117 ± 55	69 ± 50	<0.01
HGPRT	lymphocytes	178 ± 85	144 ± 89	
	granulocytes	108 ± 57	84 ± 61	
AK	lymphocytes	9 ± 4	5 ± 2	
	granulocytes	8 ± 6	6 ± 3	
5'N	lymphocytes	61 ± 27	75 ± 44	
	granulocytes	27 ± 30	41 ± 27	<0.05

DISCUSSION

The results presented demonstrate differences in PNP levels of lymphocytes and APRT and 5'N of granulocytes between normal subjects and patients with small-cell bronchogenic carcinoma. This may reflect a heterogeneity of the circulating cells. The increased levels observed of purine metabolizing enzymes appear consistent with making possible a greater flux through the purine salvage pathways in lymphocytes and granulocytes of the patients. The increased activity of PNP in the lymphocytes of the patients also provide a greater capacity for channeling pentose phosphates into phosphoribosylpyrophosphate (PRPP), which is needed for both purine- and pyrimidine nucleotide synthesis. The physiological significance of such a role for PNP in the generation of PRPP is not known.

The inherited deficiency of PNP which is associated specifically with a severe T cell deficiency indicates a specific role of PNP in T cells. Histochemical studies of lymphocytes have revealed that PNP is predominantly localized in the T cells (9). The percentage of T-cells in patients with small-cell bronchogenic carcinoma is significantly reduced (10). One would therefore expect a decrease in lymphocyte PNP and ADA activity rather than elevated PNP levels and unchanged ADA levels, as has been observed in this investigation Fig. II. An altered ratio between T and B cells would also imply alterations in ADA activity since T cells have higher ADA levels than B cells (11). In patients with hypogammaglobulinemia with subnormal levels of lymphocyte 5'N no significant alterations in the activity of PNP, ADA, HGPRT or APRT were observed (1). Lymphocytes from patients with Waldenström's macroglobulinæmia contain increased ADA activity, while the level of PNP was not elevated (12). However, in two other haematological diseases chronic lymphatic leukemia (CLL) and Hodgkin's disease (HD), which also can be viewed as immunodeficiency diseases, ADA and PNP are altered. In HD increased PNP has been measured in the peripheral lymphocytes, while decreased PNP and ADA levels were found in CLL (13). In other haematological diseases great variations in ADA levels have been seen in leukemia cells by several investigators, however, PNP levels did not vary significantly from the level of the normal lymphocytes, except for cases of CLL, where PNP activity was low (4,11).

The reason for the alterations of purine salvage enzymes is not known. Certainly variations in the stage of the immunological maturation of the circulating lymphocytes, but also a specific enzyme induction could explain these variations. Whether the immunostimulating effect of inosine (14) bears any relationship to the level of PNP in the lymphocytes is not known, neither is it known whether lymphocytes, like certain membranes of the brain (15), process binding sites for inosine.

Acknowledgments

This work was supported by Christian den X's Fond. Many thanks are due to Dr. Per Dombernowsky for permission to study patients under his care.

References

(1) Webster, A.D.B., North, M., Allsop, J., Asherson, G.L., and Watts, R.W.E., Clin. Exp. Immunol. 31:456 (1978).

(2) Uberti, J., Johnson, R.M., Talley, R., and Lightbody, J.J., Cancer Res. 36:2046 (1976).

(3) Formeister, J.F. and Tritsch, G.L., Surgery 79:111 (1976)

(4) Mejer, J. and Nygaard, P., Leuk. Res. (1979) in press.

(5) Dombernowsky, P. and Hansen, H.H., Acta Med. Scand. 204:513 (1978).

(6) BØum, A., Tissue Antigens 4:269 (1974).

(7) Mejer, J. and Nygaard, P., in Inborn Errors of Immunity and Phagocytosis, MTP Press Ltd. England 181 (1979).

(8) Lowry, O.H., Rosebrough, N.J., Farr, A.L., and Randall, R.J., J. Biol. Chem. 193:256 (1951).

(9) Borgers, M., Verhaegen, H., De Brabander, M., Thone, F., Van Reempts, J., and Geuens, G., J. Immunol. 16:101 (1977).

(10) Gross, R.L., Latty, A., Williams, E.A., and Newberne, P.M., New Engl. J. Med. 292:439 (1975)

(11) Tung, R., Silber, R., Quagliata, F., Conklyn, M., Gottesman, J., and Hirschhorn, R., J. Clin. Invest. 57-756 (1976).

(12) Sidi, Y., Boer, P., Pick, I., Pinkhas, J., and Sperling, O., Lancet. 500 (1979).

(13) Ambrogi, F., Grassi, B., Ronca-Testoni, S., and Ronca, G., Clin. Exp. Immunol. 28:80 (1977).

(14) Hadden, J.W., in The Pharmacology of Immunoregulation (ed. G.H. Werner & F. Floch) Acad. Press New York 369 (1978).

(15) Asano, T. and Spector, S., Proc. Natl. Acad. Sci. (USA) 76:977 (1979).

TREATMENT OF ACUTE LYMPHOBLASTIC LEUKEMIA WITH THE ADENOSINE DEAMINASE INHIBITOR 2'-DEOXYCOFORMYCIN

Beverly S. Mitchell, Charles A. Koller, and William N. Kelley

Departments of Internal Medicine and Biological Chemistry, Human Purine Research Center, University of Michigan Medical School, Ann Arbor, Michigan U.S.A.

Congenital deficiency of the enzyme adenosine deaminase (ADA, EC 3.5.4.4.) is associated with severe combined immunodeficiency disease characterized by marked lymphopenia and thymic involution.[1] ADA catalyzes the deamination of adenosine and deoxyadenosine to inosine and deoxyinosine, respectively, and it has been proposed that the increased levels of deoxyadenosine associated with ADA deficiency lead to the selective accumulation of dATP by lymphoid cells, inhibition of DNA synthesis and cell death.[2]

We[3] and others[4] have documented that cultured T-lymphoblasts derived from patients with acute lymphoblastic leukemia (ALL) are particularly sensitive to the toxic effects of deoxyadenosine in the presence of an ADA inhibitor. These studies provide the rationale for the use of an ADA inhibitor as a selective chemotherapeutic agent in the treatment of lymphoproliferative malignancies. We have treated a patient with refractory ALL with 2'-deoxycoformycin, a potent and specific inhibitor of ADA activity.[5]

At the time of initial diagnosis, the patient had a white blood count of 134,000/mm^3 with 96% lymphoblasts. The leukemic cells were characterized as null cells on the basis of lack of E or EAC rosette formation and lack of surface immunoglobulins; they did, however, develop the capability of forming E rosettes after 48 hrs. in culture in RPMI medium with 10% fetal calf serum.

Following multiple unsuccessful courses of chemotherapy, the patient was treated with 2'-deoxycoformycin at a dose of 0.25 mg/kg/day for a total of ten days over a two-week period. Allopurinol, 300 mg/day, was given throughout the study. The administration of

2'-deoxycoformycin was well tolerated and was not associated with any changes in hepatic or renal function. Serum uric acid remained below 6 mg %. The patient developed an initial leukopenia associated with an increasing percentage of lymphoblasts (Figure 1). The bone marrow remained hypercellular with >95% blasts. During the second week, the white count rose to 6700/mm^3 with 80% blasts and therapy was discontinued. Serum immunoglobulins remained within the normal range throughout the study. A positive PHA skin test (8 mm) turned negative at the end of the first week.

Sequential measurements of ADA activity in buffy coat cells, mononuclear cells and hemolysate showed a complete inhibition of ADA activity within 1 hr. of drug administration. The loss of activity was sustained over the subsequent fourteen days. ADA activity in the bone marrow cells was similarly inhibited, but plasma adenosine concentrations remained undetectable (<1 μM). Erythrocyte dATP levels rose from <20 pmol/ml packed rbc to 2250 pmol/ml on the fifth day of therapy.

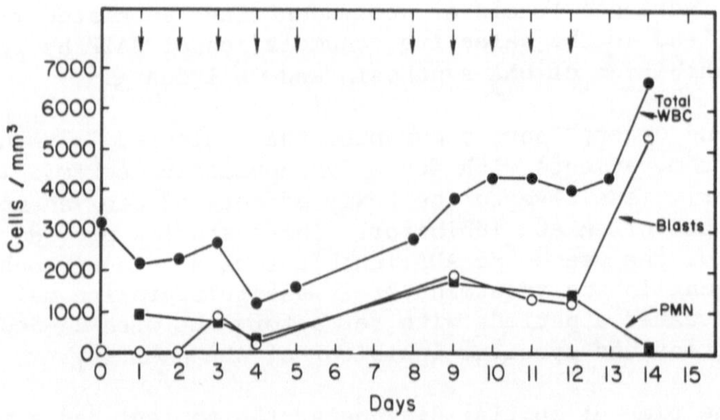

Fig. 1. Alterations in total white count, lymphoblasts and polymorphonuclear cells during 2'-deoxycoformycin treatment. Each dose of 2'-deoxycoformycin is marked by an arrow.

Table 1

Accumulation of dATP by leukemic cells before and after therapy with 2'-deoxycoformycin

Additives	dATP (picomoles/10^6 cells) Before Rx	Day 6
None	< 5	9
50 μM AdR	8	35
1 μM 2'-DCF	< 5	8
5 μM 2'-DCF	< 5	10
50 μM Adr + 1 μM 2'-DCF	31	48
50 μM AdR + 5 μM 2'-DCF	49	52

Leukemic lymphoblasts were incubated with 2'-deoxycoformycin and/or deoxyadenosine prior to and following the first week of therapy (Table 1). dATP levels were not detectable in 10^6 lymphoblasts incubated for 1 hr. alone or with deoxycoformycin. In pretreatment lymphoblasts in the presence of both 50 μM deoxyadenosine and 5 μM deoxycoformycin, dATP levels reached 49 pmol/10^6 cells. After treatment, the leukemic cell dATP level reached only 9 mol/10^6 cells, but incubation with 50 μM deoxyadenosine (without 2'-deoxycoformycin) increased dATP levels to 35 pmol/10^6 cells.

We conclude from these observations that the circulating deoxyadenosine concentrations in our patient were insufficient to result in maximal dATP accumulation in the leukemic cells. Further support for this conclusion may be derived from the slowly rising red cell dATP levels, which reached only 2,250 pmol/ml after 5 days of treatment. Alternative approaches to the use of 2'-deoxycoformycin should include more prolonged administration of the drug and/or supplemental administration of deoxyadenosine.

The in vitro accumulation of dATP by leukemic cells may be predictive of the efficacy of this form of therapy and is a potentially useful parameter by which to monitor both ADA inhibition and serum deoxyadenosine concentrations. The level of dATP necessary to result in inhibition of DNA synthesis in these cells remains to be determined. Despite the lack of response in our patient, we feel that further trials of 2'-deoxycoformycin therapy in refractory hematologic malignancies are warranted.

REFERENCES

1. E. R. Giblett, J. E. Anderson, F. Cohen, B. Pollara, and H. J. Meuwissen, Adenosine deaminase deficiency in two patients with severely impaired cellular immunity Lancet 2:1067 (1972).

2. D. A. Carson, J. Kaye and J. E. Seegmiller, Lymphospecific toxicity in adenosine deaminase deficiency and purine nucleoside phosphorylase deficiency: possible role of nucleoside kinase(s) Proc. Natl. Acad. Sci. U.S.A. 74:5677 (1977).

3. B. S. Mitchell, E. Mejias, P. E. Daddona, and W. N. Kelley, Purinogenic immunodeficiency diseases: selective toxicity of deoxyribonucleosides for T cells Proc. Natl. Acad. Sci. U.S.A. 75:5011 (1978).

4. D. A. Cason, J. Kaye, and J. E. Seegmiller, Differential sensitivity of human leukemic T cell lines and B cell lines to growth inhibition by deoxyadenosine J. Immunol. 121:1726 (1978).

5. R. P. Agarwal, T. Spector, and R. E. Parks, Tight-binding inhibitors - IV. Inhibition of adenosine deaminase by various inhibitors Biochem. Pharm. 26:359 (1977).

INCREASE OF PHOSPHORIBOSYLPYROPHOSPHATE LEVELS IN CULTURED L1210 LEUKEMIA CELLS EXPOSED TO METHOTREXATE[+]

J.M. Buesa, A. Leyva and H.M. Pinedo

Section of Chemotherapy, General Hospital of Asturias, Oviedo, Spain (J.M.B.), Section of Experimental Chemotherapy, Antoni van Leeuwenhoek Institute, (J.M.B., A.L., H.M.P.) and Department of Oncology, Free University Hospital (A.L., H.M.P.), Amsterdam, The Netherlands

INTRODUCTION

MTX[‡] is a folic acid analog that binds to the enzyme dihydrofolate reductase and inhibits its activity depleting the cells of reduced folates. It blocks the synthesis of dTMP from dUMP and de novo purine biosynthesis by decreasing the availability of reduced folates [1]. MTX effects on cell metabolism can be reversed not only by folinic acid (5-formyltetrahydrofolic acid) but also by either TdR alone or TdR plus a purine base or nucleoside [2-6] which provide for nucleotide synthesis through salvage pathways. Purine bases are converted to nucleotides by phosphoribosyltransferases in the presence of PRPP as the phosphoribosyl donor and TdR phosphorylation to dTMP by TdR kinase requires ATP. MTX has been demonstrated to cause a rapid decrease of intracellular ATP levels, which could restrict the cellular utilization of TdR [7]. The purpose of our study was to evaluate the effect of MTX on the intracellular levels of PRPP in order to assess the availability of PRPP for purine salvage during MTX treatment.

+ Supported in part by the Queen Wilhelmina Fund (Project No. UUKC 77-3)

‡ Abbreviations used are : MTX, methotrexate; dTMP, thymidylate; dUMP, deoxyuridylate; PRPP, phosphoribosylpyrophosphate; Hyp, hypoxanthine; TdR, thymidine.

METHODS

L1210 mouse leukemia cells were grown in RPMI 1640 culture medium supplemented with 10% dialyzed fetal calf serum. Cells in logarithmic growth at a concentration of 3-5 x 10^5/ml were used to test the different conditions. Before each experiment cells were counted with a hemacytometer and cell viability determined by the trypan blue exclusion test. Cell extracts were prepared by cold perchloric acid precipitation [8] and PRPP was determined by a radiochemical assay measuring the production of C^{14}-AMP from C^{14}-adenine in the presence of excess adenine phosphoribosyltransferase activity [9]. Cellular PRPP content was expressed as pmoles/10^6 cells, and values were corrected for 60% recovery of PRPP during extraction and were based on the number of viable cells. PRPP synthetase activity was determined in dialyzed cell extracts using a coupled enzyme assay similar to the method described by Fox and Kelley[9].

RESULTS

After 6 hr of exposure to 0.01, 0.1 or 1 µM MTX, cell viability was 94, 92 and 84% respectively, decreasing to 85, 67 and 60% at 12 hr and to 60, 38 and 34% after 24 hr. Less than 7% of cells were viable after 48 hr of exposure to MTX.

Untreated cells in logarithmic growth contained PRPP levels which varied during 24 hr, with values of 145 ± 65 pmoles/10^6 cells (mean ± S.D.). In cells exposed to 0.1 and 1 µM MTX, PRPP content increased and was maximal within 3 hr after MTX addition and declined to control values after 12 hr (Table 1). In cells exposed

Table 1. Methotrexate induced changes in PRPP content of L1210 cells. Cells were grown in the presence or absence of different concentrations of MTX and PRPP content determined at different time periods.

	PRPP (pmoles/10^6 cells)			
	3 hr	6 hr	12 hr	24 hr
Untreated	198 ± 11[a]	89 ± 21	166 ± 15	67 ± 12
0.01 µM MTX	161 ± 13	454 ± 29	983 ± 19	781 ± 68
0.1 µM MTX	939 ± 51	713 ± 140	628 ± 38	74 ± 11
1 µM MTX	850 ± 52	551 ± 167	413 ± 49	51 ± 9

[a] mean ± S.E.

Table 2. PRPP synthetase activity in L1210 cells after exposure to MTX. Conditions as in Table 1.

	PRPP synthetase activity (nmoles/mg protein/hr)		
	3 hr	12 hr	24 hr
Untreated	376 ± 24[a]	302 ± 49	391 ± 18
0.01 μM MTX	243 ± 31	305 ± 39	332 ± 75
0.1 μM MTX	318 ± 8	273 ± 58	244 ± 32
1 μM MTX	338 ± 25	172 ± 51	271 ± 33

[a] mean ± S.E.

to 0.01 μM MTX, cellular PRPP content reached maximal values 12 hr after exposure with values still higher than controls by 24 hr.

PRPP synthetase activity (Table 2) in untreated cells varied between 258 and 391 nmoles/mg protein/hr during 24 hr logarithmic growth. There was little change in enzyme activity in the presence of 0.01 μM MTX and with 0.1 and 1 μM MTX the activity decreased to 55-70% of control levels after 12 hr exposure to MTX.

Different conditions were tested in the presence or absence of 0.1 μM MTX. The addition of 0.1 mM folinic acid to the culture medium prevented the increase in PRPP levels after MTX addition, while folinic acid alone did not show any effect on cellular PRPP content. The addition of 0.1 mM Hyp alone decreased PRPP content to 20% of control values. When 0.1 mM Hyp was added either at 0, 3, 6, or 9 hr after MTX, cellular PRPP content also decreased to 20% or less of controls. The effect of a temporary exposure (1.5 hr) to 0.1 mM Hyp was also tested (Table 3). When Hyp was added after 1.5 hr MTX exposure PRPP values decreased and returned to that of MTX-treated controls within 3 hr. However, when Hyp was added 10 hr after MTX exposure, PRPP values decreased and only recovered to those of untreated controls.

DISCUSSION

The results of our experiments show that in L1210 mouse leukemia cells PRPP content increases after exposure to 0.01 - 1 μM MTX. Cellular PRPP content is increased most likely due to decreased utilization in de novo purine biosynthesis when the latter is blocked by MTX. The addition of 0.1 mM folinic acid prevented the PRPP increase in the presence of 0,1 μM MTX supporting that conclusion. Cells in these experiments were grown in medium supplemented with dialyzed fetal calf serum that provides purine- and pyrimidine-free

Table 3. Effect of temporary exposure to hypoxanthine on MTX enhanced PRPP levels. During continuous exposure of cells to 0.1 µM MTX, 0.1 mM Hyp was added for 1.5 hr and removed, and PRPP determined at different times.

	PRPP (pmoles/10^6 cells)	
Time (hr) after addition of MTX	0.1 µM MTX	0.1 µM MTX plus 0.1 mM Hyp
	Hypoxanthine added after 1.5 hr MTX exposure	
0	239 ± 13[a]	-
3	972 ± 73	23 ± 6
7	816 ± 41	621 ± 44
9	788 ± 15	873 ± 73
	Hypoxanthine added after 10 hr MTX exposure	
11.5	1491 ± 28	20 ± 15
14	1199 ± 11	121 ± 13
16	872 ± 39	70 ± 1
18	716 ± 32	106 ± 28

[a] mean ± S.E.

conditions. Under these conditions PRPP accumulation in the absence of purine salvage activity can be examined. When Hyp was added tc the culture medium PRPP values decreased to 20% of controls both in MTX-treated and untreated cultures indicating that the lack of purine salvage activity contributes to the MTX-induced enhancement of PRPP levels.

The specific activity of PRPP synthetase in L1210 cells did not change significantly during the first 12 hr of MTX exposure. Cellular PRPP synthesis was examined by addition of Hyp, which markedly reduced PRPP levels and measuring PRPP after subsequent removal of Hyp. Complete recovery of PRPP levels was observed after 1.5 hr but not after 10 hr of MTX exposure. These findings suggest that PRPP synthesis decreases after prolonged treatment of cells with MTX despite the continuous presence of PRPP synthetase. Although changes in the regulation of the enzyme activity can not be excluded from our data, availability of substrates, ATP and ribose 5 phosphate, could be a limiting factor for PRPP synthesis during MTX exposure. Hryniuk [7] found a 75% decrease in ATP content in L5178Y lymphoma cells, 6 hr after exposure to MTX and we have also observed a decrease of ATP in L1210 to 35% of control values, 5 hr after exposure to 0.1 µM MTX (J.M. Buesa, unpublished). Kaminskas et al.[10] observed a block in glucose consumption in Ehrlich ascites cells in vitro as a result of MTX exposure, which was believed to be a secondary effect resulting from ATP depletion of ribose 5 phosphate as well. The recovery of PRPP levels after removal of Hyp (Table 3) suggests that Hyp restores ATP cellular content. Hryniuk[7] found enhanced ATP levels in L5178Y lymphoma cells exposed to MTX plus Hyp. We also observed that ATP cellular content increased when Hyp was added 3 hr after MTX exposure of L1210 cells.

Enhanced PRPP levels could favor phophoribosyltransferase reactions. Cadman et al.[11] recently suggested that the increased uptake of 5-fluorouracil into L1210 cells after MTX exposure was a consequence of increased cellular PRPP levels. We also noted elevated PRPP levels in L1210 cells after inhibition of de novo purine biosynthesis by 0.1 mM 6-mercaptopurine riboside. The cytotoxicity of drugs depending on phosphoribosyltransferases for activation may be enhanced after treatment with MTX or other agents which block de novo purine biosynthesis. On the other hand, higher PRPP levels could favor purine salvage if purine bases are available reversing the antipurine effect and decreasing the cytotoxicity of MTX. Human purine plasma levels are probably sufficient to reverse the changes in purine metabolism induced by MTX as it has been shown that infusion of patients with TdR alone effectively prevents MTX toxicity [12].

Differential sensitivity of cell types to MTX could be in part related to their balance between de novo and salvage purine pathways and to the availability of substrates for salvage pathways. These factors should be considered in in vitro studies that are aimed to simulate in vivo conditions.

REFERENCES

1. J. R. Bertino, Folate Antagonists, in:"Antineoplastic Agents Part II," A. C. Sartorelli and D. G. Johns, ed., Springer-Verlag, Berlin (1975).
2. J. Borsa, and G. F. Whitmore, Cell Killing Studies on the Mode of Action of Methotrexate on L-cells in Vitro, Cancer Res. 29:737 (1969).
3. M. H. N. Tattersall, R. C. Jackson, S. T. M. Jackson, and K. R. Harrap, Factors Determining Cell Sensitivity of Methotrexate: Studies of Folate and Deoxyribonucleoside Triphosphate Pools in Five Mammalian Cell lines, Eur. J. Cancer 10:819 (1974).
4. W. M. Hryniuk, The Mechanism of Action of Methotrexate in Cultured L5178Y Leukemia Cells, Cancer Res. 35:1085 (1975).
5. H. M. Pinedo, D. S. Zaharko, J. M. Bull, and B. A. Chabner, The Reversal of Methotrexate Cytotoxicity to Mouse Bone Marrow Cells by Leucovorin and Nucleosides, Cancer Res. 36:4418 (1976).

6. A. Leyva, L. van de Grint, and H. M. Pinedo, Reversal of Methotrexate Toxicity to Mouse Bone Marrow and L1210 Leukemia Cells Grown in Vitro, in:"Clinical Pharmacology of Antineoplastic Drugs", H. M. Pinedo, ed., Elsevier/ North Holland Biomedical Press, Amsterdam (1978).
7. W. M. Hryniuk, L. W. Brox, J. F. Henderson, and T. Tamaoki, Consequences of Methotrexate Inhibition of Purine Biosynthesis in L5178Y Cells, Cancer Res. 35:1427 (1975).
8. T. Hisata, An Accurate Method for Estimating 5-Phosphoribosyl-1-pyrophosphate in Animal Tissues with the use of Acid Extraction, Anal. Biochem. 68:448 (1975).
9. I. H. Fox, and W. N. Kelley, Human Phosphoribosylpyrophosphate Synthetase, Distribution, Purification and Properties, J. Biol. Chem. 246:5739 (1971).
10. E. Kaminskas, and A. C. Nussey, Effects of Methotrexate and of Environmental Factors on Glygolysis and Metabolic Energy State in Cultured Ehrlich Ascites Carcinoma Cells, Cancer Res. 38:2989 (1978).
11. E. Cadman, C. Benz, and R. Heimer, Enhanced 5-Fluorouracil Nucleotide Formation Following Methotrexate is the Consequence of Increased Intracellular Phosphoribosylpyrophosphate. Proc. AACR. 20:258 (1979).
12. A. Leyva, J. Schornagel, H. M. Pinedo, High Performance Liquid Chromatography of Plasma Pyrimidines and Purines and its Application in Cancer Chemotherapy (See this volume).
13. S. B. Howell,W. D. Ensminger, A. Krishan, and E. Frei III, Thymidine Rescue of High-Dose Methotrexate in Humans, Cancer Res. 38:325 (1978).

PURINE SALVAGE PATHWAY IN LEUKEMIC CELLS

A. Goday, M.R. Grau, I. Jadraque, M.P. Rivera

Instituto de Farmacología, C.S.I.C.
Jorge Girona Salgado, s/n
Barcelona (34). España

INTRODUCTION

The availability of purine nucleotides in adequate intracellular concentrations is absolutely necessary for the survival of any animal cell. It is not known whether exogenous purines simply supplement the intracellular pools of purine nucleotides or whether they affect the proliferation in a more complex way[1].

The utilization by the cell of exogenous purines along the salvage pathway depends not only on the capacity of incorporation of these compounds into the intracellular medium but also on the kinetics in which this incorporation is carried out. Mouse leukemic lymphoblasts (L5178Y) incorporate guanine and guanosine by different uptake kinetics[2].

The present studies were undertaken to elucidate the metabolic pathway followed by guanine and guanosine, when the transport is working at initial rate. The other aim of this work is to study the manner in which the different kinetics are related to the intracellular metabolism.

METHODS

L5178Y cells were incubated with (8-^{3}H)guanine and (8-^{3}H) guanosine (0.5-100 uM) as described elsewhere[2].

The cold acid soluble (CAS) fractions were obtained with 7% TCA, 4ºC. After ether extraction of TCA, the CAS fractions were liophylized and diluted with water (50 ul).

Aliquots of cellular extracts were chromatographed by the following methods: (i) High pressure liquid chromatography[3], (ii) paper chromatography[4] and (iii) thin-layer chromatography.

Thin-layer chromatography: precoated TLC-plastic sheets of PEI-cellulose (20x20 cm. layer Thickness 0.1 mm) were used. Carrier solution, 5 ul., containing the required bases, nucleosides and nucleotides, and 20 ul. of diluted cell extracts, were applied.

Sheets were developed at room temperature with 50 mM acetic acid (60 min) followed, after drying, by a second run with destilled water (75 min), both runs being for the whole lengh of the sheet. After air drying, the nucleotides which had remained at the origin, were separated by a third run with 2.0 M sodium formiate buffer (pH 3.4) up to 6 cm.

The R_F values for the compounds separated are the following ones: GTP (0.03), GDP (0.10), GMP (0.19), Uric acid (0.34), Guanine (0.46), Xanthine (0.53), Xanthosine (0.61), Guanosine (0.74) and Inosine (0.87).

The plastic sheet was divided into rectangles containing the carrier spots, after they had been detected using a U.V. light source. They were scraped off and were shaken with 1 ml. of 1N HCl, for an hour (37ºC). The radioactivity was measured by liquid scintillation counting, and the recovery ranged from 75 to 90%[5,6].

RESULTS

As shown in Fig.1 the uptake of the base and the nucleoside follows different kinetics. From 10 uM upwards the amount

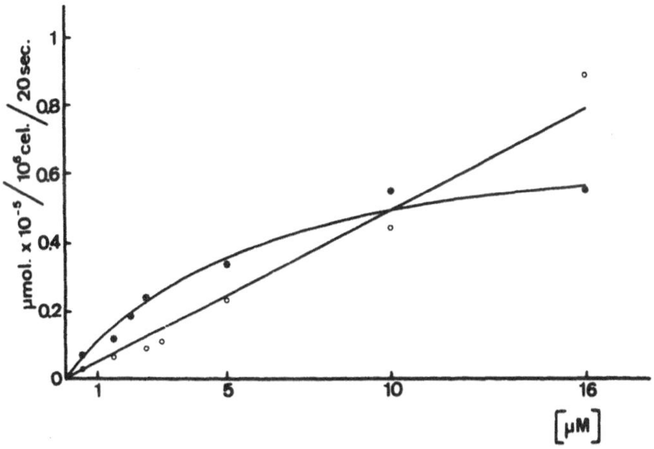

Fig.1: Incorporation of guanine (O) and guanosine (●) by L5178Y whole cell.

of guanine taken up by the cell is greater than that of guanosine. This is due to the saturation kinetics displayed by the nucleoside.

The distribution of both substrates in CAS fraction and in acid insoluble fraction is constant in all ranges of concentrations studied (Fig.2). In the acid insoluble fraction the radioactivity due to guanine is 37% and that of guanosine 9% approximately.

Fate of guanine label in CAS fraction

From 0.5 to 10 uM almost all the guanine incorporated in CAS fraction is found as nucleotides (87%). The remaining 13% is distributed as guanine (6%), guanosine (3%) and xanthine (3%). No significant counts were detected in inosine, xanthosine or uric acid. The total guanine metabolized over this concentration range is 94% (Fig.3), GTP being the predominant metabolite.

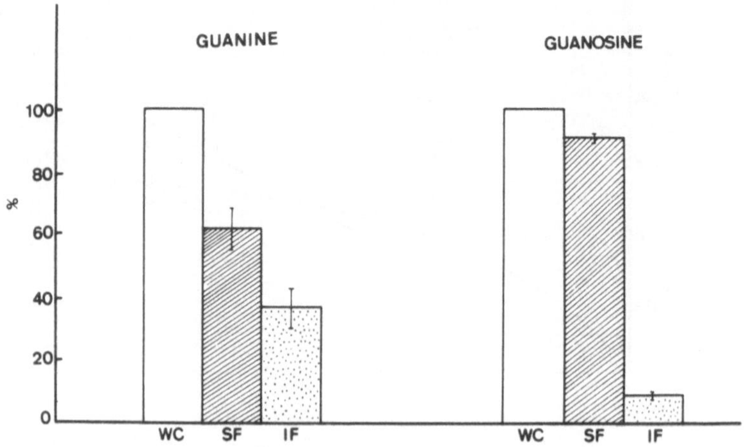

Fig.2: Distribution of label in total cell (TC), soluble fraction (SF) and insoluble fraction (IF).

At concentrations of guanine higher than 10 uM the distribution of the label in CAS fraction is different, the percentage of nucleotides decreases as the external concentration rises, being 14% at 100 uM. This decrease is accompained by a rise of the nonmetabolized base (58%) and of guanosine (11%). Traces of uric acid (2.2%) and xanthosine (6.5%) were observed when the medium concentration of guanine exceeded 10 uM.

Fate of guanosine label in CAS fraction

Between 0.5 and 16 Um the characteristics of guanosine utilization were qualitatively similar to those of the guanine in a concentration range of 0.5-10 uM.

Almost all the guanosine is metabolized (93%) and is distributed approximately as follows: nucleotides (74%), guanine (6%), xanthine (8%) and guanosine (7%). The predominant metabolite is GTP (40%), followed by GDP (19%) and GMP (15%).

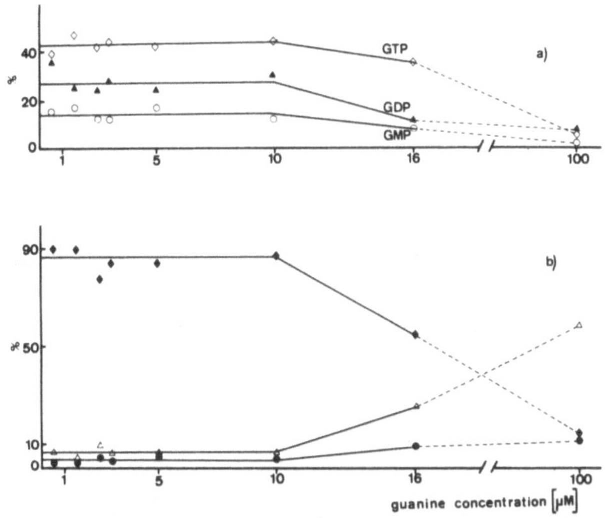

Fig.3: Label distribution in L5178Y cells incubated with (^{3}H) guanine for 20 sec. The values are expressed as percentages of CAS fraction. a) GTP (◇), GDP (▲), GMP (○). b)Nucleotides (◆), guanine (△) and guanosine (●).

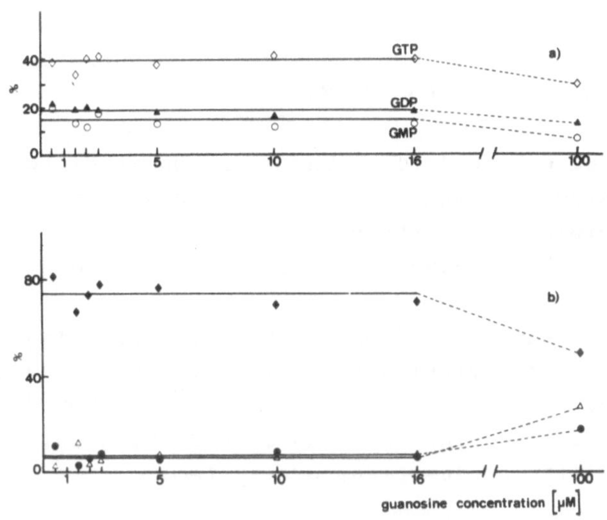

Fig.4: Label distribution in L5178Y cells incubated with (^{3}H) guanosine for 20 sec. The values are expressed as percentages of CAS fraction. a) GTP (◇), GDP (▲), GMP (○). b)Nucleotides (◆), guanine (△) and guanosine (●).

At a guanosine concentration of 100 uM, the label distribution differs quantitatively from that described for guanine at the same concentration: nucleotides (50%), guanine (27%), guanosine (17%) and xanthosine (5%) (Fig.4).

As is shown in Fig. 3 and 4, the distribution of the radioactivity is similar for guanine and guanosine. In both cases the conversion to nucleotides is reduced at the expense of guanósine accumulation. However, this effect is detected at higher concentrations of guanosine in the medium than of guanine, possible owing to the saturation mechanism of the nucleoside entry.

DISCUSSION

The metabolic distribution of guanine and guanosine was studied during short periods of incubation in order to establish its relationship with the uptake mechanisms. This involves a limitation in assesing the incorporation into nucleic acids. However, it is possible, in this conditions, to measure the relative rates of flux along the salvage pathway of purines in intact cells.

The distribution of labeling in CAS fraction is similar for guanine and guanosine, when their concentrations in the medium range from 0.5 to 10 uM.

When the extracellular concentration is higher than 10 uM, the conversion of guanine into nucleotides seems to be limited and an accumulation of the base is observed. This accumulation increases as the guanine concentration in the medium rises.

The guanosine behaviour is similar, differing in the fact that the base accumulation is noticed at higher concentrations of guanosine in the medium. This can be attributed to the lower amount of guanosine taken up by the cell, which is due to the saturability of its transport system.

It can be concluded that the processes involved in the entry of guanine and guanosine through the plasma membrane determine the total amount of substrate taken by the cell and, therefore, the quantitative distribution of the metabolites inside the cell.

REFERENCES

1- T.Hovi, A.C. Allison, K.O. Raivio, A. Vaheri, Purine metabolism and control of cell proliferation, in "Purine and Pyrimidine metabolism", Ciba Foundation Symposium 48, Elsevier Amsterdam (1977).

2- M.P. Rivera, M.R. Grau, J. Rigau, A. Goday, Purine transport and the cell cycle, 3th International Symposium on Purine metabolism in man, Madrid (1979).

3- P. Brown, The rapid separation od nucleotides in cell extracts using High Preasure Liquid Chromatography, J. Chromatog., 52:257 (1970).

4- C.E. Carter, J. Amer. Chem. Soc., 72: 1466 (1950).

5- L.N. Gonzales, S.E. Geel, Thin-Layer Chromatography of brain adenine nucleoside and nucleotides and determination of ATP specific activity, Anal. Biochem, 63:400 (1975).

6- U.E. Honegger, S.S. Bogdanov, P.R. Bally, Quantitative extraction, separation and recovery of Adenine-derived radioactivity in bases, nucleosides and nucleotides from blood platelets using PEI-Cellulose Thin-Layer Chromatography, Analyt. Biochem. 81:268 (1977).

BIOCHEMICAL CONSEQUENCES OF TREATMENT WITH THE ADENOSINE DEAMINASE INHIBITOR 2'-DEOXYCOFORMYCIN

Rosanne M. Paine, J.F. Smyth and K.R. Harrap

Institute of Cancer Research, Sutton, Surrey, England

INTRODUCTION

A congenital absence of the enzyme adenosine deaminase (ADA, EC 3.5.4.4) has been associated with profound T and B cell deficiencies in children, leading to immune dysfunction (1). Cohen *et al* (2) have demonstrated that ADA deficient children have elevated levels of 2'-deoxyadenosine triphosphate (dATP) in their erythrocytes. dATP is a known negative effector of the enzyme ribonucleotide reductase (3). In view of the finding that ADA activity is high in the blast cells of patients with T-cell acute lymphocytic leukaemia (4), an ADA inhibitor might well induce a selectively toxic event in the lymphoblast.

A Phase I clinical study of the tight-binding ADA inhibitor, 2'-deoxycoformycin (dCf) has been completed at the Royal Marsden Hospital. We report here biochemical changes, both *in vitro* and *in vivo*, following dCf treatment, and discuss these changes in relation to dATP toxicity.

MATERIALS AND METHODS

Coformycin (Cf) and dCf were supplied by the Drug Research and Development Branch, National Cancer Institute, Bethesda, USA.

<u>Cell Cultures</u>: Lymphocytes were isolated from heparinised venous blood by the method of Boyum (5) and cultured as previously described (6). Cf and dCf were prepared in 0.15M sodium phosphate buffer pH 7.1, and sterilised by filtration.

Animal Studies: BDF1 male mice, 10 to 12 weeks of age were randomly allocated into groups of ten, and received dCf 0.25mg/kg by intraperitoneal injection for 5 days. Injection volumes were 0.01ml/g body weight and control groups received vehicle only.

Patients: Patients included in the study had advanced malignant disease and had failed all previous forms of treatment. All patients reported here received dCf 0.25mg/kg by slow intravenous injection for 5 days. dCf was prepared in 4.2% sodium carbonate to maintain stability. The mean age group of the 6 male patients was 31 years (range 18-56).

Adenosine Deaminase: ADA activity in lymphocytes was assayed as described by Smyth and Harrap (7). Spleen ADA was extracted by the method of Jackson (8).

Ribonucleoside Triphosphates (rNTPs) and Deoxyribonucleoside Triphosphates (dNTPs): These were extracted and estimated in lymphocytes as previously described (6). dATP was extracted from erythrocytes by addition of 5ml of 72% methanol per ml of packed erythrocytes. After incubation for 10 minutes at 37°, the extracts were placed at -20° for 18 hours. Following 2 washes with 5ml of 60% methanol, solid material was removed by centrifugation at 40,000g at 4° for 30 minutes. The supernatant was rotary evaporated at room temperature for 5 minutes and the aqueous solution lyophilised and stored at -20°. Immediately prior to assay the extract was dissolved in 0.05M Tris HCl buffer pH 7.7 and dATP assayed according to the method of Tattersall and Harrap (9) using a calf thymus DNA template.

RESULTS

Incubation of PHA-stimulated lymphocytes with Cf 10^{-5}M or dCf 10^{-5}M for 48 hours did not affect intracellular ATP levels (Table 1). There were small increases in the other 3 ribonucleoside triphosphates, particularly UTP, compared with control cells. Exposure to Cf or dCf produced an elevation of dATP of 3.1 and 2.75 respectively (Table 2). dCTP was reduced to 62% of control in both cases. Under these conditions no inhibition of ^{3}H TdR incorporation into lymphocyte DNA was observed.

Following administration of dCf to BDF1 mice, spleen ADA activity fell to 9.9% of control on Day 4, accompanied by an elevation of erythrocyte dATP levels from 0.29nmoles (pretreatment) to 21.9nmoles/ml of packed erythrocytes (Figure 1). 7 days after termination of treatment, spleen ADA activity had recovered to 54.6% of control and dATP levels had fallen to 2.3nmoles/ml. During treatment, average spleen weights fell by 23% and had not totally recovered by Day 12.

Table 1. Ribonucleoside Triphosphate Levels in PHA-Stimulated Normal Human Lymphocytes Following 48hr Exposure to 10^{-5}M Concentrations of Cf or dCf

Additions	rNTPs % Stimulated Control			
	ATP	GTP	UTP	CTP
Cf	97.7	107.9	145	104.5
dCf	94.3	119.4	120	111.9

The results are the means of two separate experiments.

Table 2. Deoxyribonucleoside Triphosphate Levels in PHA-Stimulated Normal Human Lymphocytes Following 48hr Exposure to 10^{-5}M Concentrations of Cf or dCf

Additions	dNTPs pmoles/10^7cells			
	dATP	dGTP	TTP	dCTP
None	206	22	242	106
Cf	641	25	264	66
dCf	566	28	200	66

The results are the means of triplicate observations in two separate experiments.

The major toxicity following administration of dCf to patients was lymphocytotoxicity. A most dramatic response was seen in one patient, GMt, whose pretreatment peripheral blast count of 82,000/cumm had cleared by Day 4. Figure 2 shows graphically the inverse relationship between lymphocyte ADA activity and erythrocyte dATP levels following dCf administration to patient PF. In all patients studied a decrease in lymphocyte ADA activity was associated with a concomitant elevation of erythrocyte dATP levels both during and after treatment (Table 3).

DISCUSSION

The *in vitro* studies with PHA-stimulated lymphocytes exposed to dCf or Cf for 48 hours indicated no change in intracellular ATP but an elevation of dATP levels. This may reflect a difference in phosphorylation control of adenosine and 2'-deoxyadenosine. We have previously shown lymphocyte adenosine kinase to be substrate inhibited by adenosine (6). No reduction in DNA synthesis was observed in these cells, so clearly the elevation of dATP was insufficient to prevent lymphocyte blastogenesis.

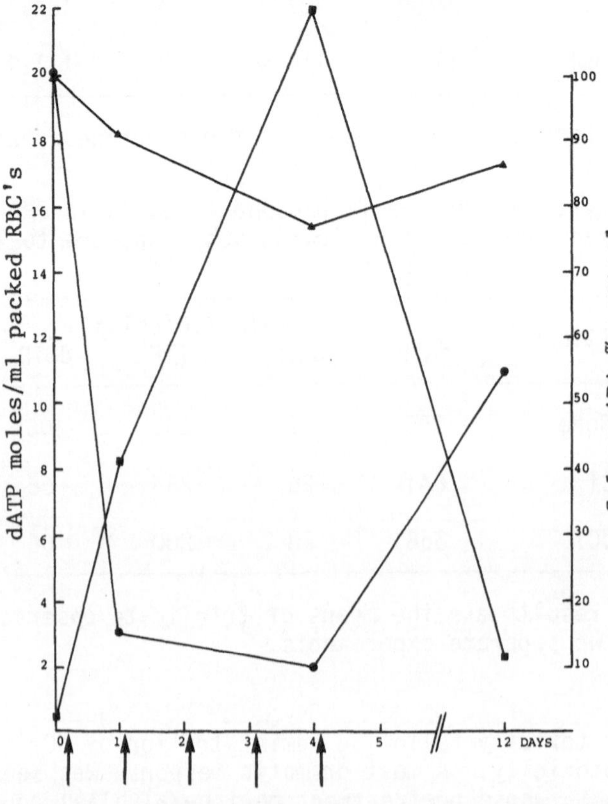

Figure 1. Spleen Weights, ADA Levels and Erythrocyte dATP Content in Mice Receiving dCf (0.25mg/kg/d x 5)

↑ dCf i.p. injection.
▲% control spleen wet weight on groups of 10 mice. Overall scatter ⟊ ± 10%.
■ Erythrocyte dATP levels, each point represents the mean of triplicate assays on pooled erythrocytes from 10 mice.
● Spleen ADA activity expressed as % control activity, each point represents the mean of duplicate assays on 10 mice. Overall scatter ⟊ ± 10%.

Table 3. Red Blood Cell dATP Levels (nmoles/ml packed RBCs) and Lymphocyte ADA Levels (units/10^7cells*)

Patient		Day Post-Treatment									
		1	2	3	4	5	6	7	8	11	14
GMg	dATP	-	-	-	-	152	-	-	-	-	-
	ADA	-	-	-	-	0.4	-	-	-	-	-
AM	dATP	ND	-	-	152	-	-	425	-	-	-
	ADA	7.1	-	-	ND	-	-	ND	-	-	-
GMt	dATP	0.5	-	975	-	-	-	-	-	-	-
	ADA	136	-	0.72	-	-	-	-	-	-	-
LT	dATP	0.9	-	-	168	-	-	350	-	-	330
	ADA	6.6	-	-	ND	-	-	ND	-	-	3.5
PF	dATP	ND	-	-	88	-	-	328	-	-	-
	ADA	22.6	-	-	3.4	-	-	0.6	-	-	-
CB	dATP	ND	ND	-	146	-	-	-	241	125	-
	ADA	16.0	1.7	-	0.8	-	-	-	1.3	0.3	-

*1 unit of activity is the amount of enzyme required to produce a decrease in absorbance of 0.01 per min under the conditions of assay (Smyth and Harrap, 1975).

The ADA results are the means of replicate assays.

dATP measurements are the means of triplicate assays on individual samples.

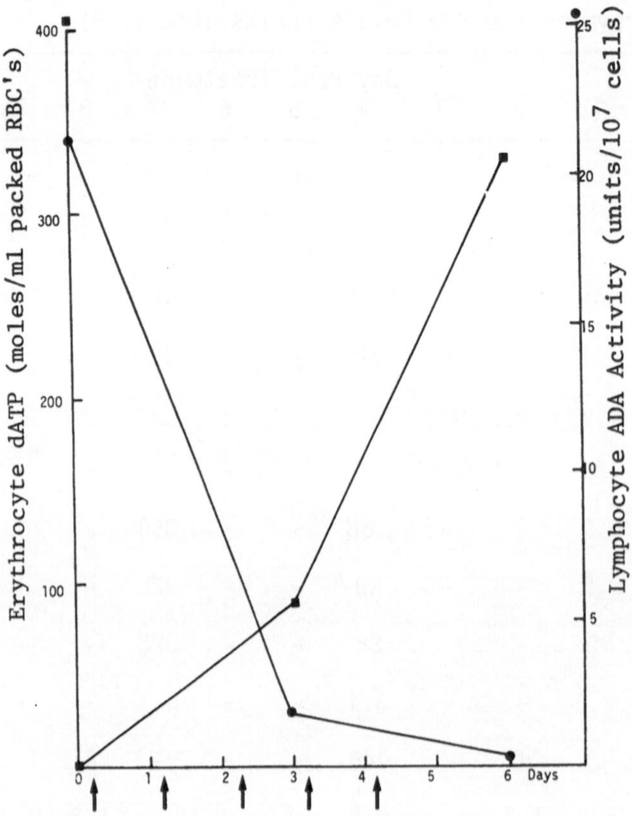

Figure 2. Erythrocyte dATP Levels and Lymphocyte ADA Levels in Patient PF Who Received 0.25mg/kg Daily x 5

↑dCf i.v. injection.

●Lymphocyte ADA activity, each point represents the mean of replicate assays.

■Erythrocyte dATP levels, each point represents the mean of triplicate assays on individual samples.

In patients dCf treatment proved to be selectively lymphocytotoxic, and in both patients and mice dCf administration resulted in large elevations of erythrocyte dATP levels. This must reflect an increase in circulating purines following tissue ADA inhibition since erythrocytes have no mechanism for synthesising deoxynucleotides *de novo*. This supports our earlier hypothesis that ADA inhibitors potentiate 2'-deoxyadenosine toxicity to lymphoid tissues through an intracellular accumulation of dATP (10).

REFERENCES

1. E.R. Giblett, J.E. Anderson, F. Cohen, B. Pollara and H.J. Meuwissen, Adenosine deaminase deficiency in two patients with severe impaired cellular immunity, Lancet 2: 1067 (1972).
2. A. Cohen, R. Hirschhorn, S.D. Horowitz, A. Rubinstein, S.H. Polmar, R. Hong and D.W. Martin Jr., Deoxyadenosine triphosphate as a potentially toxic metabolite in adenosine deaminase deficiency, Proc. Natl. Acad. Sci. USA 75: 472 (1978).
3. E.C. Moore and R.B. Hurlbert, Regulation of mammalian deoxyribonucleotide biosynthesis by nucleotides as activators and inhibitors, J. Biol. Chem. 241: 4802 (1966).
4. J.F. Smyth, D.G. Poplack, B.J. Holiman, B.G. Leventhal and G. Yarbro, Correlation of adenosine deaminase activity with cell surface markers in acute lymphoblastic leukaemia, J. Clin. Invest. 62: 710 (1978).
5. A. Boyum, Separation of leukocytes from blood and bone marrow, Scand. J. Clin. Lab. Investig. 21, supp. 97: 77 (1968).
6. K.R. Harrap and R.M. Paine, Adenosine metabolism in cultured lymphoid cells, Adv. Enz. Regln. 15: 169 (1977).
7. J.F. Smyth and K.R. Harrap, Adenosine deaminase activity in leukaemia, Brit. J. Cancer 31: 544 (1975).
8. R.C. Jackson, H.P. Morris and G. Weber, Adenosine deaminase and adenosine kinase in rat hepatomas and kidney tumours, Br. J. Cancer 37: 701 (1978).
9. M.H.N. Tattersall and K.R. Harrap, Changes in the deoxyribonucleoside triphosphate pools of mouse 5178Y lymphoma cells following exposure to methotrexate or 5-fluorouracil, Cancer Res. 33: 3086 (1973).
10. K.R. Harrap and R.M. Paine, Use of a regulatory effector as a potential antitumour agent, Excerpta Medica, Characterisation and Treatment of Human Tumours 4: 239 (1978).

ACKNOWLEDGEMENT

This work was supported by the Cancer Research Campaign.

IN VITRO AND IN VIVO EFFECT OF DEOXYCOFORMYCIN IN HUMAN T CELL LEUKEMIA

Alice L. Yu, M.D., Ph.D., Faith H. Kung, M.D.
Bohdan Bakay, D.N.S., William L. Nyhan, M.D., Ph.D.
University of California, San Diego, Department of Pediatrics
University Hospital, 225 West Dickinson Street
San Diego, California 92103

The success of anti-cancer chemotherapy is often limited by the toxicity of chemotherapeutic agents toward normal bone marrow cells and other rapidly dividing cells. It is most desirable, but it has been difficult to find anti-cancer agents that are selectively toxic to neoplastic cells. Approaches to the achievement of such chemotherapeutic selectivity were developed out of the recent elucidation of the causal association of severe combined immunodeficiency with hereditary deficiency of adenosine deaminase (ADA) (1,2). ADA, an enzyme that catalyzes the deamination of adenosine and deoxy-adenosine to inosine and deoxy-inosine, is widely distritubed in mammalian tissue (3,4). However, inherited deficiency of ADA activity is associated with selective impairment of lymphoid functions, particularly T cell functions, although the enzyme is virtually absent from all tissues examined (4). Further delineation of the preferential toxicity in the lymphoid system revealed that cultured cell lines of T lymphoblasts are far more sensitive than B lymphoblasts to the growth inhibitory effects of doexyadenosine in the presence of ADA inhibitors (6,7). It appeared possible to take advantage of these findings in the design of a chemotherapeutic regimen specific for T cell neoplasm. We therefore investigated the *in vitro* and *in vivo* effects of the artificial induction of ADA deficiency by a potent ADA inhibitor, deoxycoformycin, on the survival and metabolism of purine of T cell leukemias.

MATERIALS AND METHODS

Patient

The patient under study was a 5 year old boy who was diagnosed to have acute lymphoblastic leukemia at the age of 3½ years. He presented at that time with a white blood cell count of 716,000/μl, hepatosplenomegaly, generalized lymphadenopathy, and a large mediastinal mass. His leukemia was classified as T cell in type on the basis that 80% of his leukemic cells formed rosettes with sheep erythrocytes. Although initial remission was rapidly induced, his leukemia relapsed 8 months later. Despite aggressive treatment with different chemotherapeutic regimens, he subsequently had three more relapses with progressively shorter intervals of remission. The studies reported here were carried out three months after his last relapse. By that time, his bone marrow was so replaced by leukemic cells that he had become transfusion dependent for both RBC and platelets.
Deoxycoformycin was kindly provided by Warner-Lambert/Parke Davis Company, Detroit, Michigan.

Preparation of Cells

Leukemic cells and normal lymphocytes from peripheral blood were isolated by Ficoll-Paque technique. In some cases erythrocytes recovered from the bottom layer of Ficoll-Paque were washed with saline for further studies. Normal bone marrow cells were prepared from diagnostic bone marrow specimens of non-cancerous patients or leukemic patients in remission by dextran sedimentation. Contaminating erythrocytes were lysed by hypotonic shock.

Effects of Deoxycoformycin-Deoxyadenosine on DNA Synthesis

A micromethod was employed to test for the inhibitory effects of deoxycoformycin and deoxyadenosine on the DNA synthesis of leukemic cells. Leukemic cells or normal bone marrow cells were incubated in microtest culture plates with deoxycoformycin and deoxyadenosine. Twenty four hours later, they were pulsed with 3H-thymidine for two hours. The cells were collected on glass fiber filters with the aid of an Automatic Sample Harvester, washed with H_2O and counted in a liquid scintillation counter. The percent control uptake was calculated using uptake by cells incubated without deoxycoformycin and deoxyadenosine as 100 per cent.

Enzyme Assays

Cell lysates obtained by freezing and thawing were centrifuged at 28,000g for 20 minutes. ADA activity in cell lysates were assayed by radiochemical method as described by Snyder et al (8).

Analysis of Purine Metabolites by High Pressure Liquid Chromatography

Leukemic cells were incubated with 1μCi/ml of 8-^{14}C-labeled purine precursors; specific activity was maintained at approximately 50μCi/mM. Two hours later, the leukemic cells were extracted with 0.4M$HClO_4$, neutralized with KOH. The acid soluble fractions were analyzed both for ^{14}C radioactivity and for U.V. absorption in purine metabolites by high pressure chromatography (19).

RESULTS

In Vitro Effects of Deoxycoformycin and Deoxyadenosine

The combined effects of the ADA inhibitor deoxycoformycin and deoxyadenosine on the synthesis of DNA in freshly isolated leukemic T-cells were studied. Deoxyadenosine alone, up to 20μM or deoxycoformycin alone, up to 80μM, did not affect DNA synthesis in any experiments. However, at 5μM deoxycoformycin, DNA synthesis was inhibited 50% by 5μM deoxyadenosine and 90% by 20μM deoxyadenosine. At higher concentrations of deoxycoformycin, DNA synthesis came to a complete halt. In contrast, DNA synthesis in normal bone marrow cells was not affected by the combined presence of deoxycoformycin up to 80μM and deoxyadenosine up to 10μM. These findings indicate that in the presence of deoxycoformycin, deoxyadenosine is toxic to T leukemic cells but not normal bone marrow cells.

In order to understand the mechanism of toxicity of deoxycoformycin and deoxyadenosine toward leukemic T cells; purine metabolism was investigated in cells pre-incubated with or without deoxycoformycin and deoxyadenosine and pulse-labeled with ^{14}C-deoxyadenosine. Analysis of purine metabolites in these leukemic cells by high pressure liquid chromatography revealed a nearly tenfold increase in the level of dATP despite 2.4 fold decrease in the amount of total purine metabolites. Increased dATP content was paralled by a rise in the specific radioactivity of dATP from 1.67 nCi/nM to 6.04 nCi/nM. There was also an increase in dADP from nondetectable (<0.01/nM) to $0.19nM/10^7$. In contrase, ATP content dropped from $17.48nM/10^7$ cells to $5.10nM/10^7$ cells, which was accompanied by a reduction in specific radioactivity from 1.74nCi/nM to 0.21nCi/nM.

In Vivo Effects of Deoxycoformycin

Because of the encouraging results of these in vitro studies on this patient's leukemic T cells, we conducted a clinical trial of deoxycoformycin after the patient had failed to respond to conventional chemotherapeutic measures. From January 4 (day zero) to January 8, 1979 (day 4), he received 5 daily intravenous injections of 0.25mg/kg of deoxycoformycin. The trial had to be terminated on January 8, 1979 when the patient developed acute appendicitis and pseudomonas sepsis. Despite appendectomy, aggressive antibiotic therapy and supportive measures including multiple transfusions of RBC, platelets and granulocytes, the patient finally succumbed to sepsis on January 23, 1979.

During the clinical trial, samples of blood were collected daily, immediately before the administration of deoxycoformycin for clinical and laboratory studies. Among the changes noted, most prominent was the drop in the leukemic cell count. Initially the cell count rose from 7.2×10^3 cells/µl on day zero to 1.2×10^5 cells/µl on day 3. It then began to decline rapidly, and on day 9 it was 500 cells/µl. At this time 15% of his cells were normal leukocytes including neutrophils. This was the first appearance of neutrophils in this patient in a period of three months prior to this clinical trial. The leukemic cell count remained below 5,000 cells/µl through day 15. As measured by 3H-thymidine uptake, DNA synthesis in leukemic T cells began to decrease on day 2 and it dropped to a negligible amount on day 6.

The other prominent change was a decrease in ADA activity in both red cells and leukemic cells. On the second day, ADA activity of the red cells dropped to 12% of normal and on the 5th day it dropped to 5%. However, on the 6th day, ADA activity returned to normal levels. Deoxycoformycin had a similar effect on the ADA activity of leukemic cells.

Treatment with deoxycoformycin caused a drastic alteration of the metabolism of purines in the leukemic cells, as revealed by high pressure liquid chromatography. From day 0 to day 5, there was a progressive loss of purine metabolites in leukemic cells ($40nM/10^7$ cells to $1nM/10^7$ cells) and in red blood cells (1062nM/ml packed cells t0 476nM/ml packed cells). The concentration of ATP in leukemic cells reduced drastically from $25nM/10^7$ cells to $0.6nM/10^7$ cells. On the other hand, the concentration of dATP rose considerably from $<0.01nM/10^7$ cells to a peak of $0.8nM/10^7$ cells on day 2. The low concentrations of purine metabolites in leukemic cells harvested on day 4 and 5 suggested that these cells were in the process of disintegration. This was further supported by the finding that DNA synthesis in these cells was reduced to a negligible amount and that there was a high proportion of "smudge" cells on blood smear.

DISCUSSION

Our in vitro and in vivo studies of deoxycoformycin have indicated that the artificial induction of ADA deficiency is detrimental to the survival of human T cell leukemia and causes a profound perturbation of purine metabolism in both leukemic cells and erythrocytes. The alteration of purine metabolism that occurred after deoxycoformycin administration was evidenced by the reduction of ADA activity in both leukemic and red cells as well as by significant changes in purine metabolite profiles which were observed as early as 16 hours after the first dose of deoxycoformycin. On the other hand, cesation of DNA synthesis and the death of leukemic cells took place with some time lag. Thus, for the first three days the patient's leukemic cell count continued to quadruple daily as it had before chemotherapy. It began to fall rapidly to a very low count (500 cells/μl) although complete remission was not achieved. Complete remission might have been feasible had the patient not required transfusions of red blood cells on day 5 since erythrocytes are known to be rich in ADA activity and to restore lymphoid functions in ADA deficient children (10). These findings point to a potential therapeutic value for deoxycoformycin in T cell leukemia. They also further our understanding of the basis for lymphoid dysfunctions in patients with ADA deficiency. The return of ADA activity in leukemic and RBC cells occurring one and two days, respectively, after the last dose of dCF was unexpectedly swift. It has been shown that deoxycoformycin is a potent tight-binding inhibitor of ADA (11). Following the injection of deoxycoformycin in mice, the recovery of ADA activity was very slow in erythrocytes with 13% recovery in 48 hours, whereas, 80% recovery occurs in 48 hours in L1210 leukemic cells (12). In Rhesus monkeys, the rate of recovery of ADA in erythrocytes, after a single injection of deoxycoformycin, was in the range of 1-2% per day, which suggested the recovery was due to newly generated erythrocytes (13). Therefore, the rapid recovery of ADA activity in erythrocytes of this patient on day 6, was most likely the result of red cell transfusion on day 5.

Therapeutic selectivity of deoxycoformycin for lymphoid neoplasms was inferred from the preferential lymphoid impairment in ADA deficiency. It is further supported by our in vitro studies which indicate that DNA synthesis is not inhibited in normal bone marrow cells in the presence of deoxycoformycin and deoxyadenosine, while in leukemic T cells there is a marked inhibition of the synthesis of DNA. Whether deoxycoformycin displays similar selectivity in vivo, is difficult to ascertain because the patient studied had become transfusion dependent for both RBC and platelets before his clinical trial. However, following treatment with deoxycoformycin, his granulocytes made their first appearance in three months since his last relapse. Thus, deoxycoformycin appears to be non-myelosuppressive in vivo.

Analysis of purine metabolites following both in vitro and in vivo treatment with deoxycoformycin revealed drastic alterations in purine metabolism. These changes, which were characterized by progressive loss of the overall amount of purine metabolites, marked decrease in ATP and significant increase in dATP, were observed with both leukemic cells and red blood cells. Elevated concentrations of dATP in erythrocytes of ADA deficient patients have recently been reported (14,15). It is known that dATP exerts a potent negative feed back inhibition of ribonucleotide reductase, which leads to a reduction of the conversion of ADP, CDP, GDP and UDP to their respective deoxyribonucleotides (16). Thus, our findings are in accod with the generally held view that deoxyadenosine is the toxic substrate of ADA. A deficiency of ADA leads to accumulation of dATP, the ultimate phosphorlated metabolite of deoxyadenosine, which is in turn toxic to lymphoid cells.

References

1. Giblett, E.R., Anderson, J.E., Cohen, F., Pollara, B., Meuwissen, H.J., 1972. Lancet 2:1067

2. Meuwissen, H.J., Pollara, B., Pickering, R.J., 1975, J. Pediatrics 86:169

3. Adams, A. and Harkness, R.A., 1976. Clinical Exp. Immunol. 26:647-649

4. VanDerWeyden, M.B. and Kelley, W.N., 1976. J. Biol. Chem. 251:5448-5456

5. Hirschhorn, R., Levytska, V., Pollara, B. and Meuwissen, H.J., 1973. Nature (New Biol) 246:200-202

6. Mitchell, B.S., Mejias, E., Daddona, P.E. and Kelley, W.N., 1978. Proc. Nat. Acad. Sci. USA 75: 5011-5014

7. Carson, D.A., Kaye, J. and Seymitter, J.E., 1978. J. Immunol. 121:1726-1731

8. Snyder, F.F., Mendelsohn, J. and Seegmiller, J.E., 1976. J. Clin. Investigation 58:654-666

9. Bakay, B., Nissinen, E., Sweetman, L., 1978. Anal Biochem. 86:65

10. Polmar, S.H., Stern, R.C., Schwartz, A.L., Wetzler, E.M., Chase, P.A., Hirschhorn, R.,1976. New Eng. J. Med. 295:1337-1343

11. Agarwal. R.P., Spector, T. and Parks, R.E., 1977. Biochem. Pharmacol. 26:359-367

12. Agarwal, R.P., 1979. Cancer Res. 39:1425-1427

13. Rogles-Brown, T., Widness, J.A., Agarwal, K.C. and Schwartz, R. 1979. Proc. 7th Annual Meeting of Amer. Assoc. Cancer Res. Abstract 340.

14. Coleman, M.S., Donofrio, J., Hutton, J.J., Hahn, L., Dasud, A., Lampkin, B., Dyminski, J., 1978. J. Biol. Chem. 253:1619.

15. Cohen, A., Hirschhorn, R., Horowitz, S.D., Rubenstein, A., Polmar, S.H., Hong. R., Martin, D.W., 1978. Proc. Natl. Acad. Sci. USA. 75:472-

16. Moore, E.C. and Hurlbert, R.B., 1966. J. Biol. Chem. 241:4802-4809

UNIQUENESS OF DEOXYRIBONUCLEOTIDE METABOLISM IN HUMAN MALIGNANT T CELL LINES

Dennis A. Carson, Jonathan Kaye, Steven Matsumoto, J. E. Seegmiller and Linda Thompson

Department of Clinical Research, Research Institute of Scripps Clinic, La Jolla, California 92037 and Department of Medicine, University of California, San Diego, La Jolla, California 92093

Inherited deficiencies of the purine metabolic enzymes, adenosine deaminase (ADA) and purine nucleoside phosphorylase (PNP) specifically impair the growth and development of the lymphoid system in human beings[1,2]. We have proposed that lymphospecific toxicity in these diseases might result from the selective phosphorylation and trapping by T lymphocytes of the adenosine deaminase substrate deoxyadenosine (AdR) and the purine nucleoside phosphorylase substrate deoxyguanosine (GdR), with the subsequent formation of toxic deoxyribonucleoside triphosphates[3]. We further proposed that the phosphorylation was mediated by the lymphospecific enzyme deoxycytidine kinase, for which AdR and GdR are substrates, albeit poor ones[4]. This hypothesis was difficult to test *in vivo* because of the severe lymphopenia observed in untreated enzyme deficient patients. In an effort to find a relevant *in vitro* model system, we therefore turned to human malignant T lymphoblasts derived from patients with T cell leukemia and a mediastinal mass. These cells are conveniently available as continuous lines and share certain antigenic and biochemical characteristics with normal primitive thymocytes[5,6]. Initial studies showed that the malignant T cell lines were far more sensitive to the growth inhibitory effects of deoxyguanosine, deoxyadenosine, and thymidine, than were human B cell lines derived from normal subjects by *in vitro* infection of lymphocytes with Epstein-Barr virus[7,8].

Additional results from this and other laboratories showed that the malignant T cells were not in general more sensitive than B cells to a wide variety of antimetabolites[9,10]. Hence the increased susceptibility of the T lymphoblasts to deoxyribonucleoside toxicity appeared to be specific for this class of compounds.

The toxicity of many deoxyribonucleosides and related analogs at low concentrations requires their prior conversion to the respective triphosphate. We therefore reasoned that when incubated with deoxyadenosine, deoxyguanosine or thymidine, the T cells should accumulate more triphosphate than the B cells. This was indeed the case[11]. As shown in Figure 1 the human malignant T cell line, CEM, when incubated with 20 μM deoxyadenosine in the presence of an ADA inhibitor, accumulated much more deoxyATP than the B cell line Wil-2 which was incubated with a five-fold higher concentration of nucleoside. This difference became increasingly apparent as the time of incubation was lengthened. Identical results were obtained when T and B cells were incubated with 250 μM deoxyguanosine and when other malignant T cell and normal B cell lines were studied.

Similarly, when the T cells and B cells were incubated with 250 μM thymidine, the T cells accumulated significantly more thymidine triphosphate than the B cells at periods of up to 4 hours (Figure 2). Thereafter thymidine triphosphate levels declined in the T cells possibly as a result of a loss in cell viability.

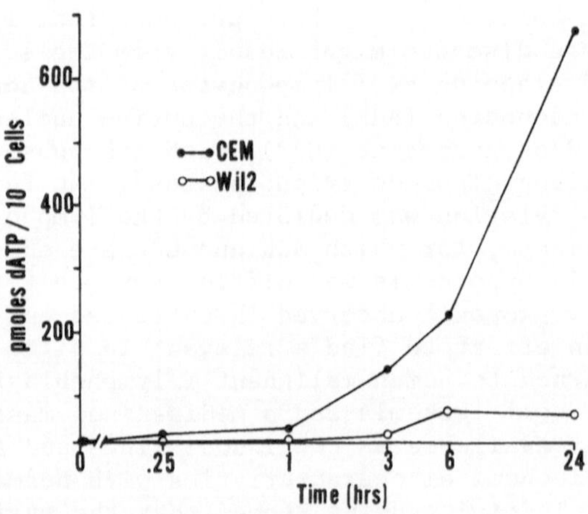

Fig. 1. Changes in dATP concentrations in T and B cell lines incubated with deoxyadenosine. To the T cell line CCRF-CEM (●——●) and the B cell line Wil-2 (o——o) at a density of 10^6 cells/ml in complete medium was added 20 μM and 100 μM deoxyadenosine, respectively, plus 5 μM of the ADA inhibitor deoxycoformycin. At the time periods indicated on the abscissa, the cells were washed, extracted, and dATP levels were assayed by high performance liquid chromatography[11].

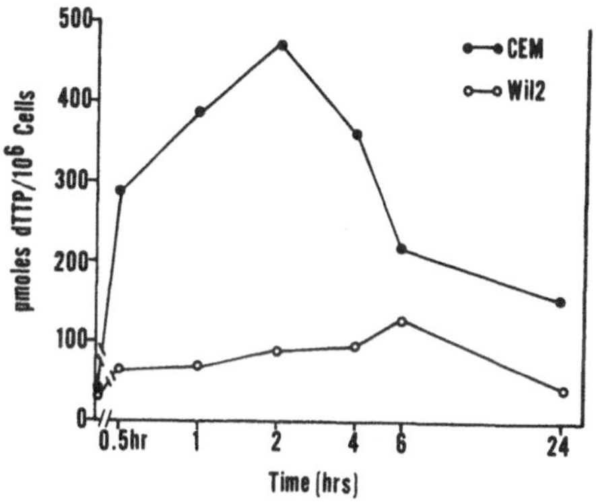

Fig. 2. Changes in dTTP concentrations in T and B cell lines incubated with thymidine. To the T cell line CCRF-CEM (●——●) or the B cell line Wil-2 (o——o) at a density of 5 x 10^5 cells/ml was added 250 μM thymidine. After varying times of incubation, the cells were washed and extracted, and dTTP levels were determined by DNA polymerase assay.

The ability of a cell to accumulate triphosphates when incubated with the respective deoxyribonucleoside is a function of several parameters. These are shown in Figure 3, and include the rate of nucleoside transport, the effectiveness of deoxyribonucleoside phosphorylation as mediated by specific kinases, the rate of deoxyribonucleotide dephosphorylation as mediated by specific nucleotidases and non-specific phosphatases, as well as the rate of cell division. We wanted to determine which one of these factors was responsible for the differential uptake of deoxyribonucleosides by T and B lymphoblasts.

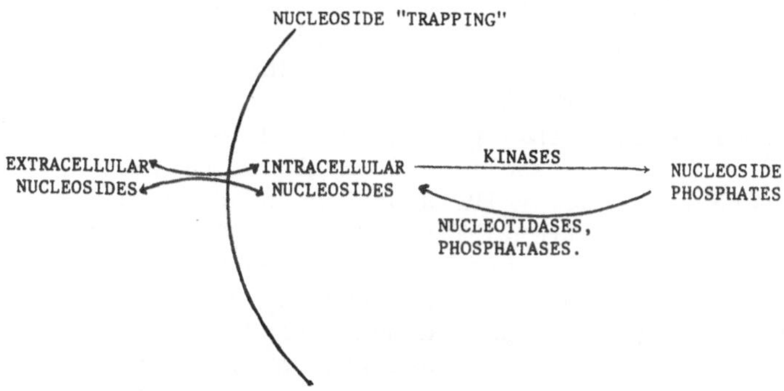

Fig. 3.

Table 1. Enzyme Levels in T and B Cell Lines

Enzyme	T Cells	B Cells	T Cell Activity / B Cell Activity
Deoxyadenosine phosphorylating activity (+deoxycoformycin)	0.74 ± 0.15	0.25 ± 0.07	2.96
Adenosine kinase	1.03 ± 0.25	1.69 ± 0.20	0.61
ADA	230 ± 90	32 ± 10	7.19
PNP	27 ± 3	74 ± 5	0.36
Thymidine kinase	0.17 ± 0.12	0.24 ± 0.06	0.71

Enzyme activities are expressed as nanomoles product/minute/mg protein ± S.E.M., as determined in extracts of logarithmically growing cells.

Table 1 shows the deoxyadenosine phosphorylating activity, adenosine phosphorylating activity and thymidine phosphorylating activity of 3 T and 3 B cell lines, as well as the levels of adenosine deaminase and purine nucleoside phosphorylase[7]. As can be seen, although the T cells had up to 3 times as much deoxyadenosine phosphorylating activity, reflecting deoxycytidine kinase levels, as the B cells, they actually had lower levels of thymidine kinase. Thus the differential ability of the T cells and B cells to take up deoxyadenosine, deoxyguanosine, and thymidine could not be attributed simply to differences in the levels of phosphorylating enzymes as measured in extracts. We therefore reasoned that the malignant T cell lines and B cell lines must differ in their rate of deoxyribonucleotide breakdown.

Figure 4 shows the rate of disappearance of deoxyATP in the T cell line CEM and the B cell line 894, each of which was pulsed with 1 mM deoxyadenosine for 3 hours in the absence of an ADA inhibitor and subsequently washed and placed in fresh medium without nucleosides. As can be seen, over a period of 30 minutes there was no decay in deoxyATP levels in the T cells. On the contrary, deoxyATP levels during this period decayed 50% in the B cells. As intact T and B cells catabolized deoxyATP at different rates, so too the breakdown of deoxyATP, deoxyGTP and deoxyTTP by broken cell preparations prepared from 3 T and 3 B cell lines differed markedly (Table 2). In this case, breakdown was measured as the conversion of radiolabelled nucleoside triphosphates to nucleosides and bases. The results are expressed as picomoles product per minute per 10^6 cells. Of note, the B cells converted deoxyribonucleoside triphosphates to nucleosides and bases from 7 to 20 fold more effectively than the T cells.

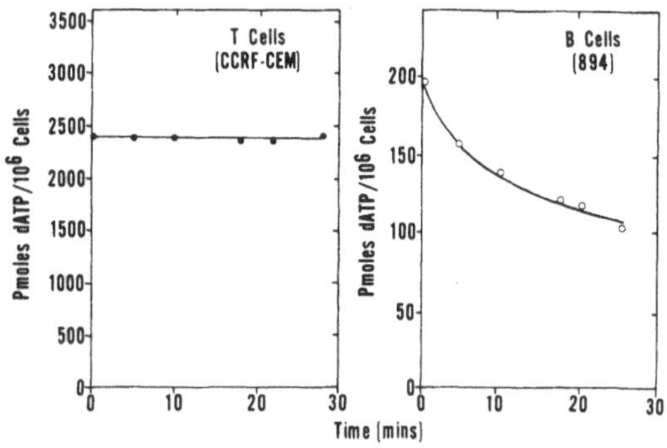

Fig. 4. Breakdown of dATP by T and B cell lines. The T cell line CCRF-CEM (●——●) and the B cell line 894 (o——o) at a density of 5 x 10^6 cells/ml were incubated for three hours with 1 mM deoxyadenosine without deoxycoformycin, then washed and placed in fresh medium. At varying time points thereafter, the cells were washed, extracted, and dATP concentrations were determined by DNA polymerase assay.

Although the rate limiting enzymes in deoxyribonucleotide catabolism are not precisely known, the 3 B lymphoblasts did have nearly 100-fold more inosine monophosphate dephosphorylating activity than the T lymphoblasts in both intact and broken cells (Table 3). The majority of the activity was accessible in intact cells, and was blocked by the specific ecto 5'-nucleotidase inhibitor, AOPCP[13]. The exact relationship of the ecto-enzyme 5'-nucleotidase to the rate of intracellular deoxyribonucleotide catabolism remains unexplained and is under current investigation. In any event, the

Table 2. Deoxyribonucleoside Triphosphate Catabolism in Broken Cells

	Substrate		
	dATP	dGTP	dTTP
T cell lines	11.1 ± 7.6	10.0 ± 5.2	4.0 ± 1.2
B cell lines	102.3 ± 32	67.3 ± 18	84.9 ± 12

Three leukemic T cell lines and three B cell lines were lysed by freeze thawing and incubated with 300 μM tritiated dATP, dGTP, or dTTP. After 30 minutes for the B cell lines and 120 minutes for the T cell lines, nucleotides were separated from nucleosides and bases by thin layer chromatography. Activities are expressed as p moles product (nucleoside + bases)/minute/10^6 cells ± S.D. (n-1 method).

Table 3. IMP Dephosphorylating Activity in T and B Cell Lines

	Whole Cells		Broken Cells	
	Without AOPCP	With AOPCP	Without AOPCP	With AOPCP
T cells	3.07 ± 1.4	3.07 ± 1.4	46 ± 8.5	42 ± 15
B cells	387 ± 319	14 ± 7.0	330 ± 225	62 ± 15

Activities are expressed as p mol product (inosine thypoxanthine)/ minute per 10^6 cells ± S.D. for three T and three B cell lines.

present experiments show unequivocally that three neoplastic human T cell lines break down deoxyribonucleoside triphosphates at a much slower rate than most B cell lines under all conditions tested.

Table 4 summarizes the levels of deoxycytidine kinase, an enzyme which, as mentioned previosly, can also phosphorylate deoxyguanosine and deoxyadenosine; adenosine kinase, an enzyme which can phosphorylate adenosine and also deoxyadenosine under certain conditions; and finally 5'-nucleotidase, in human malignant T lymphoblastoid cell lines, B cell lines, red blood cells, and other cell types. The T lymphoblasts are characterized by high levels of deoxycytidine kinase and adenosine kinase, but low levels of 5'-nucleotidase. This profile is associated with a marked accumulation of deoxyribonucleoside triphosphates by T cells incubated with the respective nucleosides. The B lymphoblasts from normal individuals, although containing appreciable levels of both deoxycytidine kinase and adenosine kinase, also have high levels of 5'-nucleotidase and hence are unable to accumulate deoxyribonucleoside triphosphates. Other experiments showed that red cells, which have low levels of deoxycytidine kinase, phosphorylated GdR poorly but could convert deoxyadenosine to deoxyATP via the normally inefficient adenosine kinase, because the cells catabolized deoxyadenosine nucleotides at a very slow rate. The pattern for other cells varied; however,

Table 4. Enzymes of Purine Metabolism in Various Cell Types

	Enzyme Level		
Cell Type	Deoxycytidine Kinase	Adenosine Kinase	Purine 5'-Nucleotidase
T lymphoblasts	High	High	Low
B lymphoblasts	High	High	High
Red blood cells	Low	High	Low
Most other cells	Low	High	Variable

none have been found with enzymatic profiles identical to the T cells. Thus, deoxyribonucleotide metabolism in the malignant T cell lines is in many ways unique. It is possible that human T cell neoplasms with similarly distinctive patterns might be effectively treated with deoxyribonucleosides and related analogs.

REFERENCES

1. E. R. Giblett, J. E. Anderson, F. Cohen, B. Pollara, and H. J. Meuwissen, Adenosine deaminase deficiency in two patients with impaired cellular immunity, Lancet 2:1067 (1972).
2. E. R. Giblett, A. J. Ammann, R. Sandman, D. W. Wara, and L. K. Diamond, Nucleoside phosphorylase deficiency in a child with severely defective T-cell immunity and normal B-cell immunity, Lancet 2:1010 (1975).
3. D. A. Carson, J. Kaye, and J. E. Seegmiller, Lymphospecific toxicity in adenosine deaminase deficiency and purine nucleoside phosphorylase deficiency: Possible role of nucleoside kinase(s), Proc. Natl. Acad. Sci. USA 74:5677 (1977).
4. T. A. Krenitsky, J. V. Tuttle, G. W. Koszalka, I. S. Chen, L. M. Beacham, III, J. L. Rideout, and G. B. Elion, Deoxycytidine kinase from calf thymus. Substrate and inhibitor specificity, J. Biol. Chem. 251:4055 (1976).
5. J, Kaplan, T. C. Shope, and W. D. Peterson, Jr., Epstein-Barr virus negative human malignant T cell lines, J. Exp. Med. 139:1070 (1974).
6. J. Minowada, T. Ohnuma, and G. E. Moore, Rosette forming human lymphoid cell lines. I. Establishment and evidence for origin from thymus derived lymphocytes, J. Natl. Cancer Inst. 49:891 1973).
7. D. A. Carson, J. Kaye, and J. E. Seegmiller, Differential sensitivity of human leukemic T cell lines and B cell lines to growth inhibition by deoxyadenosine, J. Immunol. 121:1726 (1978).
8. B. Mitchell, E. Mejias, P. E. Daddona, and W. N. Kelley, Purinogenic immunodeficiency diseases: Selective toxicity of deoxyribonucleosides for T cells, Proc. Natl. Acad. Sci. USA 75:5011 (1978).
9. G. E. Foley, and H. Lazarus, The response in vitro of continuous cultures of human lymphoblasts (CCRF-CEM cells) to chemotherapeutic agents, Biochem. Pharmacol. 16:659 (1967).
10. T. Ohnuma, J. F. Holland, and H. Arkin, Differential chemotherapeutic susceptibility of human T-lymphocytes and B-lymphocytes in culture, J. Natl. Cancer Inst. 59:1061 (1977).
11. D. A. Carson, J. Kaye, S. Matsumoto, J. E. Seegmiller, and L. Thompson, Biochemical basis for the enhanced toxicity of deoxyribonucleosides toward malignant human T cell lines, Proc. Natl. Acad. Sci. USA 76:2430 (1979).

12. A. W. Solter, and R. E. Handschumacher, A rapid quantitative determination of deoxyribonucleoside triphosphates based on the enzymatic synthesis of DNA, Biochim. Biophys. Acta 174: 585 (1969).
13. M. K. Gentry, and R. A. Olsson, A simple, specific radioisotope assay for 5'-nucleotidase, Anal. Biochem. 64:624 (1975).

Supported by National Institutes of Health Grants GM 23200, RR 05514, AM 13622, and GM 17702 and by grants from the Kroc Foundation and the Arthritis Foundation.

HIGH PERFORMANCE LIQUID CHROMATOGRAPHY OF PLASMA PYRIMIDINES AND PURINES AND ITS APPLICATION IN CANCER CHEMOTHERAPY+

A. Leyva, J. Schornagel and H.M. Pinedo

Section of Experimental Chemotherapy, Antoni van Leeuwenhoek Institute and Department of Oncology, Free University Hospital, Amsterdam (A.L., H.M.P.)
Department of Internal Medicine, State University Hospital, Utrecht (J.S.) The Netherlands

INTRODUCTION

A variety of sensitive high performance liquid chromatography (HPLC) methods have been reported for the separation of pyrimidine and purine compounds[1-4]. These methods have involved anion- and cation-exchange and reverse phase chromatography. The use of these common techniques for the analysis of plasma is sometimes hampered by the incomplete fractionation of bases and nucleosides or inadequate separation of these compounds from other plasma components. Also, in most cases elution has required more than one buffer or the use of a gradient. Eksteen et al.[5] recently demonstrated that bases and nucleosides could be rapidly and efficiently separated by HPLC using an anion-exchange resin and isocratic elution with an alcoholic phosphate buffer. Optimal results were reported achieved using 0.05 M sodium phosphate-0.005 M citric acid, pH 9.25, in 55% ethanol as the eluent and a column temperature of 70°. We initially attempted to use the same method for the analysis of plasma for pyrimidines and purines. However, it was noted that uric acid was not well separated from oxypurines. Moreover, using the conditions described above, we encountered poor column stability and inconsistent flow rates. By decreasing the phosphate and ethanol concentrations in the eluent, it was possible to obtain good column stability. With subsequent adjustments of the ethanol concentration, pH and temperature, the uric acid peak could be isolated and the complete fractionation of several common bases and nucleosides could be achieved. This paper demonstrates the effectiveness of this method

+Supported in part by the Queen Wilhelmina Fund (Project UUKC 77-3).

for analysis of pyrimidines and purines in plasma and describes its application in cancer chemotherapy with methotrexate and thymidine.

METHODS

Plasma was obtained from heparinized blood and the protein removed by cold perchloric acid (0.4 N) precipitation. The acid-soluble portion was neutralized with KOH and applied to a small DEAE-cellulose column equilibrated with 0.05 M ammonium acetate, pH 7.0. The column was washed with the same buffer eluting pyrimidines, purines and uric acid but not highly acidic compounds. The eluate was lyophilized and redissolved in a reduced volume of 0.02 N KOH yielding a pH of 7 to 8.

HPLC was performed using stainless steel columns packed with a strong anion-exchange resin (Aminex A28 or A29). Isocratic elution was carried out using an ethanol-0.01 M sodium phosphate-0.005 M citric acid buffer under two conditions, Systems A and B, differing in ethanol content and pH of eluent, particle size of resin and column temperature (Table 1). The columns were packed at pressures less than 1200 psi and at a flow rate of 0.4 ml/min. Elution was at 0.3 ml/min. Sample volumes of 10-25 µl were injected and eluted compounds were monitored with a variable wavelength detector at 260 or 280 nm or with two fixed wavelength detectors at 254 and 280 nm. Quantitation of peaks was by peak height measurements which were linear with concentration.

RESULTS AND DISCUSSION

Fig. 1 illustrates the fractionation of pyrimidines and purines using system A. Thymidine and thymine with retention times of 7 and 12.5 min, resp., were resolved from cytosine derivatives ($<$5 min) and uracil derivatives (7.5 to 14 min). Besides adenosine, eluting at 5.5 min, all other common purines followed the pyrimidines.

Table 1. HPLC Separation of Pyrimidines and Purines with Isocratic Elution

Conditions	System A	System B
Column Packing	Aminex A29 (5-9 µ)	Aminex A28 (8-12 µ)
Column Dimensions	25 cm X 3 mm	10 cm X 4.6 mm
Eluent[a]	19.0% Ethanol pH 9.15	28.5% Ethanol pH 8.35
Column Temperature	70°	60°

[a]0.01 M sodium phosphate-0.005 M citric acid

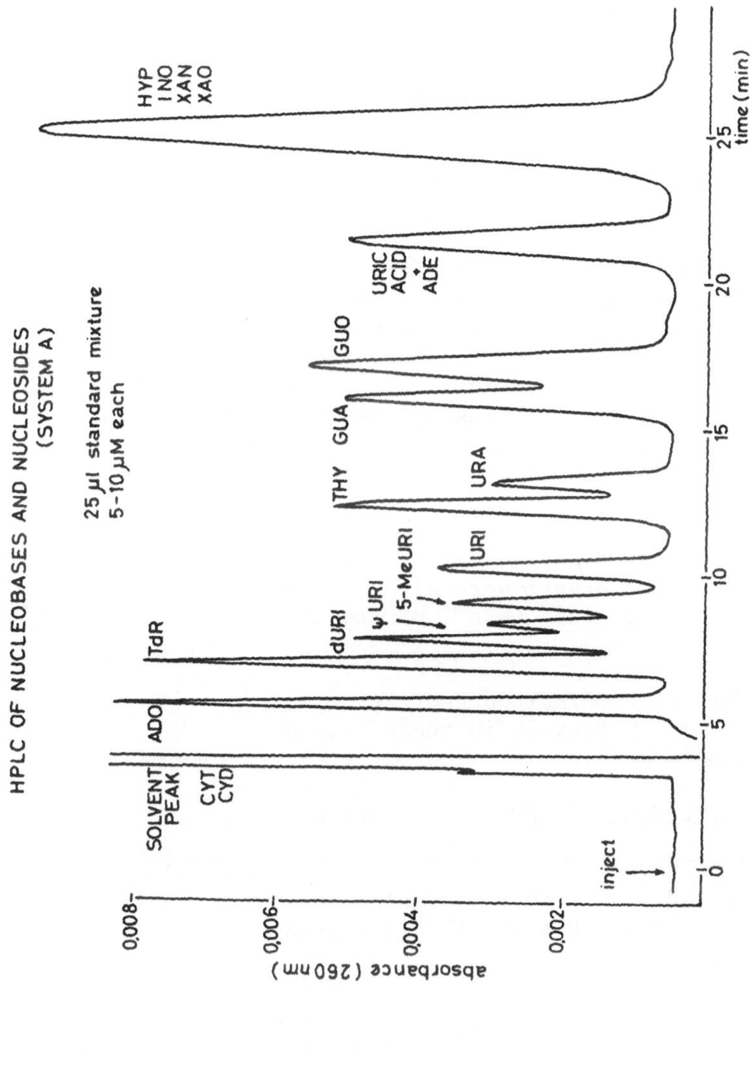
HPLC OF NUCLEOBASES AND NUCLEOSIDES
(SYSTEM A)
25 μl standard mixture
5-10 μM each
absorbance (260 nm)
0,008
0,006
0,004
0,002
inject
SOLVENT PEAK
CYT
CYD
ADO
TdR
dURI
ψ URI
5-MeURI
URI
THY
URA
GUA
GUO
URIC ACID + ADE
HYP
INO
XAN
XAO
0
5
10
15
20
25
time (min)

Fig. 1

Guanine and guanosine were resolved and were followed by co-elution of uric acid and adenine at 22 min. The final peak, eluting at 25 min, contained hypoxanthine and xanthine and their nucleosides. With system A thymidine was measurable at plasma concentrations as low as 0.2 µM, while the limit of detection for other pyrimidines and purines was 0.2 to 0.5 µM. Double wavelength detection served in confirming the identification of peaks and also allowed hypoxanthine and inosine to be distinquished from xanthine and xanthosine in the oxypurine peak due to differences in spectral properties.

System B proved more useful in the separation of purine bases and nucleosides (Fig. 2). Cytosine compounds and thymidine eluted rapidly and were not easily quantitated in plasma. Adenosine and thymine also appeared early in the elution but were resolved. Uridine and uracil eluted together at 6 min and were immediately preceeded by a peak containing pseudouridine and an unknown plasma component. Other than adenosine all purines and uric acid were well retained and well fractionated within 25 min. Notably, uric acid, normally in high concentrations in plasma, was isolated from all the common purines. Not only did this improve the measurement of purines, especially oxypurines, but also provided in the same analysis a uric acid determination as well. With system B sensitivities for bases and nucleosides ranged between 0.3 and 1.0 µM.

Table 2 gives the values for pyrimidine and purine concentrations in plasma of 19 normal individuals (10 males and 9 females, 24 to 56 years old). Adenosine, guanine, guanosine and xanthosine were not detectable (<0.5 µM) while other purines ranged between 1 and 5 µM. Thymine and uracil were also generally undetectable (<0.5 µM) while thymidine was less than 1 µM and uridine was about 4 µM.

Table 2. Normal Concentrations of Pyrimidines and Purines in Human Plasma

Pyrimidines	µM	Purines	µM
Thymine	<0.5	Adenine	1.6 ± 1.3
Thymidine	0.6 ± 0.3[a]	Adenosine	<0.5
Uracil	<0.5	Guanine	<0.5
Uridine	4.2 ± 1.2	Guanosine	<0.5
		Hypoxanthine	2.5 ± 1.0
		Inosine	0.9 ± 0.7
		Xanthine	5.2 ± 0.3
		Xanthosine	<0.5

[a]mean ± S.D.

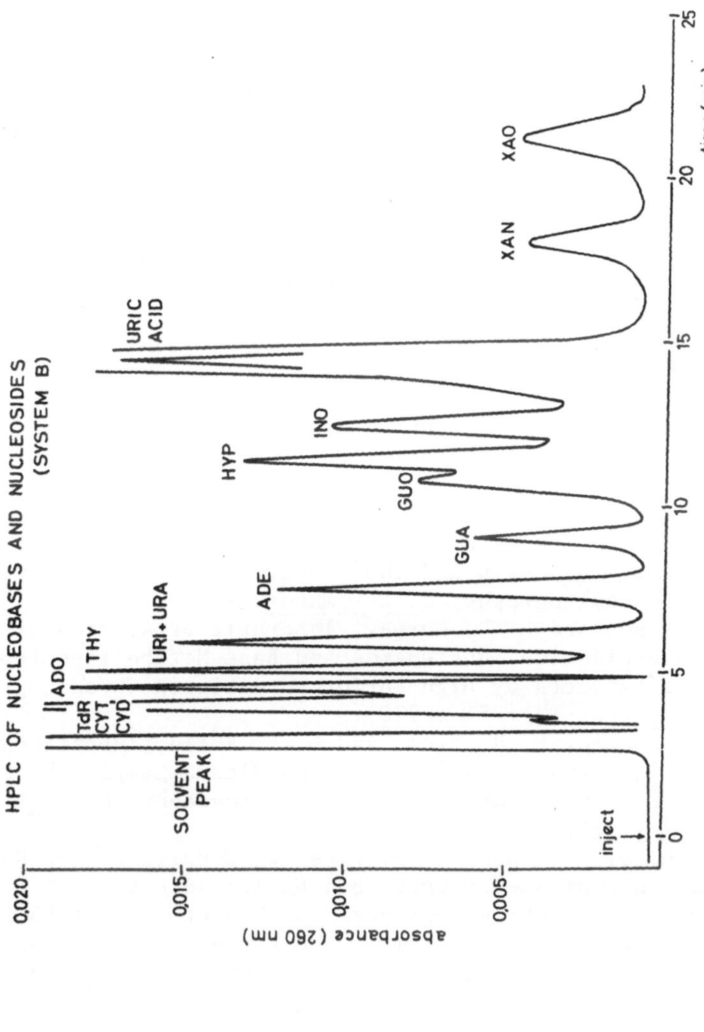
HPLC OF NUCLEOBASES AND NUCLEOSIDES
(SYSTEM B)
SOLVENT PEAK
TdR
CYT
CYD
ADO
THY
URI+URA
ADE
GUA
GUO
HYP
INO
URIC ACID
XAN
XAO
inject
absorbance (260 nm)
0,020
0,015
0,010
0,005
0
5
10
15
20
25
time (min)

Fig. 2

The levels of thymidine and purines in the plasma of patients receiving methotrexate treatment were examined to evaluate their availability for the reversal of drug toxicity. Ten patients with head and neck cancer were treated as described by Ensminger and Frei[6] with a 24-hr infusion of methotrexate, 600 mg/sqm, and a concurrrent or 24-hr delayed infusion of thymidine, 8 g/sqm/day, which lasted 72 hr. These patients received 2 or 3 courses and showed no bone marrow impairment, occasional mucositis and occasional minor toxicities. Furthermore, no antitumor response was observed. Thymidine levels in plasma increased during thymidine infusion from pretreatment levels of 0.3 μM to 0.5-10 μM, variation occuring between patients and during infusion. Plasma purine concentrations were similar to those of healthy individuals ranging between 1 and 5 μM for adenine, hypoxanthine and inosine. These micromolar levels of purines and even submicromolar levels of thymidine appear to be sufficient for the effective reversal of methotrexate toxicity presumably through their utilization in salvage pathway synthesis of nucleotides. It is yet to be determined if these low concentrations of circulating bases and nucleosides have in addition an effect on the antitumor activity of methotrexate.

REFERENCES

1. R. P. Singhal and W. E. Cohn, Cation-Exchange Chromatography on Anion Exchanges: Application to Nucleic Acid Components and Comparison with Anion-Exchange Chromatography, Biochemistry 12:1532 (1973).
2. P. R. Brown, S. Bobick and F. L. Hanley, The Analysis of Purine and Pyrimidine Bases and their Nucleosides by High-Pressure Liquid Chromatography, J. Chromatog. 99:587 (1974).
3. F. S. Anderson and R. C. Murphy, Isocratic Separation of Some Purine Nucleotide, Nucleoside and Base Metabolites from Biological Extracts by High-Performance Liquid Chromatography, J. Chromatog. 121:251 (1976).
4. A. M. Krstulovic, P. R. Brown and D. M. Rosle, Identification of Nucleosides and Bases in Serum and Plasma Samples by Reverse Phase High Performance Liquid Chromatography, Anal. Biochem. 49:2237 (1977).
5. R. Eksteen, J. C. Kraak and P. Linssen, Conditions for Rapid Separations of Nucleobases and Nucleosides by High-Pressure Anion-Exchange Chromatography, J. Chromatog. 148:413 (1978).
6. W. D. Ensminger and E. Frei III, The Prevention of Methotrexate Toxicity by Thymidine Infusions in Humans, Cancer Res. 37: 1857 (1977).

URATE-BINDING PROTEINS IN PLASMA STUDIED BY AFFINITY CHROMATOGRAPHY

M.L. Ciompi, A. Lucacchini, D. Segnini, M.R. Mazzoni

Service of Rheumatology and Department of Biochemistry

University of Pisa, Via Roma 2, 56100 Pisa, Italy

The existence in humans of one or more plasma proteins, which specifically bind uric acid (UA) is still discussed. There is no agreement on the nature of the binding protein, identified as the serum albumin[1], or as a specific $\alpha 1$-$\alpha 2$-globulin[2], nor on the extent of the bound quote. Most investigations have been till now performed by ultrafiltration[1] or column chromatography[2], but the specificity and the sensitivity of both are rather unsatisfactory; for example, recovery is lesser than 30% in the later technique[3]. On the other hand, the existence itself of a significant urate binding to human plasma proteins has been denied even recently, by means of continuous ultrafiltration and equilibrium dialysis[4].

Therefore, we have attempted a new methodological approach to the problem, utilizing a recently introduced technique. The affinity chromatography[5] is carried out by fixing, through a covalent binding, a specific ligand to an insoluble carrier, represented by an hydrophilic resin; defibrinated plasma is then filtered through a column of such resin, from which are only retained ligand-recognizing proteins; after washing, the elution is performed by means of ligand itself or other substances, which compete for the same site of the protein. In previous investigations the affinity chromatography proved to be a suitable technique for studying plasma binding mechanisms[6].

MATERIALS AND METHODS

We prepared three specific adsorbents, namely 2-amino-6,8-dihydroxy-purine, 6-amino-2,8-dihydroxy-purine and 8-amino-2,6-dihydroxy-purine, bound by the amino group to Sepharose 4B column.

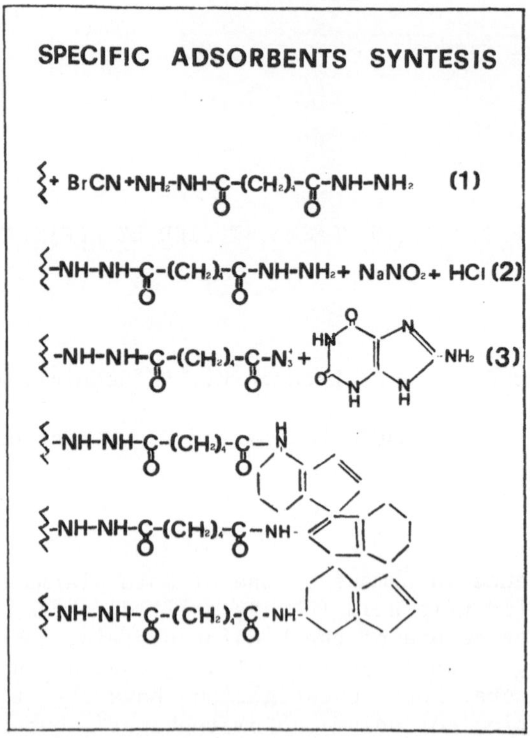

Fig. 1: Scheme of the reactions involved in the preparation of the specific adsorbants for UA binding proteins. (1) Sepharose adipic acid hydrazide was obtained by treating cyanogen bromide activated Sepharose 4B with adipic acid hydrazide, according to Wilchek and Lamed[9] . (2) Acid azide groups were generated by treating 0.2 g of dry resin, magnetically stirred, with dilute HCl (2 M, 10 ml) and 0°C cold sodium nitrite solution (4%, 4 ml); the stirring was continued for 15 min and the resin was washed with sodium borate 0.1 M, pH 8.5. (3) Then the appropriate substituted purine (1 μmole), dissolved in the borate buffer, was added; coupling to give amino-hydroxy-purine Sepharose is allowed to proceed for up to 24 hours, after which the resin was exhaustively washed with borate buffer and borate buffer plus 0.5-1 M NaCl until no absorption was detected at 260 μm.

8-amino-2,6-dihydroxy-purine was prepared according to Jones and Robins[7], 2-amino-6,8-dihydroxy-purine and 6-amino-2,8-dihydroxy-purine were purchased from Sigma Chemical (St. Louis, Mo., U.S.A.). The purine bases were bound to Sepharose by means of a spacer arm (adipic acid hydrazide), as shown in Figure 1.

Human defibrinated plasma was dialyzed against Tris buffer 20 mM, pH 7.4, to remove UA and then filtered through a ligand-containing Sepharose column (1-1.5 x 3-5 cm in size). When no otherwise specified, plasma filtration was carried out at room temperature. After plasma filtering, the column was washed with Tris buffer, to remove unspecifically bound proteins and then with the same buffer plus 0.05 M NaCl, to remove proteins, eventually adsorbed by ionic interaction. Washing was performed until complete absence of proteins, recorded at 280 μm.

The proteins specifically bound to ligand were eluted with Tris buffer plus 50 mg% UA, concentrated by dialysis against polyethyleneglycol 20.000, measured by the biuret method and analyzed by means of polyacrylamide electrophoresis, according to Davis[8]. When the native form of the protein(s) was unnecessary, the elution can be performed with 6M urea, which allows a rapid recovery.

RESULTS

The resins were previously analyzed for the amount of the specific ligand, which consistently resulted to be in the range of 10-20 μmols/g dry resin and then assayed for their specific ability to retain plasma proteins, which recognize UA. Therefore, the same amount of normal defibrinated human plasma was filtered through columns, containing the same amount of every specific ligand. After elution the protein content in the eluate was measured: the results indicate that the resin containing 8-amino-2,6-dihydroxy-purine binds consistently a major amount of plasma proteins (Table 1).

The analysis of proteins bound to 8-amino-2,6-dihydroxy-purine and specifically eluted with UA was performed by polyacrylamide electrophoresis and shows the presence of two bands: the first moving as serum albumin and the second moving in the region of α-globulins (Figure 2). The relative area of the peaks was about 7:1, being larger the albumin peak.

Table 1
Amount of plasma proteins bound by affinity chromatography with columns prepared with different amino-purines-Sepharose (5 ml)

Specific ligand	Amount of eluted proteins (mg/10 ml plasma)
2-amino-6,8-dihydroxy-purine	0.14
6-amino-2,8-dihydroxy-purine	0.17
8-amino-2,8-dihydroxy-purine	0.80

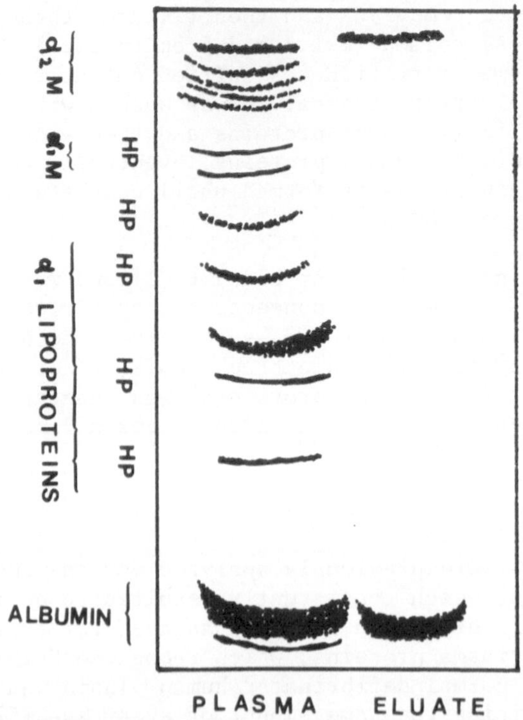

Fig. 2: Polyacrylamide electrophoresis of proteins eluted by 8-amino-2,6-dihydroxy-purine.

Temperature, in a range between 4°C and 37°C, was without any significant effect on the protein binding. On the contrary, the binding appears to be affected by the ionic strength, being maximal at a ionic strength of 0.02 and decreasing when the ionic strength rises, and by pH variations, being maximal at a pH of 5.0-5.5 and decreasing consistently when the pH falls under 4.0 or rises over 8.0.

DISCUSSION

These very preliminary results indicate that affinity chromatography is a suitable technique for studying urate protein plasma binding. Our results seem to confirm that at least two different plasma proteins recognize UA. The first is serum albumin, according to previous data of Klinenberg et al.[1], the second moves in the region of α-globulins and is likely to be identical with the α1-α2-globulin, described by Alvsaker[2]. The ability of 8-amono-2,6-dihydroxy-purine to bind a major amount of proteins seems to indicate that the hydroxyl groups in 2- and 6- positions are essential for the binding of urate to plasma proteins. We don't have attempted up to date to state the relative importance of these two positions for the binding to albumin and α-globulin. Studies are also in progress in order to quantify the amount of urate binding proteins and the quote of bound UA under normal and pathological conditions.

REFERENCES

1. D.S. Campion, R. Bluestone, J.R. Klinenberg, Uric acid. Characterization of its interaction with human serum albumin, J. Clin. Invest. 52:2238 (1973)
2. J.O. Alvsaker, Uric acid in human plasma. V. Isolation and identification of plasma proteins interacting with urate, Scand. J. Clin. Lab. Invest. 18:227 (1966)
3. J.O. Alvsaker, Urate-plasma protein interaction, Scand. J. Clin. Lab. Invest. 30:345 (1972)
4. J. Kovarsky, E.W. Holmes, W.N. Kelley, Absence of significant urate binding to human serum proteins, J. Lab. Clin. Med. 93:85 (1979)
5. P. Cuatrecasas, Protein purification by affinity chromatography. Derivatizations of agarose and polyacrylamide beads, J. Biol. Chem. 245:3059 (1970)
6. A. Lucacchini, R. Barsacchi, C. Martinelli, Isolation of human serum protein binding cardenolides by affinity chromatography, J. Solid-Phase Biochem. 2:79 (1977)
7. J.W. Jones, R.K. Robins, Potential purine antagonists. XXIV. The preparation and reactions of some 8-diazopurines, J. Am. Chem. Soc. 82:3773 (1960)
8. B.J. Davis, Disc electriphoresis. II. Method and application to human serum proteins. Ann. N.Y. Acad. Sci. 121:404 (1964)
9. M. Wilchek, R. Lamed, Immobilized nucleotides for affinity chromatography, in "Methods in enzymology", vol. XXXIV, W.B. Jakoby and M. Wilchek, eds., Academic Press, New York (1974)

CHROMATOGRAPHIC DETERMINATION OF PRPP-SYNTHETASE ACTIVITY IN HUMAN BLOOD CELLS

Per Nygaard and Kaj Frank Jensen

University Institute of Biological Chemistry B

Sølvgade 83, 1307 Copenhagen, Denmark.

The enzyme phosphoribosylpyrophosphate synthetase (EC 2.7.6.1) catalyzes a pyrophosphoryl group transfer reaction: ATP + Ribose-5-phosphate $\rightleftharpoons$ PRPP + AMP.

In blood cells 5-phosphoribosyl-α-1-pyrophosphate (PRPP) is a substrate for the _de novo_ synthesis of both purine and pyrimidine nucleotides and for purine salvage.

The activity of PRPP synthetase is most commonly measured by coupling the PRPP formation to a subsequent phosphoribosyltransferase reaction (1,2). This method suffers from the disadvantage that a brief heating step is required to terminate the PRPP synthetase reaction, which leads to some breakdown of PRPP (2). Direct isotopic assays of PRPP synthetase are available, activity may be determined by measuring the formation of radioactivity from γ^{32}P-ATP that does not absorb to charcoal. An alternative procedure is to follow ^{14}C - AMP formation from ^{14}C - ATP by separating the reaction products (2). These assays are not always well suited for enzyme measurements in crude extracts due to other ATP consuming reactions (2).

Recently we developed a method for PRPP synthetase activity measurements in cell-free extracts of bacteria (3). This method is based on the chromatographic separation of ^{32}P-labelled PRPP from γ^{32}P-ATP. This has now been applied to PRPP synthetase measurements in human red cell lysates and cell-free extracts of lymphocytes and granulocytes. The assay combines the specificity of the coupled enzymatic assays for PRPP synthetase activity with the sensitivity of the ^{32}P - transfer assay and can be used in crude extracts.

Cell Collection and Extraction

Blood was collected in heparinized tubes from normal healthy donors. Erythrocytes were isolated from 2 ml blood by centrifugation (2000 g for 5 min), washed twice with 0.9 % NaCl and resuspended in 50 mM potassium phosphate pH 7.5. Lysates were obtained by freezing (-20°) and thawing the red cells twice. Cell debris was removed by centrifugation, 6000 g for 5 min.

Lymphocytes and granulocytes were harvested from 10 ml blood after separation on a Ficoll-Isopaque gradient (4). The proportion of lymphocytes in the separated mononuclear fraction was more than 95 %. The granulocyte fraction contained more than 94 % granulocytes. The isolated cells were frozen at -15°. Storage of cells more than a few days should be avoided, because of loss of enzyme activity. The cells were disrupted by sonification for 2 x 30 sec in 50 mM potassiumphosphate pH 7.5 and centrifuged, 6000xg for 5 min. The supernatant was used for enzyme analysis.

PRPP Synthetase Assay

One enzyme unit is defined as the amount of enzyme which results in the formation of 1 nmole PRPP per min at 37°. Protein was determined by the method of Lowry (5).

The assay mixture contained in a final volume of 100 μl: 50 mM potassium phosphate pH 7.5, 20 mM NaF, 6 mM $MgCl_2$, 1 mM $\gamma^{32}P$-ATP (0.3-1.5 mCi/m mole), 1 mM ribose-5-phosphate. The mixture was prewarmed at 37° and the reaction was started by adding cell extract. At different times 10 μl samples were removed and mixed with 5 μl 0.33 M HCOOH, 2 mM ADP (marker) on a piece of parafilm. This mixture was then rapidly applied to PEI-plates, polyethyleneimine impregnated cellulose thin-layer plate on plastic sheets (Baker Chem. Comp. Philipsburg, N.J.,USA). The formic acid does not only stop the reaction but is also of importance for a proper separation of PRPP from other ^{32}P-labelled material. The plates were developped in 0.85 M potassium phosphate pH 3.4 at room temperature. If protein rich samples ($>$1 mg/ml) were applied, the plates were first developped in methanol to 5 mm above the application line and then in the above solvent. After the run, the chromatograms were thoroughly dried and subjected to autoradiography (Kodirex X-ray Film, Kodak), for about 16 hours. The nucleotide markers were localized under UV-light (265 nm). PRPP which appears between these two spots were cut out with a pair of scissors and transferred to plastic counting vials. Also ATP was counted to determine the specific activity. The samples were counted in a liquid scintillation spectrometer, using 6 ml of the following fluid in each vial, 5.5 g Permablend III, (Packard Inst. Comp. Ill. USA) per liter of toluene. PRPP could also be located by spraying with diphenylamin-anilin (6) if 10 nmoles of unlabelled PRPP were chromatographed together with the labelled sample.

Preparation of γ-^{32}P ATP

The preparation γ-^{32}P ATP is based on the exchange of ^{32}P orthophosphate into ATP catalysed by the glycolytic enzymes as described by Glynn and Chappell (7) and modified as described below:

Glyceraldehyde-3-phosphate dehydrogenase (2×10^4 units) and 3-phosphoglycerate kinase (6×10^4 units), Boehringer, Mannheim, W. Germany, were dissolved in 50 mM Tris-HCl pH 8.0 (0.55 ml) and desalted by passage through a small column (1.5 ml) of Sephadex G-25 equilibrated with the same buffer, 0.60 ml containing the enzymes was collected, 0.25 ml of low molecular weight components were added to give a final composition of: Tris-HCl pH 8.0 (50 mM), magnesium chloride (10 mM), glutathione (2 mM), nicotinamide adenine dinucleotide (0.05 mM), 3-phosphoglycerate (0.4 mM), ATP (0.2 mM), and carrierfree ^{32}P orthophosphate, A.E.C. Risø, Denmark (0.33 mCi/ml). This mixture was incubated at 37° for one minute. Usually more than 85 % of the added isotope is incorporated into ATP. The reaction was terminated by immersion in a boiling water-bath for one minute followed by cooling on ice. Then 2 % charcoal in 5 % perchloric acid (0.85 ml) was added and the mixture was left on ice for one hour with occational shaking. The charcoal was collected by centrifugation, washed twice in 5 % perchloric acid (2 ml), and once in water (2 ml). ATP was then eluted from the charcoal with 0.85 ml 50 % ethanol containing 0.2 M ammonia. After removal of the charcoal by centrifugation the supernatant was neutralized by addition of phosphoric acid and stored at -20°C. Usually a yield of more than 40 % was obtained.

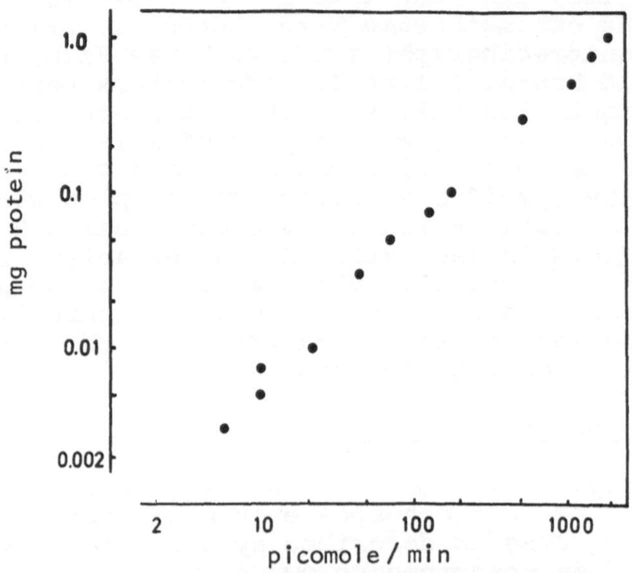

Figure I. Linearity of the PRPP synthetase assay. The assay conditions were as described in the experimental section, with the below noted modifications. Red cell lysate was diluted in 50 mM phosphate buffer pH = 7.5.

(i) At 0.25 - 1 mg protein per assay, the total incubation time was 8 min and the total cpm in ATP was 100000 (per 100 μl).

(ii) At 0.025 - 0.1 mg protein per assay, the total incubation time was 20 min and the total cpm in ATP was 100000 (per 100 μl).

(iii) At 0.0025 - 0.01 mg protein per assay, the total incubation time was 50 min and the total cpm in ATP was 600000 (per 100 μl).

RESULTS AND DISCUSSION

PRPP synthetase activity has been determined in red cell lysates and in extracts of lymphocytes and granulocytes, Table 1. The values for erythrocyte PRPP synthetase obtained with our techniques (Table 1) are slightly higher than that obtained by others, 0.7 - 0.9 units/mg protein (8,9). A linear relationship is found between the amount of lysate added and PRPP synthesized, Figure 1. A linear regression analysis of the experimental data revealed a correlation coefficient of 0.998.

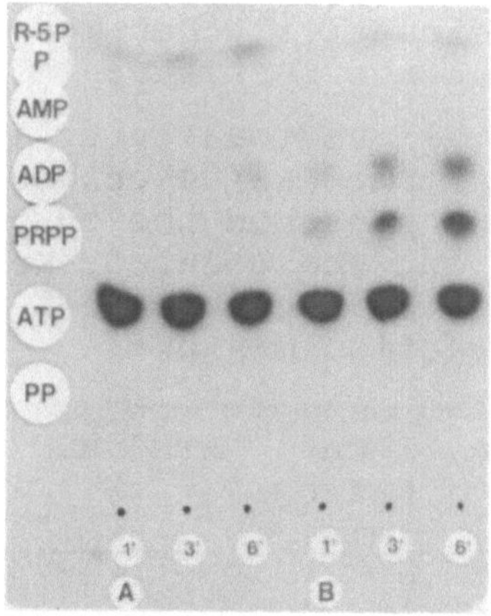

Figure II. Autoradiogram of a chromatogram used in the radiolabelled PRPP synthetase assay. The assay contained red cell lysate (1.13 mg protein) and was composed as described in Experimental Procedure. The samples were withdrawn after 1 min, 3 min, and 6 min at 37°C. (B) Complete assay mixture. (A) Ribose-5-phosphate was omitted. The total input of radioactivity in γ-^{32}P-ATP applied to the chromatogram was 6250 cpm. The rate of formation or radioactivity in the PRPP-spot in the presence of ribose-5-phosphate (B) was 158 cpm/min, without ribose-5-phosphate (A) this rate was 10 cpm/min. The specific activity of PRPP synthetase determined from these numbers is 2.2o units/mg protein.

TABLE 1

PRPP-Synthetase Activity In Human Blood Cells

Cell type	Mean (units/mg protein)	SD	In the assay (1oo µl) protein mg /ml	ATP cpm
Red cell (n = 14)	2.0	0.7	8-12	60000
Lymphocyte (n = 10)	1.9	0.7	0.4-1.2	180000
Granulocyte (n = 9)	0.8	0.3	1-2	180000

TABLE 2

Distribution Of Radioactivity Derived From γ^{32}P-ATP In The Assay For Synthetase Activity.

At time zero time, all radioactivity (100 %) was in ATP. The assays were performed as described in the experimental section, and the spots on the chromatograms (see Figure I) were cut out and counted. Numbers in parenthesis are values obtained when ribose-5-phosphate was omitted from the reaction mixture.

Cell type	Protein (mg/ml)	Incubation (min)	ATP	ADP	P_i	PRPP	Other comp.
			percentage				
Red cell (lysate)	11.3	6	74 (96)	10 (1)	1 (2)	15 (1)	<1 (<1)
Lymphocyte (extract)	0.6	30	78 (84)	7 (5)	9 (9)	3 (<1)	2 2
Granulocyte (extract)	1.1	30	80 (89)	6 (3)	11 (7)	2 (<1)	1 (1)

Figure II shows the composition of radioactive materials present in a PRPP synthetase reaction of a red cell lysate. It appears that the formation of radioactivity at the position of PRPP depends on the presence of ribose-5-phosphate. When ribose-5-phosphate is omitted in the assay mixture, negliable activity is measured. If NaF is omitted from the reaction mixture a two-fold increase in the formation of inorganic phosphate is seen. The distribution of radioactivity of the endpoint in typical assays are shown in Table 2. Such analysis may be used for the evaluation of the accumulation of the inhibitor ADP. Both in red cell lysates and in extracts of lymphocytes and of granulocytes, is the formation of ADP coupled to the PRPP synthetase reaction, most likely because of the phosphorylation of the formed AMP. In extracts of lymphocytes and of granulocytes, the formation of inorganic phosphate exceeds that of PRPP reflecting a greater influence of other ATP consuming reactions. PRPP is stable under the experimental conditions, less than 4 % can be registered as broken down to pyrophosphate.

ACKNOWLEDGMENTS

We wish to thank Inge Skibshøj and Jenny Christensen for their excellent technical assistance. We are indebted to the Bloodbank, Bispebjerg Hospital, Copenhagen, for providing blood samples.

REFERENCES

(1) Henderson, J.F. and Khoo, M.K.Y., J. Biol. Chem. 240:2349 (1965).
(2) Switzer, R.L., In The Enzymes IX:607 (Boyer, P.D. ed.) Acad. Press New York (1974).
(3) Jensen, K.F., Houlberg, U., and Nygaard, P. (1979) Analytical Biochem. (in press).
(4) Böyum, A., Tissue Antigen 4:269 (1974).
(5) Lowry, O.H., Rosebrough, N.J., Farr, A.L., and Randall, R.J., J. Biol. Chem. 193:256 (1951).
(6) Pifferi, P.G., Anal. Chem. 37:925 (1965).
(7) Glynn, J.M. and Chappell, J.B., Biochem. J. 90:147 (1964).
(8) Fox, J.H. and Kelley, W.N., J. Biol. Chem. 246:5739 (1971).
(9) Tax, W.J.M. and Veerkamp, J.H., Clin. Chim. Acta 78:209 (1977).

PURIFICATION OF MYOCARDIAL ADENOSINE KINASE USING AFFINITY AND ION-EXCHANGE CHROMATOGRAPHY

Martin P. Uitendaal, Jan W. De Jong, Eef Harmsen and Elisabeth Keijzer.

Cardiochemical Laboratory, Thoraxcenter,
Erasmus University Rotterdam,
Rotterdam (The Netherlands).

INTRODUCTION

Myocardial adenosine kinase (AK; EC 2.7.1.20) presumably plays a key role in the maintenance of adequate adenine nucleotide levels in the heart cell[1-3]. In order to study this enzyme in detail, we purified rat-heart AK to apparent homogeneity after a previous report on partial purification from this source[4]. The method presented here includes elution of AK from a 5'-AMP-Sepharose 4B column with a buffer containing adenosine. The endogenous adenosine in the fractions altered the specific activity of the radioactive substrate in the AK assay. This could be corrected for by means of HPLC adenosine measurements.

MATERIALS AND METHODS

<u>AK determination</u> was performed according to De Jong and Kalkman[1] with some modifications. The 100 µl incubation mixture contained 40 mM K-P_i (pH 7.0), 1.0 mM GTP, 1.0 mM $MgCl_2$, 10 µM [U-^{14}C]adenosine (about 50 Ci/mol) and bovine serum albumin (50 µg/ml). Samples containing up to 10 µg protein were assayed. Separation of substrate and product with DEAE-cellulose and counting were as described previously[1]. To correct for endogenous adenosine in the samples after affinity chromatography parts of the samples (2 ml) were deproteinized by heating for 3 min at 100°C and filtering through a 1.2 µm Millipore filter. Subsequently, adenosine was determined by automatic HPLC on a µBondapak C_{18} column (0.39 x 30 cm) with an elution buffer consisting of 10 mM ammonium formiate pH 5.5/methanol (87/13) at a rate of 1.0 ml/min.

Adenosine deaminase activity was determined as described previously[4]. 2 µM erythro-9-(2-hydroxy-3-nonyl)adenine (EHNA) inhibited the enzyme completely.

Protein was determined according to Bradford[5]. No interference was noted between the Coomassie Blue reagent and P_i, Tris, or GSH.

Gel electrophoresis was carried out as described by Weber and Osborn (ref. 6).

Molecular weight of AK from 5'-AMP-Sepharose 4B fractions was determined with a 2.5 x 63 cm Sephadex G-100 column with bovine serum albumin (MW 67,000), pigeon-chest muscle carnitine acetyl transferase (CAT, MW 55,000), ovalbumin (MW 45,000), soy bean trypsin inhibitor (MW 41,500), and horse-heart cytochrome c (MW 12,400) as standards. 0.1 M Tris-HCl (pH 7.5) was used as eluent.

Purification. The whole procedure was performed at 0-8°C and all elution buffers contained 1 mM GSH. A 5% homogenate of hearts from male Wistar rats (200-250 g) was prepared with a Virtis homogenizer in 0.25 M sucrose containing 10 mM Tricine-KOH buffer and 1.0 mM Na_2-EDTA (pH 7.4). The homogenate was centrifuged at 7,500·g for 10 min and the supernatant was spun for 1 h at 100,000·g. The supernatant (cytoplasmic fraction) was collected. From this fraction 10-80 ml was applied at a rate of 1 ml/min to a 2 g 5'-AMP-Sepharose 4B column (1.6 x 5 cm). The column was washed with 20 mM K-P_i buffer (pH 7.0). When the absorbance at 260 nm, monitored with a Uvicord III, was minimal, the buffer was changed to 20 mM Tris-HCl (pH 7.0). AK could be eluted with 20 mM Tris buffer containing 1 mM MgATP and 0.1 mM adenosine. The combined fractions containing AK were applied to a column of DEAE-Sephacel (1.6 x 15 cm). The column was washed with 20 mM Tris buffer and when the absorbance at 260 nm was minimal a linear gradient from 20 to 100 mM Tris-HCl (pH 7.0; total 150 ml) was started. From both columns the eluate was collected in 5 ml fractions.

RESULTS

According to the HPLC measurements, fractions from the 5'-AMP-Sepharose column contained up to 90 µM adenosine. Therefore, espe-

Table 1. Purification of adenosine kinase from rat heart.

Purification step		spec. act. (U/g)	purification factor	yield %
Homogenate		0.458	1	100
Cytoplasm		2.74	6	111
5'-AMP-Sepharose 4B		37.3	81	23
DEAE-Sephacel: fr.	39-41	8280	18100	10
	36-46	5670	12400	21

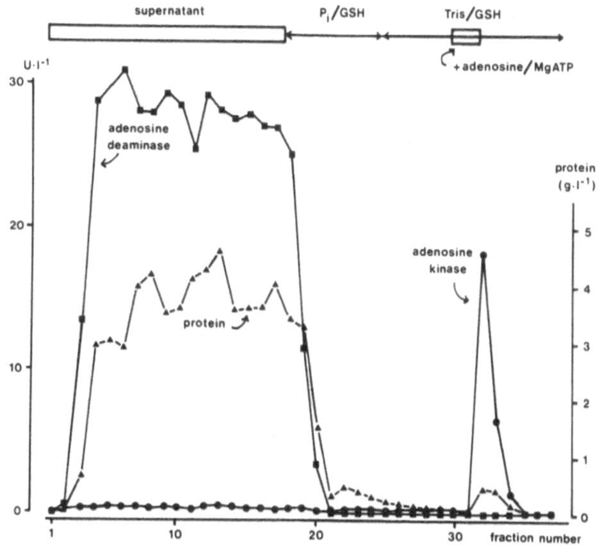

Fig. 1. Affinity chromatography of myocardial cytoplasm on a 5'-AMP-Sepharose column. Application of 79 ml cytoplasmic fraction and elution with buffers was as indicated at top of figure.

cially when large samples were added in the AK assay, correction factors up to 1.4 had to be introduced. In the first step of the AK purification - preparation of the cytoplasmic fraction - no AK activity is found in the pellet fraction, which contains most of the protein. This leads to a 6-fold purification (table 1). In the following step - affinity chromatography - AK binds to the column, while the interfering enzyme adenosine deaminase is washed off completely

Table 2. Nucleotide formation from adenosine and deoxyadenosine by rat-heart cytoplasm in the presence of inhibitors.

Substrate	Percentage of control	
	NSC 113939 (100 µM)	EHNA (2 µM)
adenosine	10	75
2'-deoxyadenosine[a]	84	0

[a]Incubation conditions were as described for adenosine, except for 10 µM [U-^{14}C]2'-deoxyadenosine in stead of adenosine. The activity was 10% of that with adenosine.

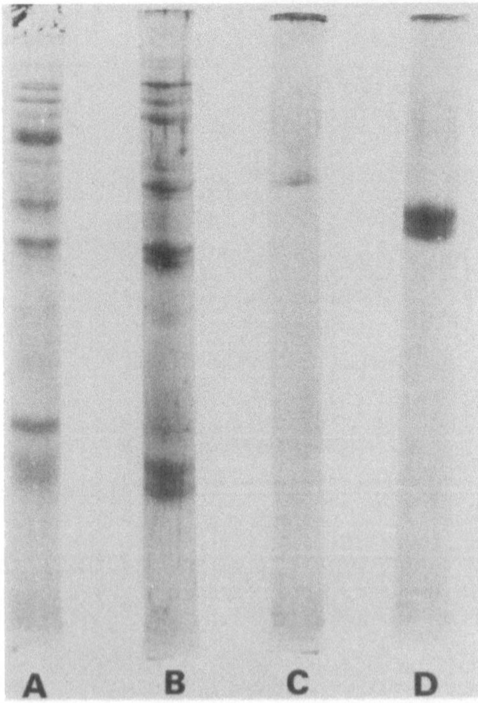

Fig. 2. SDS-polyacrylamide gel electrophoresis of samples from different purification steps. A. Cytoplasmic fraction. B. 5'-AMP-Sepharose 4B; fractions 31-36. C. DEAE-Sephacel; fractions 39-41. D. Standard: Ovalbumin.

in 20 mM P_i buffer (fig. 1). The AK can only be eluted with a buffer containing Mg^{2+}, ATP and adenosine and is stable at -25°C at this stage. The preparation, purified another 13.5 times, contains however many other proteins as shown by SDS-polyacrylamide gel electrophoresis (fig. 2). To remove these and the purine compounds used in the elution buffer, the AK-containing fractions were pooled and subjected to ion-exchange chromatography (fig. 3). The specific activity is increased by a factor 220 and the yield is 45% in the peak fractions (table 1; the fractions had to be pooled and concentrated to enable protein determination and electrophoresis). According to gel electrophoresis, the preparation contains no detectable contaminating proteins. At this stage AK is no longer stable at -25°C. Some properties of myocardial AK were investigated in partly purified fractions. The molecular weight in fractions from the affinity chromatography was determined by gel filtration and appeared to be 38,000 for rat-heart AK. Also the substrate specificity of AK has been investigated in the cytoplasmic fraction. This fraction has some capacity to

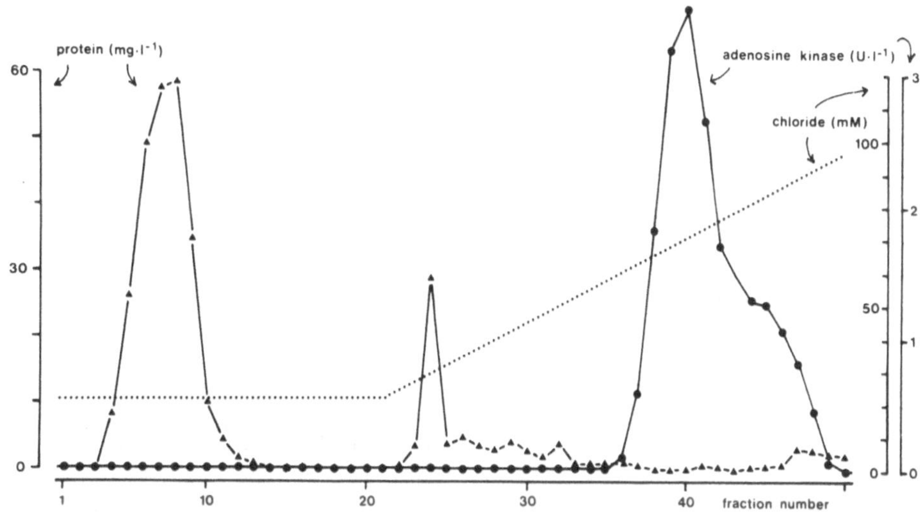

Fig. 3. Ion-exchange chromatography of partly purified AK on a DEAE-Sephacel column.

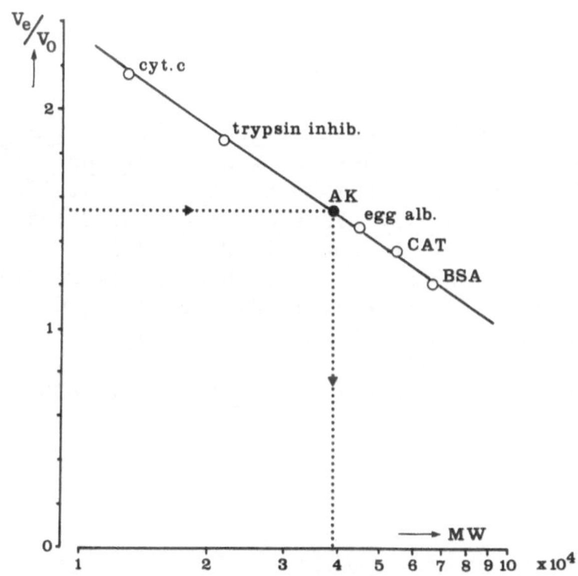

Fig. 4. Molecular weight determination of AK on Sephadex G-100.

convert 2'-deoxyadenosine to nucleotides, which bind to DEAE-cellulose (see AK determination). However, this activity is completely inhibited by 2 μM EHNA (which also inhibits adenosine deaminase completely) and is hardly affected by 100 μM NSC 113939 (4-amino-5-iodo-7-β-D-ribofuranosyl *7H*-pyrrolo(2,3-*d*)pyrimidine), which is a specific AK inhibitor[4]. Nucleotide formation from adenosine is strongly inhibited by NSC 113939 and hardly affected by EHNA (table 2).

CONCLUSION

Determination of adenosine can be done easily with HPLC and avoids underestimation of AK activity. In this study correction factors up to 1.4 had to be applied. The purification method presented gives a virtually homogeneous AK preparation which looses its activity rapidly when stored at -25°C. The yield of 10% can be increased to 21% when a somewhat lower purification factor is taken for granted (table 1). The molecular weight of rat-heart AK, 38,000, is well in agreement with results for yeast AK[7]. From table 2 it is clear that phosphorylation of deoxyadenosine by rat-heart cytoplasm is not effectuated by AK. This is confirmed by the fact that deoxyadenosine does not compete with adenosine in an AK assay[8]. Because the method described here is completed within a day, it enables further characterisation of AK in pure preparations, in spite of the lability of the enzyme.[†]

[†]Dr. De Jong is an Established Investigator, Dutch Heart Foundation.

REFERENCES

1. J. W. De Jong and C. Kalkman, Biochim. Biophys. Acta 320:388 (1973).
2. L. W. Brox and J. F. Henderson, Can. J. Biochem. 54:200 (1976).
3. J. R. S. Arch and E. A. Newsholme, Biochem. J. 174:965 (1978).
4. J. W. De Jong, Arch. Int. Physiol. Biochim. 85:557 (1977).
5. M. M. Bradford, Anal. Biochem. 72:248 (1976).
6. K. Weber and M. Osborn, J. Biol. Chem. 244:4406 (1969).
7. T. L. Leibach, G. I. Spiess, T. J. Neudacker, G. J. Peschke, P. Puchwein and G. R. Hartman, Hoppe-Seyler's Z. Physiol. Chem. 352:328 (1971).
8. D. H. Namm and J. P. Leader, Anal. Biochem. 58:511 (1974).

MICROMETHODS FOR THE MEASUREMENT OF PURINE ENZYMES IN LYMPHOCYTES

J.P.R.M. van Laarhoven, G.Th. Spierenburg,
F.T.J.J. Oerlemans and C.H.M.M. de Bruyn.

Department of Human Genetics, Faculty of Medicine,
University of Nijmegen, Nijmegen, The Netherlands.

INTRODUCTION

The involvement of purine interconversion enzyme defects in impairment of the immune system is now well documented (1-3). Although it has been suggested that deoxypurine nucleotides might be the toxic metabolites in these immune diseases (4-5), the mechanism which leads to dysfunctions of T or B cells, or both of them, is still not completely elucidated. A better understanding of purine interconversions in B and T cell subfractions might help to obtain a better view on B or T cell specificity in these immune diseases. One of the possibilities to achieve this might be a systematic analysis of purine metabolism in T and non-T lymphocytes.

The first purpose of the present study was to investigate the effect of cell destruction procedures, such as sonification and lyophilisation, on a number of purine enzyme activities. This was studied with pure lymphocyte preparations from human tonsils.

Determination of nine purine enzyme activities in lymphocyte subfractions from one peripheral blood sample using conventional methods required too much blood. Therefore, the second purpose of the present study was to develop new radiochemical micro techniques, which are based on previously described ultramicrochemical methods (6-11). This method has been applied to almost pure, unfractionated lymphocyte preparations, using 500-5000 cells per assay.

MATERIALS AND METHODS

Isolation of Lymphocytes from Tonsils.

After removal of the surrounding tissues, tonsils were homogenized (Potter; Janke & Kunkel KG) in Tris-buffered minimal essential medium (MEM; Gibco, F-14; pH 7.4) containing 15% (v/v) foetal calf serum (FCS; Gibco). After filtration over a wire mesh, the suspension was layered on lymphoprep (gravidity 1.077 gr/ml; Nyegaard. AS, Oslo). Lymphocytes were collected from the interphase after centrifugation for 20 min. at 1,000 g (room temperature) and washed once with a solution containing 155 mM NH_4Cl, 10 mM $KHCO_3$, 0.1 mM disodium EDTA (pH 7.4). Before use the cells were washed twice with 0.9% NaCl (w/v).

Isolation of Lymphocytes from Peripheral Blood.

30 ml of defibrinated (on glass beads; Ø 5 mm) venous blood was incubated for 15 min. at 37 °C. After passage through a nylon wool column (12) the effluent was diluted with MEM/Tris containing 15% FCS(v/v) to a concentration of approximately $1x10^9$ blood cells/ml. Subsequently this cellsuspension was carefully layered on top of a density gradient, that consisted of 15 ml Ficoll-Isopaque (gravidity 1.085 gr/ml; Ficoll 400, Pharmacia, Uppsala, Sweden; Isopaque 440 mg J/ml, Nyegaard & Co.AS, Oslo) and 5 ml of a lighter Ficoll-Isopaque solution (gravidity 1.055 gr/ml). The centrifugation was carried out at room temperature (20 min., 1,000 g). Lymphocytes could be collected from the interphase between the two Ficoll-Isopaque solutions. During the isolation the amounts of monocytes, granulocytes and lymphocytes were checked with a Hemalog D (Technicon Instruments Corp., Tarrytown, NY, USA). The original total leukocyte fraction in whole EDTA blood contained 29% lymphocytes, 7% monocytes, 62% granulocytes and 6% "large unstained cells" (LUC). Defibrination and filtration over a nylon wool column according to de Pauw et al. (12) yielded a fraction after Ficoll-Isopaque density centrifugation which contained over 97% lymphocytes and a very low contamination of monocytes, granulocytes and LUC.

"Macro" Enzyme Assays with Tonsillar Lymphocytes.

Lymphocytes were lysed in several ways: freezing and thawing (-20 °C, 5 cycles), lyophilisation and sonification (3 x 10 sec., output control 7; Sonifier B-12, Branson Sonic Power Co., Danbury, Ct, USA). For all enzyme assays 0.05-5 µgr protein per incubation was added (protein estimation according to Lowry et al., 13).
All enzyme assays were carried out in triplicate and in table 1 the mean values are given.

Hypoxanthine-guanine phosphoribosyltransferase (HG-PRT; E.C. 2.4.2.8) and adenine phosphoribosyltransferase (APRT; E.C. 2.4.2.7) were assayed essentially according to de Bruyn et al. (9). Purine nucleoside phosphorylase (PNP; E.C. 2.4.2.1) and adenosine deaminase (ADA; E.C. 3.5.4.4) were assayed according to previously described methods (10,11).

"Micro" Assays with Peripheral Blood Lymphocytes.

Lymphocyte suspensions containing 1,000-10,000 cells/µl were prepared in 0.9% NaCl. Aliquots of 0.5 µl were pipetted into small incubation vessels prepared from parafilm (Parafilm "M", American Can Co., Greenwich, Ct, USA). These parafilm micro cuvettes (PMC) were prepared immediately before use (6-8). The lymphocytes were frozen in the PMC's at -20 ^{o}C for 15-30 min. and subsequently lyophilised overnight. All enzyme assays were carried out in five-fold. For HG-PRT, APRT, ADA and PNP the reactions were started by adding 3 µl of the appropriate incubation mixture. The concentrations were the same as described above for the "macro" assays, except for the addition of 0.2% (v/v) triton X-100 (Sigma) in several experiments. In the HG-PRT reaction also 8-^{14}C-guanine (0.4 mM; spec. act. 55 mCi/mmol; Radiochemical Centre Amersham, UK) as a substrate; ADA was also tested with 8-^{14}C-deoxyadenosine (0.8 mM; spec. act. 45 mCi/mmol; NEN Chemicals GmbH, Dreieich, GFR). Incubation times were 1 to 4 hours at 37 ^{o}C. Quantification of enzyme activities was carried out as described elsewhere (9-11). The adenosine kinase (AK; E.C. 2.7.1.20) assay was adapted from Meyskens et al. (14). To the lyophilised lymphocytes, 3 µl of a reaction mixture was added, containing 3 µM 8-^{14}C-adenosine, 1.5 mM ATP (Boehringer Mannheim), 0.3 M trisodiumacetate/acetic acid (pH 5.7), 0.6 mM $MgCl_2$, 12.5 µM erythro-9-(2-hydroxy-3-nonyl)-adenine (EHNA), kindly supplied by Dr. H.A. Simmonds (Purine Laboratory, Guy's Hospital Medical School, London, UK) and 0.2% triton X-100. After incubation (4 hrs., 37 ^{o}C) separation of substrate and products were performed by means of high voltage electrophoresis on Whatmann 3 MM paper (0.05 M citrate buffer, pH 3.9; 70 V/cm). Purine-5'-nucleotidase (5'N; E.C. 3.1.3.5) assay; to the lyophilised lymphocytes 3 µl of a reaction mixture was added (15) containing 0.05 M Tris/HCl (pH 8.5), 0.02 M $MgCl_2$, 6.25 mM 2 -glycero-phosphate (Sigma) and 0.6 mM 8-^{14}C-AMP (spec. act. 61 mCi/mmol; Radiochemical Centre Amersham, UK). Incubation (4hrs., 37 ^{o}C) was followed by separation of substrate and products as in the AK assay.

RESULTS AND DISCUSSION

Effect of Various Lysate Preparations on Enzyme Activities.

Table 1. Effect of Different Methods of Cell Destruction on 5 Purine Enzyme Activities.

Enzyme	Procedure 1	2	3	4	5	6	7	8
HG-PRT	4.3	5.9.	4.6	6.1	2.9	5.2	4.4	4.9
APRT	7.4	8.9	7.2	9.6	7.4	8.3	8.5	7.7
ADA	49.4	85.0	52.2	84.0	54.7	69.2	51.2	81.2
PNP-Hx	44.0	90.5	71.8	110.6	55.0	33.0	51.1	57.6
PNP-Ino	15.2	21.8	23.5	26.0	20.0	15.4	20.4	24.7

Procedure 1. freezing and thawing (5 cycles, -20 ^oC)
2. freezing and thawing (5 cycles, -20 ^oC), with addition of 0.2% triton X-100
3. lyophilisation and resuspension in 0.01 M Tris/HCl (pH 7.4)
4. lyophilisation and resuspension in 0.01 M Tris/HCl (pH 7.4), with addition of 0.2% triton X-100
5. lyophilisation and sonification after resuspension in 0.01 M Tris/HCl (pH 7.4)
6. sonification preceded by lyophilisation
7. sonification
8. sonification in the presence of 0.2% triton X-100

Enzyme activities are expressed in 10^{-9} mol product formed/hour. 10^6 cells. For each enzyme the procedure which gives the highest activity is placed in a rectangle.

Tonsillar lymphocytes, suspended in 0.01 M Tris/HCl (pH 7.4), were lysed in 8 different ways in order to investigate, which method of cell destruction is to be preferred for the determination of purine enzyme activities (table 1). After preparation of the lysate, insoluble particles were removed by centrifugation (300 g, 15 min.). As can be seen in table 1 the procedures using 0.2% triton X-100 yielded highest enzyme activities, especially in the case of freezing and thawing and lyophilisation (procedures 2 and 4). This effect was seen with all enzymes tested.

In the present studies enzyme activities are expressed on a per cell basis. When expressing the activities on a protein basis it was found that, although with some procedures much more protein was released into the soluble fraction than with other procedures (e.g. lyophilisation + sonification; procedure 5), the specific enzyme activities were not higher as compared to other procedures (16). For routine determinations the method of choice for the preparation of lymphocyte lysates turned out to be lyophilisation with addition of 0.2% triton X-100 (procedure 4).

Table 2. Reproducibility of the "Micro" Assay for ADA and PNP, assayed in Five-Fold in the Same Sample.

	probe no.	net product formed (cpm)	enzyme activity (10^{-9} mol/10^{6} cells.hr)
ADA	1	6568	57.77
	2	5620	62.95
	3	6651	53.40
	4	4598	51.29
	5	7302	71.04
			59.29(7.95) mean(s.d.)
PNP	1	6591	80.00
	2	8508	78.51
	3	8549	90.16
	4	6132	80.73
	5	7302	83.72
			82.62(4.62) mean(s.d.)

ADA assay: incubation with 1500 cells for 4 hours at 37 °C, input 60,000 cpm 8-^{14}C-adenosine
PNP assay: incubation with 1500 cells for 1 hour at 37 °C, input 30,000 cpm 8-^{14}C-hypoxanthine

"Micro" chemical Enzyme Determinations.

Nine purine interconversion enzyme activities could reproducibly be assayed with relatively small numbers of cells(500-5000). Two examples are given in table 2, where the raw data and calculated

Table 3. Purine Interconversion Enzymes in Lymphocytes

Enzyme	Substrate	Spec. Act.(standard deviation)
HG-PRT	hypoxanthine	2.95(0.72)
	guanine	6.49(2.44)
APRT	adenine	8.93(1.56)
ADA	adenosine	46.27(10.94)
	deoxyadenosine	34.36(9.04)
PNP	hypoxanthine	63.87(16.70)
	inosine	12.99(1.82)
AK	adenosine	0.74(0.47)
5'N	AMP	12.26(7.08)

Activities calculated from a group of 7 healthy individuals and expressed as 10^{-9} mol/10^{6} cells.hour.

enzyme activities are given of an ADA and PNP assay; measurements were carried out in five-fold.

The mean activities in purified normal human lymphocytes of nine enzymatic reactions involved in purine interconversions are shown in table 3. ADA and PNP displayed the highest activity in pure lymphocytes whereas relatively low activities were found for HG-PRT and AK. The rather wide range of specific enzyme activities was attributed to individual variation rather than to methodological errors.

With the present "micro" methodology it becomes possible to carry out a great number of enzyme determinations with small numbers of cells. In the present study only radiochemical assays are described, but also other substrates(e.g. fluorogenic; van Laarhoven et al. unpublished) can be employed both for experimental and diagnostic purposes.

REFERENCES

1. E.R. Giblett, J.E. Anderson, F. Cohen, B. Pollara and H.J. Meuwissen, Lancet i:1067 (1972).
2. E.R. Giblett, A.J. Ammann, D.W. Ward, R. Sandman and L.K. Diamond, Lancet ii:1010 (1975).
3. N.L. Edwards, D.B. Magilavy, J.T. Cassidy and I.H. Fox, Science 201:628 (1975).
4. D.A. Carson, J. Kaye and J.E. Seegmiller, Proc. Natl. Acad. Sci. USA, 74:5677 (1977).
5. A. Cohen, L.J. Gudas, A.J. Ammann and G.E.J. Staal, J. Clin. Invest. 61:1405 (1977).
6. P. Hösli, Tecnomara A.G., Rieterstrasse 59, Postfach CH 8059, Zürich, Switzerland (1972).
7. P. Hösli, in: Birth Defects, A. Motulsky and W. Lenz, eds., Exerpta Medica, Amsterdam, p. 226 (1974).
8. P. Hösli, Clin. Chem. 23:1476 (1977).
9. C.H.M.M. de Bruyn, T.L. Oei and P. Hösli, Biochem. Biophys. Res. Commun. 68:483 (1977).
10. M.P. Uitendaal, C.H.M.M. de Bruyn, T.L. Oei, P. Hösli and C. Griscelli, Anal. Biochem. 84:147 (1978).
11. M.P. Uitendaal, C.H.M.M. de Bruyn, T.L. Oei, S.J. Geerts and P. Hösli, Biochem. Med. 20:54 (1978).
12. B.E.J. de Pauw, J.M.C. Wessels, E.J.M. Geestman, J.B.J.M. Smeulders, D.J.Th. Wagener and C. Haanen, J. Imm. Meth. 25:291 (1979).
13. O. H. Lowry, N.J. Rosebrough, A.L. Farr and R.J. Randall, J. Biol. Chem. 193:265 (1951).
14. F.L. Meyskens and H.E. Williams, Biochem. Biophys. Acta, 240:170 (1971).
15. C. Ip and T. Dao, Cancer Res. 38:723 (1978).
16. J.P.R.M. van Laarhoven, G.Th. Spierenburg, C.H.M.M. de Bruyn and E.D.A.M. Schretlen, in preparation.

A RAPID SCREENING METHOD FOR INBORN ERRORS OF PURINE AND PYRIMIDINE METABOLISM USING ISOTACHOPHORESIS

H. A. Simmonds, A. Sahota and R. Payne

Clinical Science Laboratories, Guy's Hospital, London;

LKB Instruments, Selsdon, South Croydon, U.K.

INTRODUCTION

Isotachophoresis has recently been used in the analysis of purine nucleotides and amino acids[1,2]. The speed, sensitivity and high resolving power of this relatively new technique suggested it might have potential for the separation of purines and pyrimidines in biological fluids.

In this paper we describe the use of isotachophoresis in the screening for inborn errors of purine and pyrimidine metabolism, as well as the identification of drug metabolites following allopurinol therapy in some of these situations.

MATERIALS AND METHODS

Chemicals

Purine and pyrimidine bases and nucleosides were obtained from the Sigma Chemical Company, β-alanine and Tris (hydroxymethyl) methylamine from British Drug Houses; hydroxypropylmethylcellulose (HPMC) from the Dow Chemical Company.

Instrument

The instrument used was an LKB 2127 Tachophor (LKB Instruments, South Croydon, Surrey) equipped with a UV detector capable of monitoring at 254 or 280 nm. The UV signal was recorded at a chart speed of 8 cm/min.

Determination of optimal experimental conditions

Experiments were carried out at different concentrations and pH of leading and terminating electrolyte and with varying currents and capillaries of different length. With a 43 cm capillary and a 2.5 mM leading electrolyte the zones were well separated with an analysis time of 35-45 minutes. The optimum pH finally selected for the leading electrolyte (Tris-HCl made up in 0.3% HPMC) was 7.9. The terminator was 20 mM β-alanine with solid barium hydroxide, added to precipitate bicarbonate and adjust the pH to 10.6. Using the above conditions a constant current of 20 μA was found to give the best result, following the initial period at the higher current.

Patient material and specimen handling

Urine samples from controls and patients with deficiencies of the following enzymes were analysed: adenine phosphoribosyltransferase (APRT: EC 2.4.2.7), hypoxanthine guanine phosphoribosyltransferase (HGPRT: EC 2.4.2.8), adenosine deaminase (ADA: EC 3.5.4.4), purine nucleoside phosphorylase (PNP: EC 2.4.2.1), xanthine oxidase (XOD: EC 1.2.3.2), orotate phosphoribosyltransferase (OPRT: EC 2.4.210) and orotodylate decarboxylase (ODC: EC 4.1.1.2). Details of the majority of these cases have already been published[3]. Urines were also obtained during treatment with allopurinol.

Application to biological samples

5-10 μl of urine (diluted 1:5 or 1:10 with distilled water depending on the total volume to give approximately the same pseudouridine concentration) were injected. Pseudouridine, a pyrimidine of endogenous origin excreted daily with relative constancy was found a suitable internal reference for dilution[3].

RESULTS

Calibration

A tachophor scan of a mixture of standards showed good separation[3] in the following order: orotic acid, uric acid, hippuric acid, xanthine, oxipurinol, hypoxanthine, inosine, pseudouridine, guanosine and adenine. In this system orotic acid had the highest net mobility, adenine the lowest, being of the same order as the terminator. Compounds such as adenosine and deoxyadenosine would not leave the terminator to be recorded in the tachophor scan. However, the acid lability of deoxyadenosine could be utilised for its identification following hydrolysis of the urine sample for 24 hr in HCl at pH 1 showed the appearance of a distinct adenine peak at 254 nm. Concentration of the applied sample was found to be critical. Zone mixing occurred if the capacity of the capillary was exceeded.

Application to the screening for inborn errors of purine and pyrimidine metabolism

Good resolution was obtained using appropriately diluted urine 1/5 or 1/10 from a control child, as shown in Fig. 1. The three predominant peaks in the tachophor scan of control urine were identified as uric acid, hippuric acid and pseudouridine by spiking with standards. The differential absorbance of many purines and pyrimidines at 280, as compared with 254 nm, was invaluable for their identification. At 280 nm control urines contained only one major UV absorbing peak: uric acid. Orotic acid, like uric acid, was one of the few compounds with a 280/254 ratio in excess of unity and was thus readily distinguishable.

1. Purine enzyme deficiencies

Purine 'salvage' enzymes:

(i) APRT: In the complete deficiency adenine was the only abnormal metabolite detected (Fig. 2). The adenine peak was diagnostic of the homozygous state[3].

(ii) HGPRT: In homozygotes for HGPRT deficiency the characteristic feature was the abnormally large uric acid zone, even more evident at 280 nm as compared with control scans (Fig. 2) and measurable amounts of hypoxanthine and xanthine were also present.

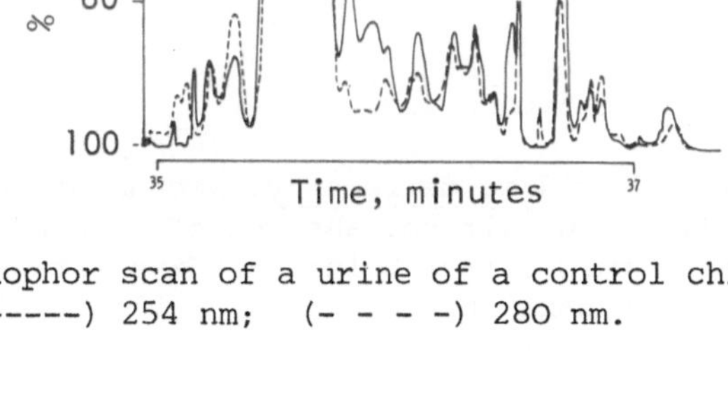

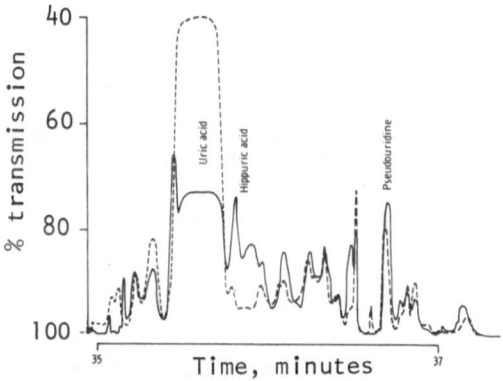

Fig. 1. Tachophor scan of a urine of a control child. (-------) 254 nm; (- - - -) 280 nm.

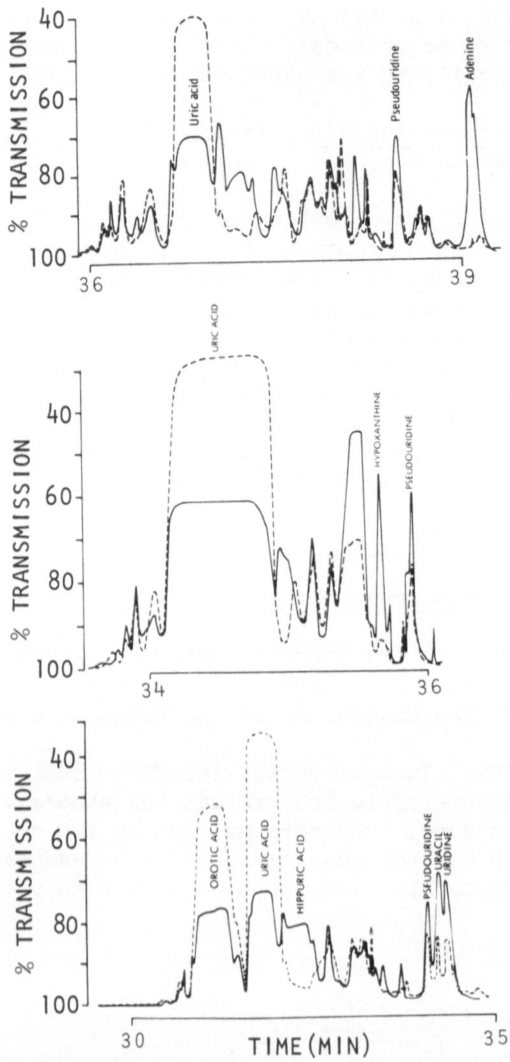

Fig. 2. Tachophor scan of the urine of: TOP - an APRT deficient child; MIDDLE - an HGPRT deficient child; BOTTOM - an OPRT/ODC deficient child.
(-------) 254 nm; (- - - -) 280 nm.

Enzymes of purine catabolism:

(i) Xanthine oxidase deficiency: Scans at 280 and 254 nm confirmed the virtual absence of uric acid, the major peak in this defect being xanthine with a small but evident hypoxanthine peak (Fig. 3).

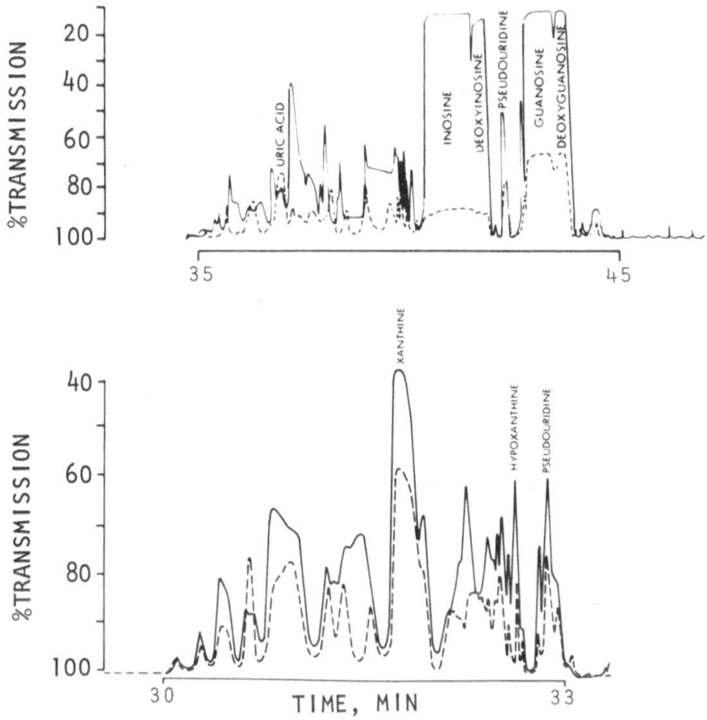

Fig. 3. Tachophor scan of the urine of: TOP - a PNP deficient child; BOTTOM - a xanthine oxidase deficient patient. (-------) 254 nm; (- - - -) 280 nm.

(ii) Purine nucleoside phosphorylase deficiency: The scan at 280 and 254 nm showed virtual absence of uric acid and its replacement by the precursors, inosine, guanosine, and the corresponding deoxyribonucleosides. This defect also shared with HGPRT deficiency a readily apparent increment in total purine end-product (Fig. 3).

(iii) Adenosine deaminase deficiency: The scan showed the presence of uric acid in normal amounts. No adenine[3] was detected by isotachophoresis. However, the deoxy-adenosine excreted in this defect[4] could be degraded to adenine by acid hydrolysis and detected by isotacho-phoresis in a repeat scan at 254 nm (Fig. 4).

2. Pyrimidine enzyme deficiencies

(i) Hereditary oroticaciduria OPRT:ODC deficiency): Urine samples from this defect showed a prominent orotic acid zone immediately before the dominant uric acid zone

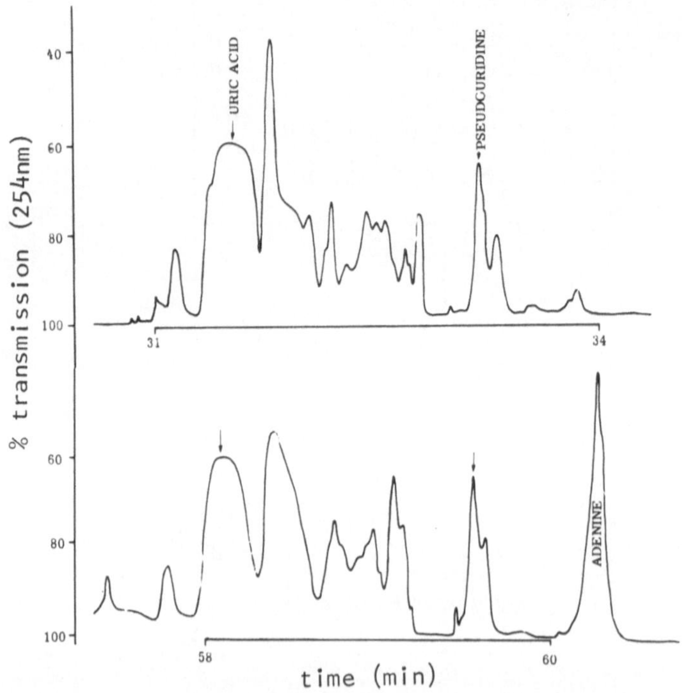

Fig. 4. Tachophor scan at 254 nm of an ADA deficient urine (TOP); after acid hydrolysis (BOTTOM)

readily identifiable because of the 280/254 ratio greatly in excess of unity (Fig. 2).

3. Pharmacological applications of isotachophoresis

Scans during allopurinol therapy reflected the results anticipated; a reduction in uric acid excretion being accompanied by a concomitant increase in urinary xanthine and hypoxanthine[3].

DISCUSSION

Specific screening procedures, based predominantly on radioisotope assays in blood cells or other tissues, have been developed for the identification of inborn errors of metabolism[3]. Their disadvantage (apart from being time-consuming) is their expense and dependence on the availability of isotopes; the activity of some of these enzymes in erythrocytes may also decrease rapidly[3].

The tachophor overcomes these difficulties, is simple to

operate and requires only a few drops of urine (often more readily obtainable than blood) which may be transported frozen without detriment to the purine or pyrimidine content. These studies have shown that isotachophoresis is applicable to the separation of urinary purines and pyrimidines and to the identification of specific inborn errors of metabolism.

Repeat runs at 254 and 280 nm could distinguish six enzyme defects by the characteristic urinary metabolites excreted: adenine in APRT deficiency, gross uricaciduria in HGPRT deficiency, xanthine in xanthine oxidase deficiency, nucleosides and deoxynucleosides of guanine and hypoxanthine in PNP deficiency, orotic acid in OPRT:ODC deficiency, deoxyadenosine (as adenine) in ADA deficiency. Isotachophoresis could also be of value to assess the effect of allopurinol therapy.

REFERENCES

1. Gower, D. C. and Woledge, R. C., 1977, The use of isotachophoresis for the analysis of nuscle extract. Sci. Tools, 24:17.
2. Everaerts, F. M. and Van der Put, A. J. M., 1970, Isotachophoresis: The separation of amino acids. J. Chromatog., 52:415.
3. Sahota, A., Simmonds, H. A. and Payne, R. H., 1979, Separation of urinary purines and pyrimidines by isotachophoresis: Usefulness in screening for inborn errors of purine and pyrimidine metabolism. J. Pharmacol. Methods, 2:303.
4. Simmonds, H. A., Sahota, A., Potter, C. F. and Cameron, J. S. 1978, Purine metabolism and immunodeficiency: Urinary purine excretion as a diagnostic screening test in adenosine deaminase and purine nucleoside phosphorylase deficiency. Clin. Sci. Mol. Med., 54:579.
5. Hoffee, P. A. and Jones, M. E. (eds), 1978, Methods in Enzymology: Purine and Pyrimidine Nucleotide Metabolism, Academic Press, New York.

ANALYSIS OF SERUM PURINES AND PYRIMIDINES BY ISOTACHOPHORESIS

F. Oerlemans[o], Th. Verheggen[*], F. Mikkers[*], F. Everaerts[*] and C. de Bruyn[o].
[o]Department of Human Genetics, Faculty of Medicine, University of Nijmegen, Nijmegen, The Netherlands.
[*]Department of Instrumental Analysis, Eindhoven University of Technology, Eindhoven, The Netherlands.

INTRODUCTION

Purine and pyrimidine metabolism receive attention from a rapidly growing number of workers in the field of inborn errors (1), hematology (2), immunology (3) and oncology (4,5).
The availability of metabolite profiles of body fluids and cell contents might attribute to a better understanding of mechanisms underlying metabolic disturbances. This enables a more direct approach for both diagnostic and experimental purposes. For identification of purines and pyrimidines thin-layer high voltage electrophoresis and chromatography can be used (6). A more rapid technique involves high performance liquid chromatography (HPLC) and is widely used at present (7,8). An alternative to HPLC for a screening of metabolite profiles might be isotachophoresis (9). This technique has recently been introduced for the separation and identification of muscle nucleotides (10) and urinary purines and pyrimidines (11). An advantage of isotachophoresis as compared to HPLC is its flexibility: buffers can be changed rapidly, no columns need to be equilibrated. In this paper two systems are presented for the separation of a number of purines and pyrimidines in serum: one low-pH system (pH 3.9) for nucleotides and one high-pH system (pH 7.75) for bases and nucleosides.

MATERIALS

The following purines and pyrimidines (analytical grade) were purchased from Merck (Darmstadt, G.F.R.): AMP, cAMP, GMP, UDP, UTP, CTP (in standard mixtures for system pH 3.9) uric acid, hypoxanthine, inosine, adenine and guanosine (in standard mixture for system pH 7.75). From Sigma (St. Louis, Ky,USA): UMP (system pH 3.9) orotic acid, xanthine and allopurinol (system pH 7.75). From Boehringer (Mannheim, G.F.R.): ADP and ATP (system pH 3.9). Finally, GTP (system pH 3.9) was obtained from Roth. Hippuric acid was purchased from Sigma, hydroxyethylcellulose (HEC) from Polysciences (Warrington, Pa, USA, cat nr. 5568). Serum was prepared from venous blood after clotting (2 hrs. at room temperature) and centrifugation for 10 min. at 1,000 g (4 °C). Storage of samples occured at -20 °C. Before analysing the samples with system pH 7.75 an ultrafiltration step (Amicon CF 25 filter) had to be introduced to remove insoluble complexes. For samples to be analysed at pH 3.9 this was not necessary.

METHODS

In isotachophoresis a discontinuous system is used consisting of two electrolytes. The leading electrolyte contains ions with a higher effective mobility than that of the metabolites to be separated (separands). The terminating electrolyte contains ions with a lower effective mobility than that of the separands. When the separation has been achieved the sample ions move in consecutive zones with a speed equal to that of the leading zone. Each zone contains only ionic species with equal effective mobilities (9). The separation takes place in narrow capillaries (I.D. 0.2-0.8 mm), according to the column coupling system described

System I (pH 3.9) for separation of purine and pyrimidine nucleotides

	leading electrolyte	terminating electrolyte
Anion	Cl^-	$Caproate^-$
Concentration	0.01 M	0.005 M
Counter ion	$GABA^+$	$Tris^+$
pH	3.9	6.5
Additive	0.25% HEC	-

System II (pH 7.75) for separation of purine and pyrimidine bases and nucleosides

	leading electrolyte	terminating electrolyte
Anion	Cl^-	OH^-
Concentration	0.01 M	saturated $Ba(OH)_2$
Counter ion	0.01 M $Tris^+/Li^+$	Ba^{++}
pH	7.75	11.5
Additive	0.25% HEC	-

by Everaerts et al. (12). The volume of the injected sample is 3 µl. The equipment has been described in a preceeding paper (13) and elsewhere (12,14). The zones are detected by means of a conductivity detector and a U.V. detector.

RESULTS AND DISCUSSION

With the low-pH system (pH 3.9) a rapid (10 min. analysis time) and reproducible separation was obtained of a standard solution containing 11 purine- and pyrimidine nucleotides (fig. 1, A; UV trace). The UV trace of a standard solution consisting of 9 bases, nucleosides and some other metabolites shows that with the high-pH system (pH 7.75) also these metabolites can be separated (fig. 1B). Both systems have been applied to the analysis of serum.

As could be anticipated not many nucleotides were detected with the pH 3.9 system (data not shown). Bases and nucleosides were present at higher concentrations. A number of them were detected in the serum of a hypouricemic individual (fig. 2A). A preliminary identification of several U.V. absorbing compounds was attempted with standard solutions. An example of this is shown in fig. 2B,

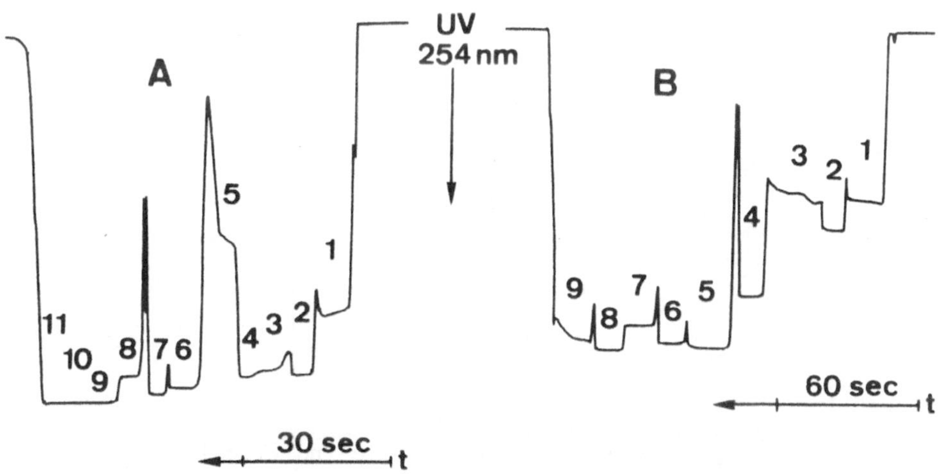

Fig. 1. Analysis of standard solutions.

A. UV trace obtained with the low-pH system (pH 3.9) of nucleotides. 1=UTP; 2=GTP; 3=ATP; 4=UDP; 5=CTP; 6=GDP; 7=ADP; 8=UMP; 9=GMP; 10=cAMP; 11=AMP.

B. UV trace obtained with the high-pH system (pH 7.75) of bases, nucleosides and some other metabolites.
1=orotic acid; 2=uric acid; 3=hippuric acid; 4=xanthine; 5=hypoxanthine; 6=inosine; 7=allopurinol; 8=guanosine; 9=adenine.

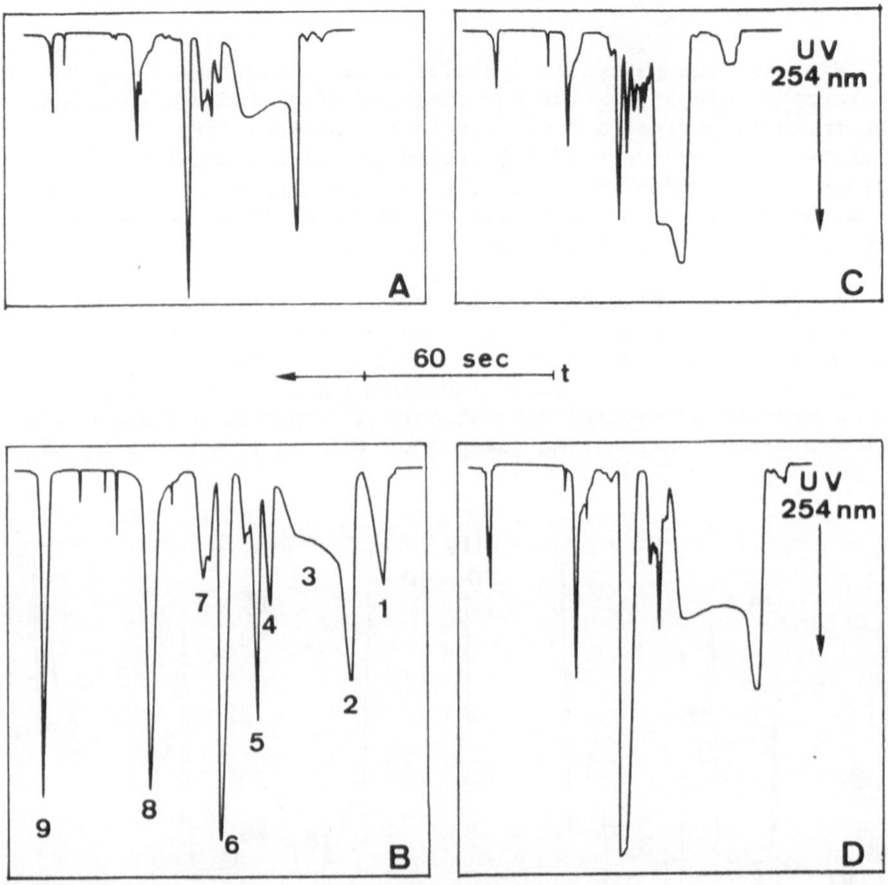

Fig. 2. Analysis of serum at pH 7.75.
A. UV trace of a serum from a hypourecemic individual.
B. Same serum, but "spiked" with a standard solution (see fig. 1B).
C. UV trace of a serum from a healthy control individual.
D. UV trace of a serum from a patient with the Lesch-Nyhan syndrome (not under allopurinol treatment).

where the analysis of a mixture of the same hypourecemic serum and a standard solution is given. Further identification of this sample was not attempted. The same holds for the U.V. trace of a normal serum (pooled from several controls) and of a serum from a Lesch-Nyhan patient (not under allopurinol treatment), shown in fig. 3C and 3D, respectively.

A UV trace of the electrolyte system showed some minor impurities. These peaks will also feature in the metabolic profiles. A possibility to differentiate between the electrolyte impurities and the separands might be to increase the volume of the sample injected: the metabolite zones will increase whereas the interfering electrolyte zones will increase much less. It should be noted that the U.V. absorption of the hippuric acid zone is not constant (fig. 2). This is due to the fact that a steady-state mixed-zone (9, page 168; 15) is formed with a constituent which has an effective mobility equal to that of hippurate. Fig 2B shows that several metabolites can be identified directly. However, to be absolutele sure about the identity of a certain metabolite "spiking" is not sufficient. Further possibilities to identify compounds include:

- The step height of the conductivity signal (universal detector; 9), which gives qualitative information regarding the constituents involved.
- The enzymatic conversion of a metabolite by purified enzymes which is a sensitive and specific way to identify the metabolite.
- Information regarding extinction, e.g. the E_{280}/E_{254} ratio.
- Changes in the operational electrolyte systems (such as complexing agents, solvents, mixtures of solvents, pH; see also ref. 9). For this it is especially attractive to use isotachophoresis because of its flexibility in rapidly changing the operational conditions.

We intend to carry out the identification of metabolite profiles along the above described lines.

REFERENCES

1. V.A. McKusick, no. 10260, 10280, 20160, 24275, 26612, 13890, 30620, 30800, 25890, 25892, 13894 and 27830, Mendelian inheritance in man, John Hopkins Univ. Press, Baltimore (1975).
2. W.N. Valentine, K. Fink, D.E. Paglia, S.R. Harris and W.S. Adams, J. Clin. Invest. 54:866 (1974).
3. J.E. Seegmiller, H. Bluestein, L. Thompson, R. Willis, S. Matsumoto and D. Carson, in: Models for the study of inborn errors of metabolism, p.p. 153-170. F. Hommes ed., Elsevier/ North Holland Biomedical Press, Amsterdam (1979).
4. E.M. Scholar and P. Calabresi, Cancer Res. 33:94 (1973).
5. C. Ip and T. Dao, Cancer Res. 38:723 (1978).
6. H.A. Simmonds, Clin. Chim. Acta 23:353 (1969).

DETERMINATION OF URIC ACID IN SERUM:

COMPARISON OF A STANDARD ENZYMATIC METHOD AND ISOTACHOPHORESIS

F. Oerlemans[°], Th. Verheggen[*], F. Mikkers[*],
F. Everaerts[*] and C. de Bruyn[°].
[°]Department of Human Genetics, Faculty of Medicine,
University of Nijmegen, Nijmegen, The Netherlands.
[*]Department of Instrumental Analysis, Eindhoven
University of Technology, Eindhoven, The Netherlands.

INTRODUCTION

Analytical techniques such a high-performance liquid chromatography (HPLC) and isotachophoresis can be applied for determination in biological fluids of a series of metabolites, including uric acid. Methods using HPLC have been reported for serum uric acid (1), blood and urine (2). Isotachophoresis has been employed in the analysis of nucleotides in muscle extracts (3) and urinary purines and pyrimidines, including uric acid (4).

Unlike other available methods, such as the colorimetric (5) and the enzymatic (6,7), the determination of uric acid with HPLC and isotachophoresis are much less hampered by interfering substances such as drugs and biological metabolites. In contrast to the HPLC procedure, where pre-treatment of biological samples is often necessary (e.g. removal of proteins), the samples can mostly be applied directly in isotachophoresis. Moreover in the case of uric acid, the ratio of free and protein-bound urate can be determined conveniently using a simple ultrafiltration step.

Untill now isotachophoresis has not been used for rapid determination of serum uric acid. Isotachophoresis is an electrophoretic separation method, taking advantage of the non-diluting phenomenon of the sample zone in the steady-state (8). In this paper an isotachophoretic system is given for the quantification of uric acid in serum (injected volume 1-3 μl).

MATERIALS AND METHODS

Materials

Uric acid (Na-salt), HCl (titrisol), Tris, ε-aminocaproic acid (EACA) and 2-morfolinoethanesulfonic acid (MES) (all analytical grade) were purchased from Merck (Darmstadt, G.F.R.). Hydroxyethyl-cellulose (HEC) was obtained from Polysciences (Warrington, Pa., USA, cat. no. 5568): a 0.5% (w/v) stock solution was purified by ionexchange. Uricase was purchased from Løvens Kemiska Fabrik (Ballerud, Denmark). Serum was prepared from venous blood after clotting (2 hrs. at room temperature) and centrifugation for 10 min. at 1,000 g (4 ^{o}C). Eventual storage of samples was at -20 ^{o}C. CF 25 centriflow filters (M.W. cut off 25,000) were purchased from Amicon (Oosterhout, The Netherlands).

Isotachophoretic Methods

Isotachophoresis is an electrophoretic separation method which employs a discontinuous system of electrolytes. The leading electrolyte contains an ionic species with a high effective mobility and a counter ion with buffering capacity. The terminating electrolyte contains an ionic species with a low effective mobility. The sample must be introduced on the boundary of leading electrolyte and terminating electrolyte. The separation is based on the difference in effective mobilities between the different ionic species involved. In the steady-state, the sample ions move in consecutive zones with a speed equal to that of the leading zone. Each zone only contains ionic species with equal effective mobility and the concentration is adjusted to that of the leading electrolyte according to the Kohlrausch regulation functions (8,9).

System, pH 5.0, for the isotachophoretic determination of uric acid.

	leading electrolyte	terminating electrolyte
Anion	Cl^-	MES^-
Concentration	0.01 M	0.005 M
Counter ion	$EACA^+$	$Tris^+$
pH	5.0	6.5
Additive	0.25% HEC	-

In this study an isotachophoretic equipment is used in which two teflon capillaries with different internal diameters are mounted (10, 11). All these experiments can also be carried out with commercially available equipment, e.g. with the Tachophor (LKB, Sweden), although more sample and analysis time is needed to obtain comparable results.

Enzymatic Methods

The assays were performed at the laboratory of the Department of Neurology (Faculty of Medicine, Nijmegen) with an ABA 100 bichromatic analyser (ABBOTT). The determination of uric acid is based on the succesive action of three purified enzymes which are added to the reaction mixture: uricase, katalase and aldehyde dehydrogenase (7). The formation of NADPH from $NADP^+$ in the latter reaction (measured both at 380 and 340 nm) is used for the quantification of uric acid. As standards sera with known concentrations of uric acid were used.

RESULTS

To serum, extensively dialysed against 0.9% NaCl, a standard amount of uric acid was added, giving a final concentration of 474 µM. The isotachophoretic analysis yielded recoveries of 99.0-100.5%.

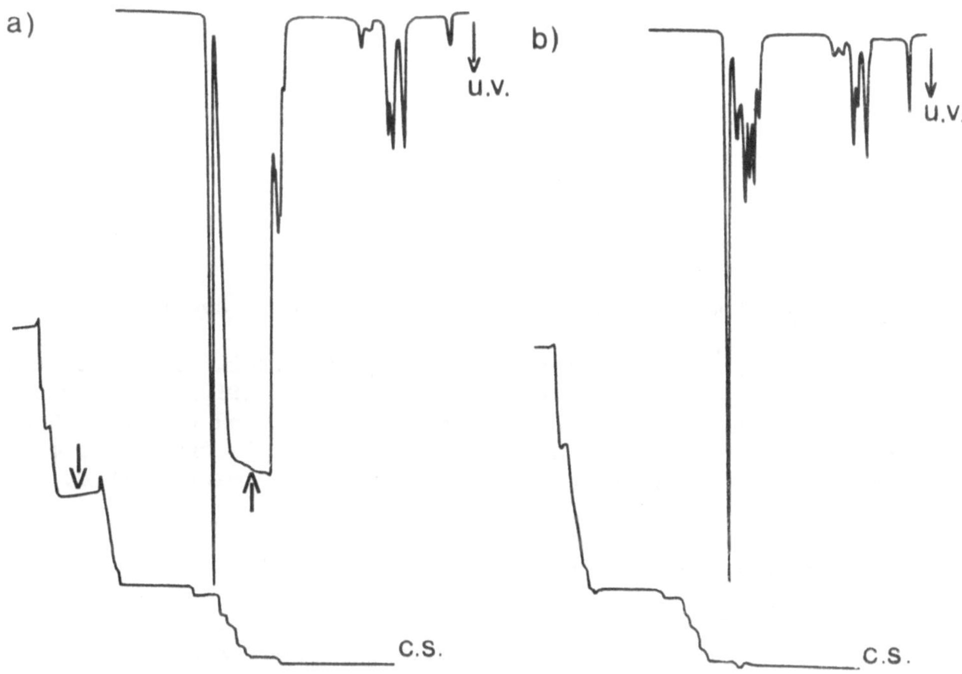

Figure 1. Effect of incubation of serum with purified uricase.
a. before incubation with uricase; the arrows indicate the uric acid zone as determined by the UV detector(u.v.) and the conductivity detector(c.s.).
b. after incubation with uricase.

To estimate the amount of uric acid bound to serum proteins under our experimental conditions the recovery from ultrafiltered and nonfiltered samples was compared. When undialysed serum was passed through an Amicon CF 25 filter (M.W. cut off: 25,000) 85.1% of the total serum uric acid was recovered in the ultrafiltrate, indicating that approximately 15% was bound to protein with a M.W. exceeding 25,000. The lower amounts of uric acid in the ultrafiltrate as compared to nonfiltered samples was not due to the CF 25 filter: when a standard solution of uric acid (474 μM in in water) was passed through it, the recovery was 99.4%.
In addition, the effect of high pH on the binding of urate to serum protein was studied. The pH of normal serum samples (pH 7.2-7.4) was adjusted to pH 10.0 and after ultrafiltration still approximately 7% (instead of 15%-20%) of total urate was bound. A small part of the sera showed some turbidity, as judged from visual inspection. Those samples were rapidly passed through a Millipore filter (Millex: 0.22 μ). This did not affect the recovery of uric acid.

Identification

The identity of uric acid was confirmed in several ways. In fig. 1 an experiment is shown which demonstrates that the uric acid zone is abolished by pre-incubation of the sample with uricase. Injection of an extra small amount of uric acid gave an increased length of the uric acid zone. Furthermore, the conductivity signal (step height) was specific for uric acid.

Comparison between Enzymatic and Isotachophoretic Results

Table 1 shows that there exists a good correlation between the data obtained with sera (not filtrated and ultrafiltrated over CF 25) from 4 controls analysed with the enzymatic and the isotachophoretic method. The day to day variance was about 2% with the isotachophoretic and approximately 10% with the enzymatic method.

Effects of Metabolites and Drugs

It is well known that several metabolites and drugs can interfere with the enzymatic determination. When homogentisic acid was added to serum samples (0.5-5.0 gr/l) higher values for uric acid were obtained with the enzymatic mathod, whereas no effect was seen with isotachophoresis.

Table 1. Results of Enzymatic and Isotachophoretic Uric Acid Determinations in Serum from 4 Healthy Control Individuals.

Sample	enzymatic method			isotachophoresis		
no.	NF	UF	% bound	NF	UF	% bound
1	383	283	26%	392	282	28%
2	292	233	20%	294	224	24%
3	483	400	17%	483	415	14%
4	375	317	15%	361	298	17%

Uric acid concentrations in μM.
NF: not filtrated; UF: ultrafiltrated (CF 25 filter.

DISCUSSION

The present isotachophoretic method for the determination of uric acid levels in serum is quantitative, reliable and reproducible. In contrast to some HPLC procedures (2) there is no need for deproteinisation: the samples can be applied directly, but also a HPLC system without deproteinisation has been described (1). However, the advantage of isotachophoresis over HPLC is the flexibility of the system. Once an electrolyte system has been chosen the analyses can be done with very low coefficients of variation (2%). In the present stage the isotachophoretic serum uric acid determination is more accurate than the enzymatic method, although the latter is faster when automatised.

Homogentisic acid is a metabolite which occurs in increased quantities in the urine of alcaptonuria patients. It interferes with the enzymatic uric acid determination at 340 nm by giving lower values than really present (7). When homogentisic acid was added to serum (0.5-5.0 gr/l) no effect was seen with isotachophoresis, whereas increased levels were read with the enzymatic procedure carried out with the bichromatic (380 and 340 nm) analyser ABA-100.

A decisive advantage of isotachophoresis (and also of HPLC) is the possibility to run whole metabolite profiles. In contrast to HPLC, there is no need to equilibrate columns enabling a fast switching from one electrolyte system to another when different conditions have to be tested.

The usefulness of isotachophoresis in the screening for inborn errors of purine and pyrimidine metabolism by analysing urinary bases and nucleosides has already been shown (4). We have recently developed operational systems for the analysis of purine- and pyrimidine nucleosides and bases in serum (12). Both for experimental and clinical purposes alternative possibilities are opened up, such as for the pharmacokinetic analysis of drug metabolism, or for studies on the binding of metabolites, such as uric acid, to serum proteins (table 1; refs. 13,14).

ACKNOLEDGEMENTS

The authors thank Mrs. Gerrie Steenbergen (Dept. of Neurology, University Hospital, Nijmegen), for the enzymatic analysis of uric acid on the ABA-100 analyser.

REFERENCES

1. W.D. Slaunwhite, L.A. Pachla, D.C. Wenke and P.T. Kissinger, Clin. Chem. 21:1427 (1975).
2. J.A. Milner and E.G. Perkins, Anal. Biochem. 88:560 (1978).
3. D.C. Gower and R.C. Woledge, Science Tools, 24:17 (1977).
4. A. Sahota, H.A. Simmonds and R.H. Payne, J. Pharm. Meth. 2:303 (1979).
5. N. Kageyama, Clin. Chim. Acta, 31:421 (1971).
6. L. Liddle, J.E. Seegmiller and L. Laster, J. Lab. Clin. Med. 54:903 (1959).
7. R. Haeckel, J. Clin. Chem. Clin. Biochem. 14:101 (1976).
8. F.M. Everaerts, J.L. Beckers and Th.P.E.M. Verheggen, Isotachophoresis. Theory, instrumentation and applications, Elsevier Scientific Publishing Company, Amsterdam (1976).
9. F. Kohlrausch, Ann. Phys. Chem. (Leipzig), 62:209 (1897).
10. F.M. Everaerts, Th.P.E.M. Verheggen and F.E.P. Mikkers, J. Chromatogr. 169:21 (1979).
11. Th.P.E.M. Verheggen, F.E.P. Mikkers and F.M. Everaerts, in: Protides of the biological fluids. Proceedings of the 27th colloquim, H. Peters, ed., Pergamon Press, New York, in press.
12. F. Oerlemans, Th. Verheggen, F. Mikkers, F. Everaerts and C. de Bruyn, Adv. Exp. Med. Biol., this volume.
13. J.R. Klinenberg and I. Kippen, J. Lab. Clin. Med. 75:503 (1970).
14. P.C. Farrell, R.P. Popovich and A.L. Babb, Biochim. Biophys. Acta 243:49 (1971).

ADDITION PRODUCTS OF URIC ACID AND FORMALDEHYDE

Peter A. Simkin and Qwihee P. Lee

Division of Rheumatology, Department of Medicine

University of Washington, Seattle, Washington

In order to calibrate colorimetric determinations of serum and urine uric acid, standard dilutions are routinely prepared from a stock solution of uric acid in 2% Formalin. We made a similar solution by dissolving 100 mg of uric acid in 100 ml of 1% formaldehyde in water and applied it to a polyacrylamide column (BIO GEL P-2) monitored spectrophotometrically at 280 nm. Uric acid, dissolved in water or in lithium carbonate solution, characteristically adsorbs to the column matrix and ultimately elutes as a single peak after 33 hours (1). In contrast, the solution of uric acid with formaldehyde unexpectedly resulted in three distinct elution peaks. A small peak was seen with the characteristic slow mobility of free uric acid, but two much larger peaks, designated F_2 and F_1, were eluted earlier after 19 and 25 hours. This simple chromatographic technique thus demonstrated the presence of two new reaction products and isolated them for further characterization.

Uric acid and formaldehyde were measured independently in each column fraction. The elution profiles corresponded precisely with each other in a pattern indicating a constant association of uric acid and formaldehyde in each of the two new peaks. In the ten tubes under the center of the F_2 peak, the mean molar ratio of formaldehyde to uric acid was 2.00 ± 0.07 (SD). In the ten tubes under the center of the F_1 peak, the mean molar ratio was 1.03 ± 0.04. We therefore concluded that the F_2 peak consists of an addition product comprised of one molecule of uric acid with two molecules of formaldehyde. Similarly, the F_1 adduct consists of one molecule of uric acid with one molecule of formaldehyde.

All column fractions adsorbing ultraviolet light contained uric acid as determined by the enzymatic spectrophotometric method.

Uric acid was responsible for each of the three peaks, since their absorbency at 293 nm could be entirely eliminated by the specific enzyme, uricase. These experiments were conducted in 0.15 M borate buffer at pH 9.3 since glycine buffer, commonly used in this procedure (2), itself reacts with formaldehyde. The amount of uric acid in each fraction was determined directly by measuring UV absorption at 285 nm (the Maximal wavelength of uric acid at pH 4.0).

Each column fraction was analyzed for formaldehyde content by the chromotropic acid method (3). For those fractions containing uric acid, appropriate uric acid control solutions were employed in the determination. The elution peak for free formaldehyde occurred after 10 hours, indicating that this small molecule does not adsorb significantly to the polyacrylamide resin.

In 1% formaldehyde the concentrations of F_1 and F_2 were essentially equal. At lower concentrations F_1 and free urate were predominant while at formaldehyde concentrations greater than 1%, F_2 was the principal but never the sole product observed.

The isolated F_1 and F_2 adducts remained quite stable on the column as evidenced by the consistent symmetry of their elution peaks. Dissociation would have caused a significant skewing of the peaks which was never seen. Stability of the adducts in the absence of excess formaldehyde was studied further by rechromatography of the isolated F_1 and F_2 peaks. Repeated column runs over an 8-day period showed progressive dissociation into the constituent molecules. Thus, the F_2 adducts (at 4° in pH 4.0 acetate buffer) decayed ($t_{½} \cong 30$ hours) into the F_1 adduct and free formaldehyde. The F_1 adduct was considerably more stable, but dissociated in turn ($t_{½} \cong 120$ hours) into uric acid and formaldehyde.

The site and the chemical nature of adduct formation is uncertain but the effects resemble those observed by previous workers studying related compounds. Several prior studies have examined the interactions in solution of formaldehyde with purine and pyrimidine compounds (4,5). This work has primarily employed UV spectrophotometry and pH titration to demonstrate reactions thought to be N-hydroxymethylation of amino groups or of acidic imino nitrogen. The reaction with ring imino nitrogen was very rapid, had a low equilibrium constant and was characterized by small spectral changes. Each of these characteristics resemble those we observed in the reactions of formaldehyde with uric acid. In previous studies, final definition of the reaction has been hampered by inability to isolate the reaction products and by their presumed instability in the absence of excess formaldehyde. With our simple column method, we have been able to isolate two stable adducts of xanthine as well as of uric acid and a single adduct of theobromine. We were unable to isolate adducts of formaldehyde with hypoxanthine, theophylline, or caffeine. We believe that further applications of this column technique will make it possible to define the sites and the nature of many of these interactions.

Adduct formation presumably explains the markedly enhanced solubility of uric acid in formaldehyde solutions. Pathologists have long known that urate crystals rapidly dissolve from gouty tissued fixed in Formalin (37% formaldehyde). To illustrate the extent of this effect, we were able to dissolve 1 gram of uric acid in 10 ml of formalin, a concentration higher by more than three orders of magnitude than that attainable in an aqueous solution of the same pH. This massive enhancement of solubility is the most striking effect of formaldehyde on the physical properties of uric acid.

Formation of urate crystals from hyperuricemic body fluids is the most crucial step in the pathogenesis of gout, but only a small percentage of hyperuricemic individuals ever develop clinical disease. Factors affecting the solubility of uric acid are thus of obvious relevance to the problem of gout. Factors currently known to affect the solubility of sodium urate include temperature, ionic strength, pH, and the glycosaminoglycans of connective tissue. After extensive investigation of the question, it seems unlikely that protein binding materially enhances urate solubility.

Our studies show that formaldehyde dramatically increases the solubility of uric acid. It may be that, as in the "activation" of tetrahydrofolic acid (6), excess formaldehyde _in vitro_ duplicates a metabolic step performed _in vivo_ by a specific enzyme. It may also be that physiologic molecules such as glucose may weakly react with uric acid in a way that is qualitatively similar to the reaction with formaldehyde. In view of these considerations, the possible interactions of uric acid with other small molecules deserves further study.

In summary, uric acid and formaldehyde interact in aqueous solution to form addition products (adducts). They have formaldehyde to uric acid molar ratios of 1:1 and 2:1, are readily isolated on polyacrylamide columns and are relatively stable in the absence of excess formaldehyde. Adduct formation explains the markedly enhanced solubility of uric acid in formaldehyde solutions.

REFERENCES

1. Simkin, P.A. J. Chromatogr. 1970, 47, 103.
2. Liddle, L., Seegmiller, J.E., Laster, L. J. Lab. Clin. Med. 1959, 54, 903.
3. Eyring, E.J., Ofengand, J. Biochem. 1967, 6, 2500.
4. Lewin, S. J. Chem. Soc. 1964, 1, 792.
5. Lewin, S., Barnes, M.A. J. Chem. Soc. (B) 1966, 1 , 478.
6. Osborn, M.J., Talbert, P.T., Huennekens, F.M. J. Am. Chem. Soc. 1960, 9, 4921.

AUTOMATED RETRIEVAL OF PURINE LITERATURE

L. Ferreiro and A. Rey

Instituto de Información y Documentación en Ciencia y Tecnología (CSIC)
Madrid (6) (Spain)

INTRODUCTION

Over the last few years, the published literature relating to purines from a wide variety of disciplines, has almost duplicated. This fact represents a challenge for bibliographical automated searching services.

As is known, the computer storage is supplied with a number of characterizing words (keywords) corresponding to the nuclear concepts of each paper published in the primary sources used. These keywords are recognised by the machine as graphic structures, only when they are identical to those which form part of the profiles or searching expressions. That is to say, the computer does not set up associatives, ideological, or semantic relations between the terms which it receives for recognition through the profiles and those which its memory contains.

In the present paper a searching strategy is considered, which uses an independent profile series for retrieving as many references as posible of papers implicitly connected with the central subject, taking as an example: "Cellular toxicity of the adenosine and immunosuppression".

METHODS

The Chemical Abstracts Search (CA), Biosis Previews (B Pr) and Excerpta Medica (E Med) bases were used via the Dialog System (1973-1978)

The first profile (A) mentioned above was developed through four subprofiles, taking as starting point the biochemical basis for two cited biological phenomena of adenosine.

All profiles were traced according to the boolean logic relationships (AND, OR, and NOT) expressed in the search strategy (Fig. 1)

```
          1      1898  ADENOSINE
          2        34  C AMP
          3      7960  CYCLIC (W) AMP
          4     10695  CELLULAR
          5      6496  CONCENTRATION
          6      3909  ACCUMULATION
          7      5568  LEVEL
? C (2+3)    (4+5+6+7)

      8  715(2+3)    (4+5+6+7)
? C   8
      9    5   1 8
```

Fig. 1. Search Strategy example (a.3 profile)
Set Items Descriptors
(+=OR; =AND;- =NOT)

RESULTS

Table 1.- Independent profile series used

Profiles	Valid References Obtained		
	CA.	P Pr	E Med
A	64	125	20
a.1	23	49	60
a.2	30	17	41
a.3	52	24	-
a.4	32	25	-
Total valid refs.	201	241	121
Total references	(545)	(503)	(278)
Precision Index	36.8%	47.9%	43.5%
Duplicate references (not included in the total above)	34	93	20

Profiles. (A):Adenosine. 2'-Deoxyadenosine. Toxicity Immunosuppression. (a.1): ADA Inhibition. (a.2): Pyrimidine. Synthesis. Inhibition. (a.3): c AMP Accummulation. (a.4): PP-ribose-P. Intracellular. Concentration. Decrease. (CA). Purine nucleoside phosphorylase. Deficiency. (B Pr)

As is shown in Table 1, from the total number of references obtained, the precision index (% valid references with respect to the total references obtained) was calculated for each data base. All these precision indexes were in normal range.

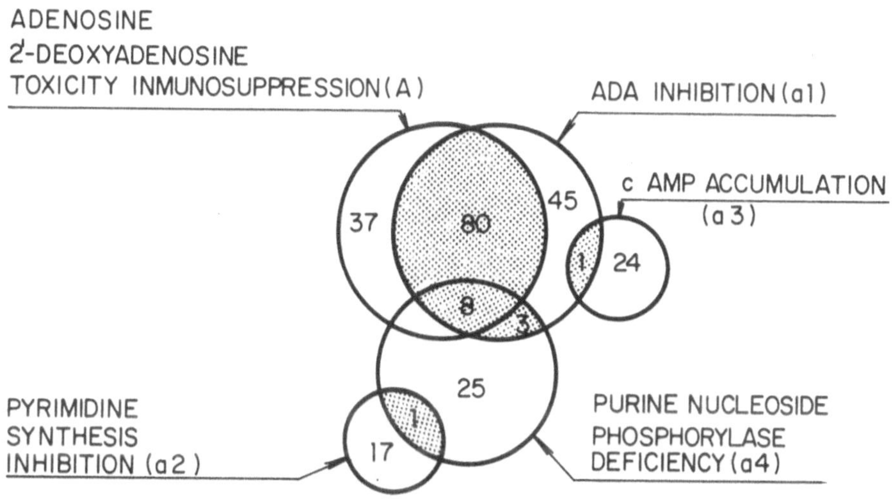

Fig. 2 Venn's Diagram (Biosis Previews)

Venn's diagram traced (Fig. 2) shows the number of valid references, individual and shared, obtained for each profile in Biosis Previews data base. The greatest number of references (88) which would have been obtained by the A–a.1. profiles intersection, would have fallen to eight if the a– a.1– a.4 profiles intersection had been used, and the system yield would have fallen to nil with more complex intersections.

CONCLUSIONS

The scientific papers of the revised literature are not a unit when considered terminologically for automated retrieval. A 60–80% of the total valid references obtained through each data base was stored in the computer memory in virtually closed compartments although they belonged to the same material complex.

The retrieval of these papers requires independent profiles series which although they give rise 14–28% duplicate references, the final yield can be increased by 100–200% of the yield obtained through the basic profile of the central subject considered.

AUTHOR INDEX

SUBJECT INDEX